AF356594

Nanostructural Materials with Rare Earth Ions: Synthesis, Physicochemical Characterization, Modification and Applications

Nanostructural Materials with Rare Earth Ions: Synthesis, Physicochemical Characterization, Modification and Applications

Editor

Rafal J. Wiglusz

MDPI • Basel • Beijing • Wuhan • Barcelona • Belgrade • Manchester • Tokyo • Cluj • Tianjin

Editor
Rafal J. Wiglusz
Institute of Low Temperature
and Structure Research Polish
Academy of Science
Poland

Editorial Office
MDPI
St. Alban-Anlage 66
4052 Basel, Switzerland

This is a reprint of articles from the Special Issue published online in the open access journal *Nanomaterials* (ISSN 2079-4991) (available at: https://www.mdpi.com/journal/nanomaterials/special_issues/nano_rare_earth).

For citation purposes, cite each article independently as indicated on the article page online and as indicated below:

LastName, A.A.; LastName, B.B.; LastName, C.C. Article Title. *Journal Name* **Year**, *Volume Number*, Page Range.

ISBN 978-3-0365-3457-2 (Hbk)
ISBN 978-3-0365-3458-9 (PDF)

Contents

About the Editor

Rafal J. Wiglusz (Prof. dr. hab.) received degrees from the Chemistry Department of the Wroclaw University (Poland) (M.Sc. and Ph.D.) and completed his postdoctoral studies at the Department of Chemistry of the University of Cologne (Germany). In 2008, he moved to the Institute of Low Temperature and Structure Research of the Polish Academy of Sciences, Poland, where he is presently Full Professor and Head of the Division of Biomedical Physicochemistry.

Rafal J. Wiglusz has published over 170 papers in the fields of Physics, Chemistry, Materials Science, Biochemistry, Medicine, Nanotechnology and Engineering. His works have been quoted over 2094 times in the ISI Index, at an average of over 28 citations per publication, while his h-index is currently 28.

His research goals lie in the preparation of nanometer-sized materials (oxides, phosphates, metallic, and magnetic particles), followed by the creation of periodically ordered nanostructures based on single nanoparticles. A small particle size implies high sensitivity and selectivity. These new effects and possibilities are mainly a result of quantum effects that are a result of the increasing ratio of surface-to-volume atoms in low-dimensional systems. An important factor in this context has been the design and fabrication of nanocomponents with/displaying new functionalities and characteristics for the improvement of existing materials, including inorganic–organic materials, polymers, hydrogels and composites, with their production, characterization and application as a biomaterial for bio-medical applications for various 3D-printing technologies.

 *nanomaterials*

Editorial

Nanostructural Materials with Rare Earth Ions: Synthesis, Physicochemical Characterization, Modification and Applications

Rafal J. Wiglusz

Institute of Low Temperature and Structure Research, PAS, Okolna 2, 50-422 Wroclaw, Poland; r.wiglusz@intibs.pl; Tel.: +48-071-395-42-74; Fax: +48-071-344-10-29

Citation: Wiglusz, R.J. Nanostructural Materials with Rare Earth Ions: Synthesis, Physicochemical Characterization, Modification and Applications. *Nanomaterials* **2021**, *11*, 1848. https://doi.org/10.3390/nano11071848

Received: 7 July 2021
Accepted: 13 July 2021
Published: 16 July 2021

Publisher's Note: MDPI stays neutral with regard to jurisdictional claims in published maps and institutional affiliations.

The success of nanotechnology in the field of physical, chemical and medical sciences has started revolutionizing the drug delivery science and theranostics (therapy and diagnostics) [1,2]. The specific advantages include superior pharmacodynamics, pharmacokinetics, reduced toxicity and improved targeting capability. This approach has great potential to produce novel diagnostics and therapeutics—theranostic—because of nanomaterials show unexpected and interesting chemical and physical properties different from those of the original in the micro-sized scale [3]. Therapies combining the use of bioactive materials and progenitor cells or an active substance become clinical reality, increasing the prospects for the development of engineering and regenerative medicine [4]. One of these perspectives is a diagnostics and personalized therapy, i.e., theranostics [5].

In this case of the active substance, the drug-delivery vehicle, as a critical quality attribute in the drug delivery science, needs special attention for the formulation development, which can be successfully achieved via nanotechnology. Drugs incorporated in nanocarriers, either physically entrapped or chemically tethered, have the potential to target the physiological zone of the disorder sparing normal cells from collateral consequences. Targeting several molecular mechanisms, for either treatment or prevention of difficult-to-treat diseases, for the design of various nanotechnology-based drug delivery systems is one of the prime focuses of the formulation scientist at the present juncture.

Much attention has been devoted to developing new drug-delivery systems with many advantages compared with the conventional forms of dosage, such as, among others, enhanced bioavailability, greater efficiency, lower toxicity, controlled release [6–12]. An ideal drug-delivery system should be characterized by: (1) maximum biocompatibility and minimal antigenic properties [13]; (2) appropriate particle size, which is important for the particles to reach a particular location in the body due to the size of the vessels of the human circulatory system [14]; (3) the ability to transport the desired drug molecules to the targeted cells or tissues and release them in a controlled manner [15]. So far, different types of drug-delivery systems have been developed, such as, i.a. biodegradable polymers [16], xerogels [17], hydrogels [18], mesoporous materials [4,11]. Among different drug-delivery systems, mesoporous materials (such as SBA-15, MCM-41 and mesoporous silica nanoparticles) have gained increasing interest, particularly as drug storage and release hosts due to their unique surface and textural properties [14,19,20].

Materials designed for biomedical applications should be characterized by a high sensitivity and specificity, a lack of functional interference with the sample, photochemical stability, non-toxicity, long time of storage and, as far as possible, detection of a substance in the presence of others. Moreover, nanoscale materials have been exploited as active components in a wide range of technological applications in the biomedical field [21–25]. Particularly, in the field of biomedicine, nanoparticles can be used as drug-delivery vehicles that can target tissues or cells [13,24] and can be functionalized with special characteristics (such as magnetization, fluorescence and near-infrared absorption) for qualitative or quantitative detection of tumor cells [23,25–27].

1

It is well-known that nanoscale fluorescent materials have attracted much interest due to the increasing demand for efficient photosensitive materials not only for sophisticated optoelectronic and photonic devices but also for a broad range of biomedical applications [28–34]. In biomedical areas, luminescent materials, mainly including fluorescent organic molecules [35,36] and semiconductor nanoparticles [37,38], have been widely investigated in biological staining and diagnostics. However, some serious problems of photobleaching and quenching of fluorescent organic molecules and the toxicity of semiconductor quantum dots are critically evident and have seriously limited their applications in biomedical areas [38,39]. Furthermore, high performance in function-specific biological applications requires that the composites possess some unique characteristics, such as uniform morphology, large surface areas, good dispersion, etc. [39]. Recently, a class of stable, efficient and self-activated luminescent materials whose emission is induced by the defects or impurities in host lattices, has been prepared by various synthesis routes [40–43]. These novel self-activated inorganic materials may be a promising fluorescent material for biodetection due to their good optical properties and nontoxicity.

Apatites are inorganic compounds with a general formula $M_{10}(XO_4)_6Y_2$, where M represents divalent cations (e.g., Ca^{2+}, Sr^{2+}, etc.), $XO_4 = PO_4^{3-}$, VO_4^{3-}, etc. and Y represents anions: F^-, OH^-, Cl^-, Br^-, etc. The hexagonal structure in apatites belongs to $P6_3/m$ space group and allows the cations to localize in the 4(f) and 6(h) positions [44] and is able to accommodate a variety of univalent cations as substituents. In that case, charge compensation, proposed by P. Martin and et al. [45], allows explaining the substitution of divalent calcium ions to trivalent lanthanide ions in apatite with a simple mechanism. It is worth mentioning that apatites themselves, such as calcium apatites $Ca_{10}(PO_4)_6(Y)_2$, are biocompatible and are natural building blocks for bones and teeth [46]. This feature combined with highly photostable luminescent properties of rare-earth dopants, makes nanocrystalline apatites highly attractive as luminescent bio-labels [47]. However, these materials have not been extensively synthesized or examined in the nanocrystalline form [48] which is a prerequisite for being internalized by cells for bio-imaging or sensing applications [49].

Several strategies have been developed in the synthesis of nanoparticles so far, involving such techniques as microemulsion, precipitation, thermal decomposition, chemical vapor deposition and others. However, the best results and control over particle size, crystallinity and purity can be ensured using microwave technology [50]. For instance, our group was able to obtain highly crystalline, phase pure, bio-compatible uniform and low agglomerated nano-apatites such as $Ca_{10}(PO_4)_6(OH)_2$ for bio-applications [51,52]. Another important feature is that the materials were produced in environmentally friendly conditions in ethylene glycol solution that is non-toxic for living organisms. Thus, this strategy seems to be very attractive for the synthesis of luminescent or multifunctional materials offering the possibility of bio-imaging measurement. The proposed synthesis technique allows for thorough control over the desired composition (as it was shown in the article [53]) which cannot be simply achieved using other techniques. Moreover, the proposed compounds can be considered as non-toxic due to their insolubility in body fluids and high chemical stability. It is well known that the solubility of oxide nanoparticles is one of the most important factors of their toxicity related to their chemical composition [54]. For example, the toxic effect of iron oxide nanoparticles originates mainly from the catalytic production of free radicals through Fenton type reaction [55]. To date, quantum dots (QD) characterized by high absorbance, high quantum yield, narrow emission bands and high resistance to photobleaching were considered as the most promising materials for FI applications in medicine. Currently, the main issue regarding QDs and their biomedical applications is their extreme toxicity (semiconductors—derivatives of highly toxic heavy metals such as Cd or Pb) [56]. One of the promising alternatives is offered by the application of inorganic compounds such as apatites doped or co-doped with optically active rare earth metals for bio-imaging [57].

Furthermore, calcium is the fifth most abundant element by mass in the human body (1.4–1.66%) where it is a common cellular ionic messenger with many functions and serves also as a structural element in bones (hydroxyapatites—99%) [58]. Calcium and its compounds play an important role in controlling numerous biological processes in living systems. Concentrations of free Ca^{2+} in biological cells are widely studied with fluorescent probes. The probes have a high selectivity for free calcium and exhibit marked changes in their photophysical properties upon binding. In particular, changes in fluorescence intensity (intensity probes) or spectral shift (ratio probes) upon binding to Ca^{2+} are monitored. The main drawback of intensity probes is that the intensity of fluorescence is affected by both the probe concentration and the free Ca^{2+} concentration. Consequently, a quantitative determination of Ca^{2+} distributions requires the probes to be distributed homogeneously in the sample. Conventional quantitative determinations of Ca^{2+} concentration with ratio probes overcomes the dependence on local probe concentration by exploiting ratiometric procedures using excitation or detection at two wavelengths [59,60]. The advent of fluorescence lifetime imaging techniques [1,61–64] opens new horizons for the quantitative determination for bio-imaging, in particular using intensity probes [65]. Fluorescence lifetime imaging is determined by factors such as the chemical environment of a fluorescent molecule and thus provides valuable information about its ion binding states. Importantly, since the lifetime is independent of fluorescence intensity, such measurements have wide-ranging applications to samples in which the probes have an inhomogeneous distribution. An additional advantage of the lifetime imaging technique is that the images are not compromised by photobleaching and absorption effects.

Nanocrystalline probes doped with lanthanide ions based on apatites meet these requirements [66]. Their narrow emission lines as well as long life-times render them suitable for use as luminescent markers in biology and medicine [67]. Therefore, this strategy seems to be very attractive for the complete elimination of the effects associated with local concentration of ions in the sample. Moreover, the surface functionalization of nanomaterials with biologically active organic ligands results in a better stability of the colloidal dispersion. It will contribute to measurable progress in the possible extension of bio-imaging techniques. Independently of the scientific goal related to theranostics, the synthesis and study of spectroscopic properties of lanthanide-ion doped apatites could also be an important area of research.

Funding: The research was funded by the National Science Centre (NCN) within the projects "Preparation and characterisation of biocomposites based on nanoapatites for theranostics" (No. UMO-2015/19/B/ST5/01330).

Institutional Review Board Statement: Not applicable.

Informed Consent Statement: Not applicable.

Data Availability Statement: Not applicable.

Acknowledgments: Rafal J. Wiglusz would like to thank all authors who submitted their research to this Special Issue, the referees who reviewed the submitted manuscripts and the Assistant Editor, Tina Tian, who made it all work.

Conflicts of Interest: Author declares no conflict of interest.

References

1. Szyszka, K.; Targońska, S.; Lewińska, A.; Watras, A.; Wiglusz, R.J. Quenching of the Eu^{3+} luminescence by Cu^{2+} ions in the nanosized hydroxyapatite designed for future bio-detection. *Nanomaterials* **2021**, *11*, 464. [CrossRef]
2. Zienkiewicz, J.A.; Strzep, A.; Jedrzkiewicz, D.; Nowak, N.; Rewak-Soroczynska, J.; Watras, A.; Ejfler, J.; Wiglusz, R.J. Preparation and Characterization of Self-Assembled Poly(l-Lactide) on the Surface of β-Tricalcium Diphosphate(V) for Bone Tissue Theranostics. *Nanomaterials* **2020**, *10*, 331. [CrossRef]
3. Saiz, E.; Zimmermann, E.A.; Lee, J.S.; Wegst, U.G.K.; Tomsia, A.P. Perspectives on the role of nanotechnology in bone tissue engineering. *Dent. Mater.* **2013**, *29*, 103–115. [CrossRef]
4. Zakrzewski, W.; Dobrzynski, M.; Rybak, Z.; Szymonowicz, M.; Wiglusz, R.J. Selected Nanomaterials' Application Enhanced with the Use of Stem Cells in Acceleration of Alveolar Bone Regeneration during Augmentation Process. *Nanomaterials* **2020**, *10*, 1216. [CrossRef] [PubMed]

5. Watras, A.; Wujczyk, M.; Roecken, M.; Kucharczyk, K.; Marycz, K.; Wiglusz, R.J. Investigation of Pyrophosphates KYP_2O_7 Co-Doped with Lanthanide Ions Useful for Theranostics. *Nanomaterials* **2019**, *9*, 1597. [CrossRef] [PubMed]

6. Fischer, K.E.; Alemán, B.J.; Tao, S.L.; Daniels, R.H.; Li, E.M.; Bünger, M.D.; Nagaraj, G.; Singh, P.; Zettl, A.; Desai, T.A. Biomimetic Nanowire Coatings for Next Generation Adhesive Drug Delivery Systems. *Nano Lett.* **2009**, *9*, 716–720. [CrossRef]

7. Sobierajska, P.; Serwotka-Suszczak, A.; Szymanski, D.; Marycz, K.; Wiglusz, R.J. Nanohydroxyapatite-Mediated Imatinib Delivery for Specific Anticancer Applications. *Molecules* **2020**, *25*, 4602. [CrossRef]

8. Vivero-Escoto, J.L.; Slowing, I.I.; Wu, C.W.; Lin, V.S.Y. Photoinduced intracellular controlled release drug delivery in human cells by gold-capped mesoporous silica nanosphere. *J. Am. Chem. Soc.* **2009**, *131*, 3462–3463. [CrossRef] [PubMed]

9. Wei, W.; Ma, G.H.; Hu, G.; Yu, D.; Mcleish, T.; Su, Z.G.; Shen, Z.Y. Preparation of hierarchical hollow CaCo3 particles and the application as anticancer drug carrier. *J. Am. Chem. Soc.* **2008**, *130*, 15808–15810. [CrossRef] [PubMed]

10. Yang, Q.; Wang, S.; Fan, P.; Wang, L.; Di, Y.; Lin, K.; Xiao, F.S. pH-responsive carrier system based on carboxylic acid modified mesoporous silica and polyelectrolyte for drug delivery. *Chem. Mater.* **2005**, *17*, 5999–6003. [CrossRef]

11. Zhao, W.; Chen, H.; Li, Y.; Li, A.; Lang, M.; Shi, J. Uniform rattle-type hollow magnetic mesoporous spheres as drug delivery carriers and their sustained-release property. *Adv. Funct. Mater.* **2008**, *18*, 2780–2788. [CrossRef]

12. Rösler, A.; Vandermeulen, G.W.M.; Klok, H.A. Advanced drug delivery devices via self-assembly of amphiphilic block copolymers. *Adv. Drug Deliv. Rev.* **2001**, *53*, 95–108. [CrossRef]

13. Chen, F.H.; Gao, Q.; Ni, J.Z. The grafting and release behavior of doxorubincin from $Fe_3O_4@SiO_2$ core-shell structure nanoparticles via an acid cleaving amide bond: The potential for magnetic targeting drug delivery. *Nanotechnology* **2008**, *19*. [CrossRef]

14. Ritter, J.A.; Ebner, A.D.; Daniel, K.D.; Stewart, K.L. Application of high gradient magnetic separation principles to magnetic drug targeting. *J. Magn. Magn. Mater.* **2004**, *280*, 184–201. [CrossRef]

15. Sobierajska, P.; Dorotkiewicz-Jach, A.; Zawisza, K.; Okal, J.; Olszak, T.; Drulis-Kawa, Z.; Wiglusz, R.J. Preparation and antimicrobial activity of the porous hydroxyapatite nanoceramics. *J. Alloys Compd.* **2018**, *748*, 179–187. [CrossRef]

16. Uhrich, K.E.; Cannizzaro, S.M.; Langer, R.S.; Shakesheff, K.M. Polymeric Systems for Controlled Drug Release. *Chem. Rev.* **1999**, *99*, 3181–3198. [CrossRef] [PubMed]

17. Yang, H.H.; Zhu, Q.Z.; Qu, H.Y.; Chen, X.L.; Ding, M.T.; Xu, J.G. Flow injection fluorescence immunoassay for gentamicin using sol-gel-derived mesoporous biomaterial. *Anal. Biochem.* **2002**, *308*, 71–76. [CrossRef]

18. Caliceti, P.; Salmaso, S.; Lante, A.; Yoshida, M.; Katakai, R.; Martellini, F.; Mei, L.H.I.; Carenza, M. Controlled release of biomolecules from temperature-sensitive hydrogels prepared by radiation polymerization. *J. Control. Release* **2001**, *75*, 173–181. [CrossRef]

19. Yang, P.; Huang, S.; Kong, D.; Lin, J.; Fu, H. Luminescence functionalization of SBA-15 by YVO_4:Eu^{3+} as a novel drug delivery system. *Inorg. Chem.* **2007**, *46*, 3203–3211. [CrossRef]

20. Vallet-Regi, M.; Rámila, A.; Del Real, R.P.; Pérez-Pariente, J. A new property of MCM-41: Drug delivery system. *Chem. Mater.* **2001**, *13*, 308–311. [CrossRef]

21. Slowing, I.I.; Trewyn, B.G.; Giri, S.; Lin, V.S.Y. Mesoporous silica nanoparticles for drug delivery and biosensing applications. *Adv. Funct. Mater.* **2007**, *17*, 1225–1236. [CrossRef]

22. Manna, L.; Milliron, D.J.; Meisel, A.; Scher, E.C.; Alivisatos, A.P. Controlled growth of tetrapod-branched inorganic nanocrystals. *Nat. Mater.* **2003**, *2*, 382–385. [CrossRef]

23. Shi, D. Integrated multifunctional nanosystems for medical diagnosis and treatment. *Adv. Funct. Mater.* **2009**, *19*, 3356–3373. [CrossRef]

24. Farokhzad, O.C.; Langer, R. Impact of nanotechnology on drug delivery. *ACS Nano* **2009**, *3*, 16–20. [CrossRef] [PubMed]

25. Walt, D.R. Miniature analytical methods for medical diagnostics. *Science* **2005**, *308*, 217–219. [CrossRef] [PubMed]

26. Wang, W.; Shi, D.; Lian, J.; Guo, Y.; Liu, G.; Wang, L.; Ewing, R.C. Luminescent hydroxylapatite nanoparticles by surface functionalization. *Appl. Phys. Lett.* **2006**, *89*, 183106. [CrossRef]

27. Burns, A.; Ow, H.; Wiesner, U. Fluorescent core–shell silica nanoparticles: Towards "Lab on a Particle" architectures for nanobiotechnology. *Chem. Soc. Rev.* **2006**, *35*, 1028–1042. [CrossRef] [PubMed]

28. Diamente, P.R.; Burke, R.D.; Van Veggel, F.C.J.M. Bioconjugation of Ln^{3+}-Doped LaF_3 nanoparticles to avidin. *Langmuir* **2006**, *22*, 1782–1788. [CrossRef]

29. Dong, C.; Van Veggel, F.C.J.M. Cation exchange in lanthanide fluoride nanoparticles. *ACS Nano* **2009**, *3*, 123–130. [CrossRef]

30. Syamchand, S.S.; Sony, G. Europium enabled luminescent nanoparticles for biomedical applications. *J. Lumin.* **2015**, *165*, 190–215. [CrossRef]

31. Jeong, J.; Cho, M.; Lim, Y.T.; Song, N.W.; Chung, B.H. Synthesis and characterization of a photoluminescent nanoparticle based on fullerene-silica hybridization. *Angew. Chem. Int. Ed.* **2009**, *48*, 5296–5299. [CrossRef] [PubMed]

32. Wang, M.; Mi, C.C.; Wang, W.X.; Liu, C.H.; Wu, Y.F.; Xu, Z.R.; Mao, C.B.; Xu, S.K. Immunolabeling and NIR-excited fluorescent imaging of HeLa cells by using NaYF 4:Yb,Er upconversion nanoparticles. *ACS Nano* **2009**, *3*, 1580–1586. [CrossRef] [PubMed]

33. Wang, F.; Tan, W.B.; Zhang, Y.; Fan, X.; Wang, M. Luminescent nanomaterials for biological labelling. *Nanotechnology* **2006**, *17*, 1–13. [CrossRef]

34. Ninjbadgar, T.; Garnweitner, G.; Boärger, A.; Goldenberg, L.M.; Sakhno, O.V.; Stumpe, J. Synthesis of luminescent ZrO_2:Eu^{3+} nanoparticles and their holographic sub-micrometer patterning in polymer composites. *Adv. Funct. Mater.* **2009**, *19*, 1819–1825. [CrossRef]

35. Kim, J.; Kim, H.S.; Lee, N.; Kim, T.; Kim, H.; Yu, T.; Song, I.C.; Moon, W.K.; Hyeon, T. Multifunctional Uniform Nanoparticles Composed of a Magnetite Nanocrystal Core and a Mesoporous Silica Shell for Magnetic Resonance and Fluorescence Imaging and for Drug Delivery. *Angew. Chemie* **2008**, *120*, 8566–8569. [CrossRef]

36. Zumbuehl, A.; Jeannerat, D.; Martin, S.E.; Sohrmann, M.; Stano, P.; Vigassy, T.; Clark, D.D.; Hussey, S.L.; Peter, M.; Peterson, B.R.; et al. An amphotericin B-fluorescein conjugate as a powerful probe for biochemical studies of the membrane. *Angew. Chemie Int. Ed.* **2004**, *43*, 5181–5185. [CrossRef]

37. Gao, X.; Nie, S. Molecular profiling of single cells and tissue specimens with quantum dots. *Trends Biotechnol.* **2003**, *21*, 371–373. [CrossRef]

38. Yong, K.T.; Ding, H.; Roy, I.; Law, W.C.; Bergey, E.J.; Maitra, A.; Prasad, P.N. Imaging pancreatic cancer using bioconjugated inp quantum dots. *ACS Nano* **2009**, *3*, 502–510. [CrossRef]

39. Yang, P.; Quan, Z.; Hou, Z.; Li, C.; Kang, X.; Cheng, Z.; Lin, J. A magnetic, luminescent and mesoporous core-shell structured composite material as drug carrier. *Biomaterials* **2009**, *30*, 4786–4795. [CrossRef]

40. Green, W.H.; Le, K.P.; Grey, J.; Au, T.T.; Sailor, M.J. White phosphors from a silicate-carboxylate sol-gel precursor that lack metal activator ions. *Science* **1997**, *276*, 1826–1828. [CrossRef]

41. Zawisza, K.; Strzep, A.; Wiglusz, R.J. Influence of annealing temperature on the spectroscopic properties of hydroxyapatite analogues doped with Eu3+. *New J. Chem.* **2017**, *41*, 9990–9999. [CrossRef]

42. Ogi, T.; Kaihatsu, Y.; Iskandar, F.; Wang, W.N.; Okuyama, K. Facile synthesis of new full-color-emitting BCNO phosphors with high quantum efficiency. *Adv. Mater.* **2008**, *20*, 3235–3238. [CrossRef]

43. Angelov, S.; Stoyanova, R.; Dafinova, R.; Kabasanov, K. Luminescence and EPR studies on strontium carbonate obtained by thermal decomposition of strontium oxalate. *J. Phys. Chem. Solids* **1986**, *47*, 409–412. [CrossRef]

44. Sudarsanan, K.; Young, R.A. Significant precision in crystal structural details. Holly Springs hydroxyapatite. *Acta Crystallogr. Sect. B Struct. Crystallogr. Cryst. Chem.* **1969**, *25*, 1534–1543. [CrossRef]

45. Martin, P.; Carlot, G.; Chevarier, A.; Den-Auwer, C.; Panczer, G. Mechanisms involved in thermal diffusion of rare earth elements in apatite. *J. Nucl. Mater.* **1999**, *275*, 268–276. [CrossRef]

46. Zhang, H.; Ye, X.J.; Li, J.S. Preparation and biocompatibility evaluation of apatite/wollastonite-derived porous bioactive glass ceramic scaffolds. *Biomed. Mater.* **2009**, *4*, 045007. [CrossRef] [PubMed]

47. Doat, A.; Pellé, F.; Gardant, N.; Lebugle, A. Synthesis of luminescent bioapatite nanoparticles for utilization as a biological probe. *J. Solid State Chem.* **2004**, *177*, 1179–1187. [CrossRef]

48. Lebugle, A.; Pellé, F.; Charvillat, C.; Rousselot, I.; Chane-Ching, J.Y. Colloidal and monocrystalline Ln^{3+} doped apatite calcium phosphate as biocompatible fluorescent probes. *Chem. Commun.* **2006**, *6*, 606–608. [CrossRef]

49. Lubojanski, A.; Dobrzynski, M.; Nowak, N.; Rewak-Soroczynska, J.; Sztyler, K.; Zakrzewski, W.; Dobrzynski, W.; Szymonowicz, M.; Rybak, Z.; Wiglusz, K.; et al. Application of Selected Nanomaterials and Ozone in Modern Clinical Dentistry. *Nanomaterials* **2021**, *11*, 259. [CrossRef]

50. Wiglusz, R.J.; Grzyb, T.; Lis, S.; Strek, W. Hydrothermal preparation and photoluminescent properties of $MgAl_2O_4$:Eu^{3+} spinel nanocrystals. *J. Lumin.* **2010**, *130*, 434–441. [CrossRef]

51. Wiglusz, R.J.; Kedziora, A.; Lukowiak, A.; Doroszkiewicz, W.; Strek, W. Hydroxyapatites and Europium(III) doped hydroxyapatites as a carrier of silver nanoparticles and their antimicrobial activity. *J. Biomed. Nanotechnol.* **2012**, *8*, 605–612. [CrossRef] [PubMed]

52. Targonska, S.; Rewak-Soroczynska, J.; Piecuch, A.; Paluch, E.; Szymanski, D.; Wiglusz, R.J. Preparation of a New Biocomposite Designed for Cartilage Grafting with Antibiofilm Activity. *ACS Omega* **2020**, *5*, 24546–24557. [CrossRef] [PubMed]

53. Pazik, R.; Wiglusz, R.J.; Strek, W. Luminescence properties of $BaTiO_3$:Eu^{3+} obtained via microwave stimulated hydrothermal method. *Mater. Res. Bull.* **2009**, *44*, 1328–1333. [CrossRef]

54. Brunner, T.J.; Wick, P.; Manser, P.; Spohn, P.; Grass, R.N.; Limbach, L.K.; Bruinink, A.; Stark, W.J. In vitro cytotoxicity of oxide nanoparticles: Comparison to asbestos, silica, and the effect of particle solubility. *Environ. Sci. Technol.* **2006**, *40*, 4374–4381. [CrossRef] [PubMed]

55. Núñez, M.T.; Garate, M.A.; Arredondo, M.; Tapia, V.; Muñoz, P. The cellular mechanisms of body iron homeostasis. *Biol. Res.* **2000**, *33*, 133–142. [CrossRef]

56. Sousanis, A.; Poelman, D.; Smet, P.F. SmS/EuS/SmS Tri-Layer Thin Films: The Role of Diffusion in the Pressure Triggered Semiconductor-Metal Transition. *Nanomaterials* **2019**, *9*, 1513. [CrossRef]

57. Cao, H.; Zhang, L.; Zheng, H.; Wang, Z. Hydroxyapatite nanocrystals for biomedical applications. *J. Phys. Chem. C* **2010**, *114*, 18352–18357. [CrossRef]

58. Tordoff, M.G. Calcium: Taste, intake, and appetite. *Physiol. Rev.* **2001**, *81*, 1567–1597. [CrossRef]

59. Grynkiewicz, G.; Poenie, M.; Tsien, R.Y. A New Generation of Ca^{2+} Indicators with Greatly Improved Fluorescence Properties. *J. Biol. Chem.* **1985**, *260*, 3440–3450. [CrossRef]

60. Tsien, R.Y.; Poenie, M. Fluorescence ratio imaging: A new window into intracellular ionic signaling. *Trends Biochem. Sci.* **1986**, *11*, 450–455. [CrossRef]

61. Buurman, E.P.; Sanders, R.; Draaijer, A.; Gerritsen, H.C.; van Veen, J.J.F.; Houpt, P.M.; Levine, Y.K. Fluorescence lifetime imaging using a confocal laser scanning microscope. *Scanning* **1992**, *14*, 155–159. [CrossRef]

62. Lakowicz, J.R.; Szmacinski, H.; Nowaczyk, K.; Johnson, M.L. Fluorescence lifetime imaging of free and protein-bound NADH. *Proc. Natl. Acad. Sci. USA* **1992**, *89*, 1271–1275. [CrossRef] [PubMed]
63. Morgan, C.G.; Mitchell, A.C.; Murray, J.G. Fluorescence decay time imaging using an imaging photon detector with a radio frequency photon correlation system. In *Proceedings of the Time-Resolved Laser Spectroscopy in Biochemistry II*; SPIE: Bellingham, WA, USA, 1990; Volume 1204, pp. 798–807. [CrossRef]
64. Wiatrak, B.; Sobierajska, P.; Szandruk-Bender, M.; Jawien, P.; Janeczek, M.; Dobrzynski, M.; Pistor, P.; Szelag, A.; Wiglusz, R.J. Nanohydroxyapatite as a biomaterial for peripheral nerve regeneration after mechanical damage—in vitro study. *Int. J. Mol. Sci.* **2021**, *22*, 4454. [CrossRef]
65. Zakrzewski, W.; Dobrzynski, M.; Nowicka, J.; Pajaczkowska, M.; Szymonowicz, M.; Targonska, S.; Sobierajska, P.; Wiglusz, K.; Dobrzynski, W.; Lubojanski, A.; et al. The influence of ozonated olive oil-loaded and copper-doped nanohydroxyapatites on planktonic forms of microorganisms. *Nanomaterials* **2020**, *10*, 1997. [CrossRef]
66. Wiglusz, R.J.; Bednarkiewicz, A.; Strek, W. Synthesis and optical properties of Eu^{3+} ion doped nanocrystalline hydroxya patites embedded in PMMA matrix. *J. Rare Earths* **2011**, *29*, 1111–1116. [CrossRef]
67. Pazik, R.; Tekoriute, R.; Håkansson, S.; Wiglusz, R.; Strek, W.; Seisenbaeva, G.A.; Gun'ko, Y.K.; Kessler, V.G. Precursor and solvent effects in the nonhydrolytic synthesis of complex oxide nanoparticles for bioimaging applications by the ether elimination (Bradley) reaction. *Chem. A Eur. J.* **2009**, *15*, 6820–6826. [CrossRef] [PubMed]

Review

Selected Nanomaterials' Application Enhanced with the Use of Stem Cells in Acceleration of Alveolar Bone Regeneration during Augmentation Process

Wojciech Zakrzewski [1], Maciej Dobrzynski [2], Zbigniew Rybak [1], Maria Szymonowicz [1] and Rafal J. Wiglusz [3,*]

[1] Department of Experimental Surgery and Biomaterial Research, Wroclaw Medical University, Bujwida 44, 50-345 Wroclaw, Poland; wojciech.zakrzewski@student.umed.wroc.pl (W.Z.); zbigniew.rybak@umed.wroc.pl (Z.R.); maria.szymonowicz@umed.wroc.pl (M.S.)

[2] Department of Conservative Dentistry and Pedodontics, Wroclaw Medical University, Krakowska 26, 50-425 Wroclaw, Poland; maciej.dobrzynski@umed.wroc.pl

[3] Institute of Low Temperature and Structure Research, Polish Academy of Sciences, Okolna 2, 50-422 Wroclaw, Poland

* Correspondence: r.wiglusz@intibs.pl; Tel.: +48-071-395-42-74; Fax: +48-071-344-10-29

Received: 31 May 2020; Accepted: 16 June 2020; Published: 22 June 2020

Abstract: Regenerative properties are different in every human tissue. Nowadays, with the increasing popularity of dental implants, bone regenerative procedures called augmentations are sometimes crucial in order to perform a successful dental procedure. Tissue engineering allows for controlled growth of alveolar and periodontal tissues, with use of scaffolds, cells, and signalling molecules. By modulating the patient's tissues, it can positively influence poor integration and healing, resulting in repeated implant surgeries. Application of nanomaterials and stem cells in tissue regeneration is a newly developing field, with great potential for maxillofacial bony defects. Nanostructured scaffolds provide a closer structural support with natural bone, while stem cells allow bony tissue regeneration in places when a certain volume of bone is crucial to perform a successful implantation. Several types of selected nanomaterials and stem cells were discussed in this study. Their use has a high impact on the efficacy of the current and future procedures, which are still challenging for medicine. There are many factors that can influence the regenerative process, while its general complexity makes the whole process even harder to control. The aim of this study was to evaluate the effectiveness and advantage of both stem cells and nanomaterials in order to better understand their function in regeneration of bone tissue in oral cavity.

Keywords: stem cells; nanomaterials; bone augmentation; nanohydroxyapatite

---

1. Introduction

Nowadays, the progress that has been made in dental surgery allows for far developed tissue regeneration in oral cavity and is expected to expand in the nearest future. Because implant dentistry has become a desirable option for replacement of missing teeth, the effectiveness of this technique is mostly dependent on the proper quality and quantity of alveolar bone [1]. The excessive bone loss forbids the placement of dental implants in the ideal prosthetic position [2]. Among many undesirable conditions, bone may be compromised owing to tumour, trauma, periodontal disease, and so on. It was confirmed by Neophytos D. et al. [3] that alveolar bone with a width of 5 mm requires augmentation procedure before successful implant placement.

Bone is biologically privileged tissue, because it has the capacity to undergo regeneration as a part of repair process [4].

There are several bone manipulation techniques that are used to achieve a predictable long-term success for dental implants, and as is later on explained in this study, autologous bone graft still remains the "gold standard" for the process of bone augmentation, as it is characterised by the most effective osteogenic, osteoconductive, osteoinductive, and immunogenic properties [5].

Lack of dental tissue in the alveolar ridge is eventually destructive for either maxilla or mandible, which will be discussed later in this study. When it comes to optimal implant and periodontal aesthetics, preservation of the labial appearance of the alveolar process in frontal region of maxilla and mandible is crucial [6].

It is important to underline that, when there is an insufficient amount of bone to sustain primary and/or secondary implant stability, the alveolar ridge needs to be augmented before placement of the implant. Inadequate alveolar bone height and width often require bone manipulation before, at the time of, or even after the implant surgery. In the case the bone is reduced, but there is enough of it for primary stability of implant, it is possible to directly cover the parts of implant that are still exposed after implant placement with bone graft [7]. In the case of implants, primary stability is crucial, and inability to acquire such a status is one of the most important contraindications for patient implantation.

Among grafts, it is possible to distinguish among the following: autologous grafts, allografts, and xenografts, which will be thoroughly explained in the study.

Alveolar process is a bony ridge that is present on both maxillary and mandibular bone.

The aim of this study is to summarise current research on bone tissue engineering for the clinician with a focus on stem cells, and to review the success of bone augmentation with their help. This work also attempts to confirm the utility of nanomaterials and stem cell-based therapies in acceleration of bone augmentation processes.

2. Changes of the Alveolar Process Following Extraction

2.1. Degradation Period

Generally, in order to avoid the degradation processes of bone after extraction, and preserve a proper extraction socket architecture, it is recommended to apply the technique of immediate implant placement at the time of extraction. It is commonly known that the bone modelling process occurring in alveolar process after tooth extraction should be avoided. The bone develops together with teeth, which influence its volume and shape. The alveolar bone supporting the teeth is characterised by distinctive features like rapid and continuous remodeling in response to stimuli by force [8–10]. According to Bodic F. et al. [11], the alveolar bone is subjected to mechanical loads for only 15–20 min per day. Because both maxilla and mandible are tooth-dependent tissues, after loss of tooth, it reacts with a reduction of alveolar ridge in both apicocoronal and buccolingual dimensions [12,13] owing to compromised blood supply and resorption of thin bundles of bone during healing [14]. The process is called atrophy. Changes in the alveolar ridge quickly result in alterations of soft tissue (gingiva), which is attached directly to the former structure. Major changes in the extraction site tend to occur within the first 12 months after the extraction [12,15], while the loss of height of the alveolar bone occurs during the first 3 months. These processes occur because of the fact that tissue must adapt its mass and structure to changing mechanical demands. In the absence of stimuli, such as forces derived from swallowing and mastication, the alveolar bone undergoes resorption [16]. After a tooth extraction, there is a cascade of inflammatory reactions that are activated, while the extraction socket is temporarily closed by the blood clot. Although tissue integrity is quickly restored, the residual ridge is being formed, which has life-long catabolic remodeling effects on either maxilla or mandible. There is no major difference in bony tissue degradation development when it comes to different regions of extraction sockets in oral cavity.

2.2. Healing Period

The alveolar bone healing process is complex, and its effectiveness depends on the efficacy of hosts' inflammatory response [16]. The bone microarchitecture analysis allows to evolve healing with trabecular thickness, and progressively increase the number over time [10]. The healing process starts within 12 months after tooth extraction [12], while reorganisation of lamina dura takes place throughout its duration. Because oral tissues are located in an environment rich in microorganisms, the healing process is always impaired by the lack of sterility [17]. The alveolar bone healing process usually occurs without histological cartilage formation, while long bone healing is a process of endochondral ossification [18].

Fracture healing is a process in which the restored bone shows a lack of scar tissue and formation of blood clot is a crucial step in order to begin such a healing process [19]. Cascade of reactions following blood clot formation allows effective tissue healing in alveolar bone, because platelets in the clot carry specific growth factors (GFs). Both osteoblasts and osteoclasts have direct contact with lymphocytes, which suggests a regulatory role of the immune system, especially in later stages of the healing process [19]. Non-alveolar and alveolar bones differ in nature of the cells surrounding them. For example, the latter lacks muscle stem cells which play essential role in fracture healing [20,21].

In the later stages of the healing process, the blood clot is being replaced with granulation tissue [22], which then leads to the development of newly formed vessels. When modelling processes commence, at first, apical and lateral walls of the alveolus are restored. Later, the healing process goes toward the centre and the coronal region of the alveolus. According to Scala A et al. [23], it takes around one month to close the extraction socket with a newly formed bone. The process eventually ends with corticalization of the socket and formation of bone marrow.

Residual ridge is an alveolar process that is formed after healing of soft tissues and bone, followed by extractions. Although it is a life-long process, the reduction of bone is most aggressive during the first 6 months. Residual ridge resorption (RRR) depends mostly on the site of the ridge and occurs differently among individuals. The basic structural change in RRR is about reduction of the size of the ridge under mucoperiosteum.

When a structure, for instance, maxillary bone, undergoes stress, it becomes deformed. The mechanical aspect of bone remodelling is mostly associated with Wolffs's law [21] of bone transformation, which simply says that bone remodels in response to the forces applied, although this explanation describes very briefly such a complex physiological process like bone remodelling.

The biggest amount of bone loss occurs in the horizontal dimension and happens mainly on the facial aspect of the ridge. On the other hand, vertical ridge height occurs most intensely on the buccal aspect [14]. Long-term lack of tooth in the bone results in increased narrowing and shortening of the ridge, which generally relocates palatally/lingually. Stages of alveolar ridge reduction after tooth loss are presented in Figure 1.

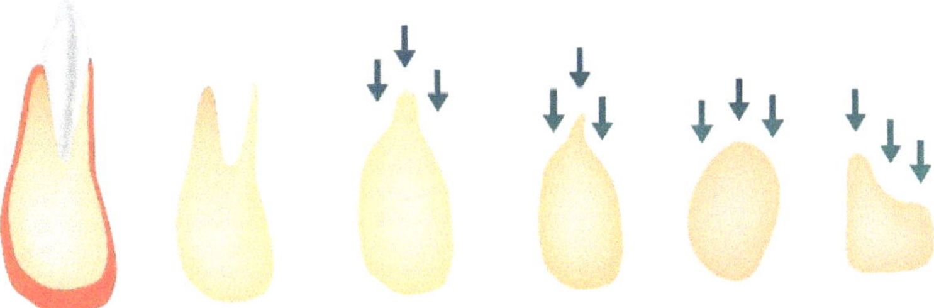

Figure 1. Stages of alveolar ridge reduction after tooth loss, order 1—pre extraction, order 2—post extraction, order 3—high, well-rounded, order 4—knife edge, order 5—low, well-rounded, order 6—depressed.

3. Factors Influencing Alveolar Bone Loss and Regeneration

3.1. Bone Loss Factors

After tooth extraction, the alveolar ridge undergoes uneven atrophy processes [22,25]. Although loss of teeth results in naturally irreversible alveolar bone resorption [26], the destructive process of the bone may start even before extraction of the tooth. It may be complicated with gingivitis present, which leads to periodontopathy, endodontic lesions, or trauma injury. After such a situation, further loss of bony tissue owing to extraction may result in severe complications that occur more quickly than in a case in which the bone stays intact before tooth removal.

Bone degradation can be also caused by several metabolic bone diseases, like vitamin D-resistant rickets (VDRR), focal infections, hyperparathyroidism, age-related parietal bone atrophy, or Paget's disease. VDRR for instance, is a disease affecting mainly dentin, while enamel remains unchanged [27]. Spontaneous pulpal abscesses, which are formed without carious lesions, are detectable. Big tubular clefts in the region of pulpal horns are visible, with submicroscopic defects in the enamel layer, leading to facilitated invasion of bacterial toxins [28]. Increased bacterial invasion of teeth results in accelerated teeth loss, eventually causing accelerated bone resorption.

3.2. Bone Regeneration Factors

Despite bone having a mineral nature, it is a vital and dynamic organ. The histogenesis of bone is directly from mesenchymal connective tissue in the intramembranous bone formation process, and from pre-existing cartilage in endochondral bone formation. Following tooth removal, the normal healing process takes approximately 40 days, starting with clot formation and culminating in a socket filled with bone covered by connective tissue and epithelium [29,30]. The biological principles of bone regeneration comprise the following: osteoinduction, osteogenesis, and osteoconduction. Optimization of these processes has been the goal of new materials used in hard tissue engineering. Osteoinduction process allows migration, followed by proliferation of unspecialised connective tissue cells into bone-forming cell lineage [31]. It induces osteogenesis [32] and GFs determine its action [33].

During osteogenesis, a formation of new bone from both Haversian systems and osteoblastic cells of the grafted bone takes place [34,35]. A direct transfer of vita cells to the area that will regenerate new bone occurs.

Osteoconduction, which is the last of the aforementioned principles, focuses on bone growth on a surface. It implies recruitment of non-adult cells and their stimulation to preosteoblasts [36]. The osteoconduction process means bone growth on a surface. It is a phenomenon often seen in the case of bone implants. Because its function is to provide space and substrates for the biochemical and cellular event progressing the bone formation process, it eventually results in osteogenesis [32,37]. After successful implant insertion, proper biologic width and aesthetics should allow for remodelling of the soft tissue and bone, occurring between 6 months and 1 year [38].

3.3. Osteoinductive Factors

Bone repair is a multistep process that involves migration, differentiation, and activation of considerable amount of cell types [39,40]. Taking into consideration that bone tissue is highly vascularised, it requires both bone tissues and blood vessels to be formed in a tight integrity [41]. Current bone regenerative strategies pursue mimicking natural bone regeneration. Bone morphogenetic proteins (BMPs) and vascular endothelial growth factors (VEGFs) are two key regulators of osteogenesis and angiogenesis, acting by promoting osteogenic and endothelial differentiation of stem cells, respectively [42,43]. Both factors act synergistically during bone regeneration.

BMPs are one of the most researched and crucial morphogenetic signals coordinating tissue architecture in the whole organism. Having the appropriate concentration and being placed on specific scaffold, they are capable of inducing new bone formation by turning mesenchymal stem cells into chondroblasts and osteoblasts [44]. BMPs are being increasingly used in surgeries.

They belong to the transforming growth factor (TGF) beta superfamily [45]. There are currently at least 20 members of the aforementioned family and, among them, BMP-2 is one of the most common factors in use.

Among the most useful functions of BMPs, one can distinguish among the following: induction of cell replication, chemotaxis, induction of differentiation, anchorage-dependent cell attachment, osteocalcin synthesis/mineralisation [46], and alkaline phosphatase activity [47]. Recent studies confirmed that using recombinant BMP in order to correct bony defects, furcations, and fenestration leads to periodontal regeneration with ankylosis [48]. On the contrary, when BMP-7 was used in the augmentation process, it resulted in serious periodontal increase without ankylosis. BMPs can also be used to alleviate implant wound healing. Rutherford et al. [49] have shown that application of Osteogenic protein-1 (OP-1) around the extraction socket escalated bone growth measured histologically at 3 weeks.

VEGF is the signal protein produced by cells, having the ability of vasculogenesis and angiogenesis. It additionally mediates osteogenesis [50]. Street et al. [40] prove, that localised VEGF delivery is beneficial even for osteoblasts migration and bone turnover. This means that, delivered to a bone defect, it is an effective strategy to accelerate bone healing [51]. Additionally, VEGF has been successfully used to improve maturation of newly formed bone. VEGF belongs to a sub-family of GFs, that is, the platelet-derived growth factor family of cystine-knot GFs. The serum concentration of VEGF increases in chronic hypoxic conditions like diabetes mellitus [52], because it is a part of the system responsible for restoring oxygen supply in the case of inadequate oxygenation of tissues.

4. Augmentation Techniques

The augmentation procedure in dentistry is aimed to increase the volume of alveolar bone, particularly when placement of intrabony implant would otherwise be considered problematic [53]. In order to regenerate a sufficient amount of bone to allow successful implant placement, a ridge augmentation technique is recommended. The intramembranous bone formation pathway is used when intraoral bone augmentation techniques are applied by the dental surgeon. The bone augmentation technique, which is used in order to reconstruct different alveolar ridge defects, depends on the horizontal and vertical extent of the defect. Predictability of corrective procedures is influenced by the span of the edentulous ridge and amount of attachment on neighbouring teeth [54].

4.1. Guided Bone Regeneration (GBR)

GBR is similar to guided tissue regeneration, but focuses on development of hard tissues. It is a surgical procedure based on using barrier membranes, with or without bone graft/bone substitutes. Bony regeneration by GBR depends on the migration of osteogenic and pluripotential cells to the defect site in bone and exclusion of cells impeding bone formation [55,56]. It is important to underline that, in order to accomplish successful regeneration of a bone defect, the rate of osteogenesis must exceed the rate of fibrogenesis from the surrounding soft tissue [30,57]. The GBR technique requires four principles in order to successfully fill osseous defect-space maintenance for bone in-growth, stability of the fibrin clot to make the uneventful healing possible, exclusion of epithelium and connective tissue to allow space to be filled with bony tissue, and primary wound closure to promote undisturbed healing [58].

The mechanism of GBR is focused on selective in-growth of bone-forming cells into a bone defect region, which is enhanced when adjacent tissue is kept away with a membrane [59]. This additionally allows to protect the wound from both salivary contamination and mechanical disruption.

There are several techniques used in GBR regarding tri-dimensional bony tissue reconstruction. They are all based on packing bone substitutes into the bony defect and covering it with resorbable or non-resorbable membranes.

4.2. Bone Augmentation Methods and Precise Implant Placement

The goal of successful augmentation following tooth loss is to allow performing effective prosthetic replacement that is in harmony with the rest of the adjacent natural dentition.

Resorption of alveolar bone is a natural consequence of tooth loss. This process causes clinical problems, especially in terms of aesthetics. In order to overcome the changes in the oral cavity, it is required to carry out treatment that makes it possible to preserve the natural tissue shape, in order to prepare for prosthetic appliance like an implant [60]. The clinical outcome of implant treatment is challenged especially in compromised bones of elderly patients [61].If more alveolar ridge is preserved, it will guarantee optimal implant placement and proper functioning of prosthetic appliance. Nevertheless, nowadays, clinicians are usually faced with the necessity to place implants in the alveolar bone of smaller volume. Such a situation requires the clinician to carry out a proper pre-treatment with augmentation techniques that will promote a more predictable regenerative outcome [54]. As Figures 2 and 3 show, the properties of alveolar bone can be assisted by several methods.

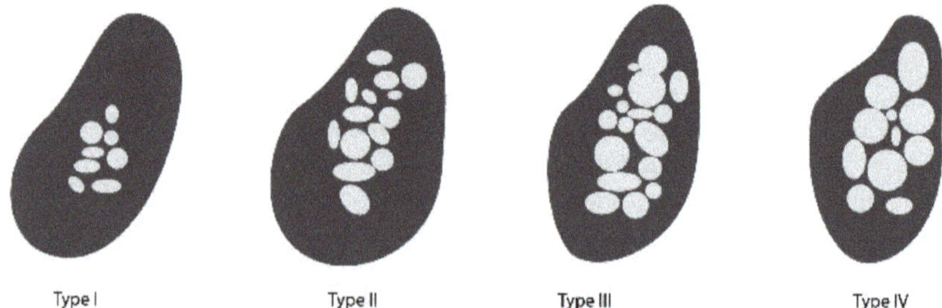

Figure 2. Lekholm and Zarb classification: Type I, whole bone is built of very thick cortical bone; Type II, thick layer of cortical bone surrounds the core of dense trabecular bone; Type III, thin layer of cortical bone surrounds the core of trabecular bone of good strength; Type IV, very thin layer of cortical bone with low density trabecular bone of poor strength.

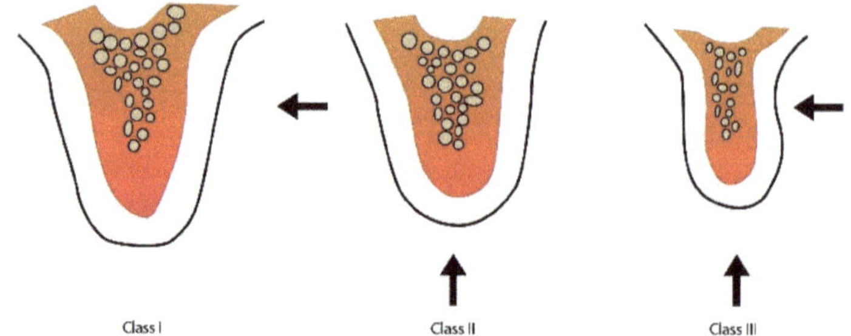

Figure 3. Ridge defect classification of edentulous patients according to Seibert (1983).

The ideal alveolar ridge width and height make the placement of a natural appearing pontic possible, which provides maintenance of a plaque-free environment [62]. The structural loss of the residual alveolar ridge can occur as a result of tooth extraction, surgical procedures, periodontal disease, or congenital defects [63,64]. In a situation with a bone missing, the overlying soft tissue tends to collapse into the bone defect, making it difficult to recreate oral cavity aesthetics after application of functional prostheses. Alveolar deformities classification is based on quantity of volumetric horizontal and vertical tissue loss within the alveolar process. Such a classification was established to standardise communication between clinicians in the selection and sequencing of reconstructive procedures [65]. It is crucial to thoroughly evaluate the contour of the partially edentulous ridge before starting the process of fixed partial denture fabrication. As presented in Figure 3, according to Seibert, Class I represents bucco-lingual loss of tissue with normal height of ridge. Class II defect is represented by the loss of alveolar height in apico-coronal axis. On the other hand, Class III has a combination of bucco-lingual and apico-coronal loss of tissue. According to this classification, the bone augmentation technique is dependent on the horizontal and vertical extent of the defect.

- Class I: bucco-lingual loss of tissue with normal ridge height in an apicocoronal direction;
- Class II: apico-coronal loss of tissue with normal ridge width in a bucco-lingual direction;
- Class III: combination of bucco-lingual and apico-coronal loss of tissue resulting in loss of height and width.

Immediate implant placement can be achieved without additional surgical treatment. However, slight hard tissue augmentation may be needed to add support to periimplant mucosa. There are situations that require soft tissue addition in order to aid maintenance [66,67]. There are several important factors that need to be taken into consideration during planning optimal placement of implants in the alveolar bone, that is, soft and hard tissue management, aesthetic factors, and proper quality of prosthetic restoration.

Conventionally, the placement of dental implants sacrifices much bone tissue during the drilling procedure. However, there are several implantation technique ideas that allow limited bone removal, especially in the case of patients with a limited amount of alveolar bone [68,69].

In order to perform a successful implantation, the dentist has to remember the periimplant values of hard and soft tissues. If the implants are placed too tightly, it will result in insufficient vertical blood supply to the papillae. Angulation of implants is crucial for a proper papilla development afterwards. There is no sufficient support for the papillae in two divergent crowns, while convergent crowns do not allow soft tissue to develop naturally. Tarnow et al. [70] demonstrated a proper relationship with regard to both implant to natural tooth and implant to implant. Regarding the former, in order to avoid horizontal bone loss, which will affect adjacent tooth, the distance should be about 2 mm. The latter requires a distance of at least 3 mm, which, when avoided, creates accelerated bone loss patterns in such areas [71]. It is important to underline that each implant loses periimplant bone within the first year and then stabilises [72].

The crown-to-implant ratio should be 1:1 or less, while the minimum height of the implant is 10–12 mm. The lower height of implant has already been proven to show a high failure rate [73]. In general, the actual height of bone is required to be 12 mm of bone actual height for a macroretentive screw-type implant to properly support occlusal forces [71].

4.3. Membranes in GBR

A GBR membrane acts as a barrier preventing fast-growing soft tissue from invading space required to be filled with a new bone [74]. Membrane materials for bone tissue engineering are usually divided into natural biomaterials [75], like chitosan, inorganic materials represented by nanohydroxyapatite [76], and synthetic polymer materials with polylactide-co-glycolide (PLGA) [77] as an example. Current approaches on graft materials exclusively face serious limitations [78]. Membranes themselves are not able to recover the defects in bone tissues completely, owing to their lack of satisfactory osteoinduction. The results change definitely with the incorporation of osteoinductive factors into semisolid or porous membranes-scaffolds [79,80].

Providing adequate space for bone regeneration is one of the fundamental principles of GBR. Various animal studies have proven that excluding the epithelium and connective tissue makes it possible to create that space, permitting slow migration of osteoblasts to the wound, which results in new bone formation [58,81]. Reinforced membranes allow the space maintenance by preventing membrane collapse that can occur owing to increased pressure from neighbouring tissues.

4.4. Resorbable Membranes

Degradable membranes can be made from collagen (natural), or poly (l-lactic acid) (PLLA) and PLGA [82] (polymeric). Many of these membranes are formulated with antibiotics, usually tetracyclines [83]. There are two important challenges in osteoinduction process that need to be overcome. The first one is to retain the osteogenic factor for a sufficient amount of time to generate the desired biological response, while the second concerns the biocompatibility of the material.

Achieving a desired tissue response is strictly dependent on both degradation components of the extracellular scaffold and concentration of inductive factors released from the matrix.

It is important to underline that bone can only grow when provided with space to do so. Thin, polymeric membranes allow to hold back soft tissues, but provide no significant mechanical support during bone healing. Polymeric adhesive or calcium phosphate cement are required for greater mechanical strength. In such a case, the bone is repaired alongside cement degradation, which can be a challenge.

One of the biggest advantages of resorbable membranes is the fact that they do not require a second surgical procedure over time. They undergo disintegration. On the other hand, it is important to underline, that such materials also have their limitations. Neiva R. et al. [84] have confirmed that using membranes to produce similar gains of keratinized tissue formation was failed in comparison with connective tissue grafts. Collagen material is resorbed by the host with the use of neutrophils and macrophages, causing no inflammation, while expanded polytetrafluoroethylene (ePTFE) material undergoes hydrolysis reaction, which, in some cases, may cause it.

4.5. Autogenous Bone Grafts

The "gold standard" bone graft material in traditional augmentation techniques is autologous bone graft. At the same time, allografts avoid donor site issues, but can cause a higher risk of infection and immune reaction of host tissue [85,86]. Autologous material is better than allograft, because it maintains bone structures, such as minerals, collagen, viable osteoblasts, and BMPs. Although autografts are a gold standard, they still present several significant limitations. Their harvesting is connected with the second concurrent surgical procedure, high donor site morbidity, and resorption [87,88].

Soft-tissue grafts are free gingival grafts that can be harvested from several sites in the patient oral cavity. In the reconstruction of minor alveolar defects, bone grafts from the retromoral region are one of the best intraoral sources possible [89–91]. Surgical operation concerning this region of oral cavity causes minimal discomfort for the patient, has relatively uncomplicated surgical access, and is in proximity of donor and recipient sites, which can lower the anesthesia doses required and cause only minor complications [92]. Its downside poses an unnecessary risk of complications owing to the involvement of the second surgical site. However, when the patient has inadequate thickness of palatal tissues, it is difficult to harvest a sufficient amount to place the graft properly.

Microvascular flap use is one of the states of the art and effective techniques for the repair of significant bone and soft tissue defects. While the vascularised pedicle allows proper perfusion to the harmed area, it also results in complete osseointegration of the bone graft [93]. On the other hand, even this technique represents several disadvantages—it requires a long intraoperative time for the patient, and may result in a permanent deficit, when a muscle or bone are included in the flap [94–96].

4.6. Allografts

Allograft is a tissue graft between individuals of the same specimen, but of nonidentical genetic composition. Generally, cadaver bone is a source for allografts, as it is available in large quantities [97]. One should realise and remember that the aforementioned bone has to undergo multiple treatment sequences in order to make it neutral for the host's immune system and to avoid cross-contamination of disease. Allografts can be used as an alternative, but they have very limited osteogenicity and resorb more rapidly than autogenous bone. It is a useful material in case of patients requiring a non-union type grafting, who have inadequate autograft bone quantity, or when it is hard to obtain tissues from the donors' site.

The main disadvantage of allografts is related to the relatively poor capacity for osteoconduction and osteoconduction, when compared with autologous graft. Another disadvantage of allografts is connected with an improper rate of resorption. It has to be clearly emphasised that bone tissue tends to be resorbed quickly, even after augmentation support, unless loading is provided with dental implant. Implants can be placed at the time of surgical procedure or 6 months later, after the stabilisation of the graft, which minimizes the resorption process.

One of the possible alternatives for soft-tissue grafts is acellular dermal matrix allografts (Alloderm). Alloderm is a donated human dermis, composed of a structurally integrated complex that constitutes basal membrane and an extracellular matrix [98]. Their use reduces the likelihood of cross-infection. Wagshall et al. [99] claim that if a graft material could be used to replace the palatal grafts, then all the possible complications connected to donor site would be immediately eliminated. This would result in alveolar ridge augmentation, being more acceptable for the dental patients.

4.7. Xenograft

Xenografts are a graft specimen from the inorganic portion of animal bones. Bovine is one of the most common sources for their extraction. In order to remove their antigenicity, the removal of the organic component is processed, whereas the remaining inorganic components both provide a natural matrix and serve as an excellent source of calcium. The disadvantage of xenografts is that they are only osteoconductive, and the resorption rate of bovine cortical bone is slow [15].

4.8. Bone Substitute Materials and Genetic Engineering

Genetic engineering can be done in two different ways. The first idea focuses on direct in vivo delivery of genes. There are reported cases in which recombinant human BMP-2 was mixed with an absorbable collagen sponge to treat open long-bone fractures [100]. Govender et al. [101] explained in their study that there was 44% reduction in the risk for failure in healing with less secondary invasive interventions and reduced healing time. The factor used during this study was recombinant BMP-2 called rhBMP-2/ACS. The second idea focuses on using autologous bone alternatives of either animal, human, or synthetic origin [102,103]. Such a technique has its limitations, for example, the risk of bacterial contamination [104], or effectiveness limited mainly to reconstruction of small bony defects [105].

5. Nanomaterials Application in Alveolar Bone Regeneration

5.1. Nanohydroxyapatite (n-HAp)

Nanomaterials, when compared with bulk materials, possess features like macroscopic quantum tunnelling or quantum size, causing altered physiochemical properties [106,107]. In oral biology, nanotechnology applications are mainly focused on augmentation procedures for osseous tissue regeneration and implants osseointegration enhancement [106]. Even though supraphysiological doses are necessary to combat the poor pharmacokinetics of this compounds, the nanocarriers can overcome such limitations by stabilizing the bioactive molecules.

Although bone autografts are considered the "gold standard" in clinical bone repair, they still have several limitations owing to the amount of bone that can be used, as well as an increased risk of donor site morbidity. Because of that, a considerable amount of study has been undertaken in order to develop effective regenerative strategies, leading to bone augmentation with limited side effects [108,109]. The n-HAp has a hierarchical architecture at multiple levels, including macrostructure (cancellous and cortical bone), microstructure (trabeculae), sub-microstructure (lamellae), nanostructure (embedded minerals and fibrillary collagen) [110], and sub-nanostructure (proteins and minerals). The presence of nanotubes or nanocrystals in the composite materials allows for enhancing the mechanical properties of the scaffolds. The nanosized materials present enhanced characteristics, like wettability, charge, roughness, and adsorption of proteins. Moreover, the nanotextured surfaces enhance in vitro osteogenesis and promote mineralization. Furthermore, in the case of the nanomaterials, aqueous contact angles become three times smaller, leading to increased adhesion of the osteoblasts in comparison with micro-sized materials.

Special composition and architecture allow them to have self-regenerative and self-remodelling ability in response to damaging signals and mechanical stimuli [111]. This makes nanomaterial an ideal candidate for bone graft development, as it is capable of recapitulating the organisation of the natural extracellular matrix, in order to regulate bone forming cells activity [112], as can be seen in Figure 4.

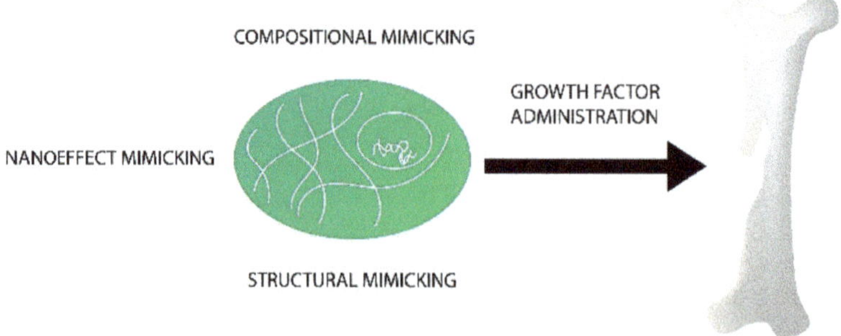

Figure 4. Schematic representation of how a synergistic combination of compositional, nanoelements, well-defined structure, and growth factor administration may endow nanomaterials with a "self-regenerative" capacity for the regeneration of large critical defect bone in a natural bone-healing way, especially at an in vivo level.

Synthetic biomaterials for bone repair should provide mechanical support and biological compatibility in order to promote bone tissue regeneration based on healing. Hydroxyapatite is an interesting inorganic mineral with potential dental [113], maxillofacial [114], and orthopaedic applications [115] that has a typical lattice structure as $(A_{10}(BO_4)_6C_2)$ which defines A, B, and C by Ca, PO_4^{3-}, and OH^- [116]. Hydroxyapatite $(Ca_{10}(PO_4)_6(OH)_2$ is the principal inorganic mineral component of animal and human bones and teeth, and is difficult to dissolve in a solution where the ration of the calcium-to-phosphorus is 1:67 [117]. There are other forms of calcium phosphate present in nature, but HAp is the least soluble of them. The enamel is the hardest substance consisting of relatively large HAp and fluorapatite (FAp) crystals that are 25 nm thick, 40–120 nm wide, and 160 to 1000 nm long. In contrast to enamel, hydroxyapatite is present in bone as plates or needles, while its dimensions range from 40–60 nm long, 20 nm wide, and 1.5 to 5 nm thick.

Nano-HAp is a nanoform of hydroxyapatite with a range of unique properties and diameters ranging in size between 1 and 100 nm [118]. From these dimensions derives a distinct activity of the particles. It has been one of the most studied biomaterials in the medical fields, and has also proven to have strong biocompatibility [119], stability, and nontoxicity. Material possesses an ability of intense ion-exchange against various cations, causing HAp to have high bioactivity [120]. Diversity of n-HAp utility can be seen in Figure 5 and Table 1.

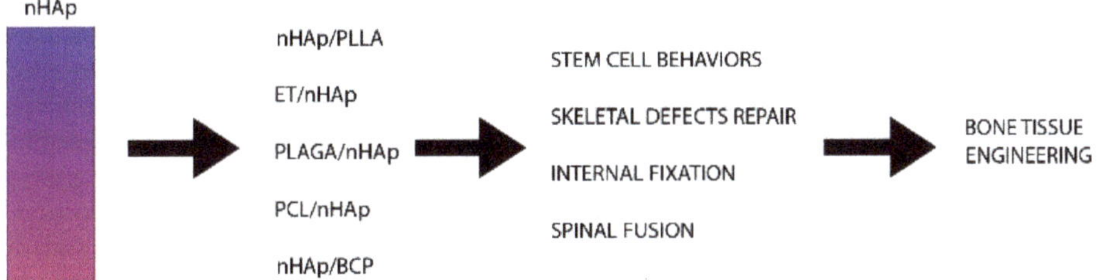

Figure 5. The application of nanohydroxyapatite (n-HAp) scaffolds in bone tissue engineering. PLGA, polylactide-co-glycolide.

5.2. Examples of Bone Regeneration Using Nanohydroxyapatite and Stem Cells in Published Studies

Use of stem cells for bone healing and regeneration still remains in its infancy [121,122]. Composite grafts are able to incorporate osteogenic, osteoconductive and osteoinductive properties onto a compound [123,124]. For instance, local autogenous bone marrow can be harvested in order to combine with a bioceramic material. Also, use of antibiotic-loaded or antimicrobial bone graft substitutes has advantages over nonresorbable antibiotic carriers due to its biodegradability [125]. Nowadays, Nowadays, bone is the second most common transplanted tissue in comparison with blood tissue [126]. Implant bone graft-carrier allows to release the incorporated growth factor at the desirable rate and concentration. Additionally, it can be formed to be structured for facilitate cellular infiltration and growth [127].

Table 1. The application of n-HAp scaffolds in bone tissue engineering. BMP, bone morphogenetic protein.

Material	Application	Merit	Reference
Autograft	Spine fusion	Gold standard	[124,125]
Allograft	Craniofacial bone injury	Osteoinductive, osteoconductive	[126]
BMP	Open tibial fractures	Osteoinduction	[127]
Bioactive glass	Osteomyelitis	Anti-infective carrier	[128]
Composites	Femoral or cancellous bone defects	Biocompatible, tunable physiochemical properties	[129,130]
Synthetic polymers	Spine fusion, loading-bearing sites	Controlled degradation, mechanical strength	[131,132]
Natural polymers	Spinal fusion	Flexible, biocompatible, and biodegradable	[133]
Ceramic	Craniofacial bone defect	Biodegradable, osteointegrative, osteoconductive	[134]
Glass-ceramic	Femoral	Osteogenic	[135]

Dahabreh et al. [128] check in their studies influence of bone graft substitutes on osteoprogenitor cells in terms of proliferation, differentiation and adherence. Moreover, Bojar et al. [129] confirmed effectiveness of alloplastic materials as an alternative for autologous transplants and xenografts in oral surgery and dental implantology.

The chemical composition most of them is hydroxyapatite. However, there is still a doubt to be successfully used in regenerative medicine. Although, the treatment using the allograft tissue is preceded by tissue freezing and freeze-drying as well as sterilization, there is always a risk of disease transmission from a donor [130] or rejection [131]. Furthermore, Kattimani et al. [132] proved that the limitations of allograft's use with stem cells and nanohydroxyapatite has resulted in new alternatives.

Nanosized bioceramics highly active surfaces and size make them a promising platform for bone regeneration. Such ceramics present increased osteoblast adhesion when compared with regular sized ceramics [133]. Their nanometre grain size is responsible for the increased osteoblast functions, like adhesion, proliferation, and differentiation. One of the examples of nanophase ceramics is n-HAp. It has been successfully used, among other applications, as a coating of orthopaedic implants, filler of composites, and bone filler [134]. According to several authors, the aforementioned material, in its pure form, is limited owing to its brittleness [135,136]. Wide ranges of solutions have been proposed in order to compensate problems with nanohydroxyapatite use, like incorporation of

cells are distributed directly into the defect. Nano-sized scaffolds are believed to better control the differentiation process, owing to their interaction with extracellular matrix (ECM) [162].

Loaded GFs, as well as adsorbed drugs on graphene and its derivatives, were able to increase osteogenic differentiation owing to increased local concentration. At the same time, the bone morphogenetic proteins (BMPs) are the most potent osteoinductive proteins for bone tissue regeneration [163].

5.7. Structural Effects of Nanomaterials on Bone Regeneration

At the time of bone regeneration, the porous architecture of scaffolds provides sufficient microenvironments for nutrient/waste exchange, cell proliferation, differentiation, and angiogenesis. The special structure of natural nanomaterials enhances bone with dynamic biological functions and mechanical durability. Nowadays, studies confirm that nanostructures at different dimensional levels play different roles in the regulation of bone regeneration [112]. For instance, during the initial period of implantation, biomaterials must provide structural support in the defect site for bone regeneration. Their nanofeatures act as a good enhancer to acquire proper mechanical properties and stability of osseous regeneration [165]. Nanoparticles can be incorporated into materials to form nanomaterials with adjustable mechanical strength, which can be used to induce stem cells' osteogenic differentiation [166]. Lastly, nanoparticles alone can have the ability to improve osteogenesis for bone regeneration, for instance, Laponite [112]. In such materials, there is a possibility to adjust the conformation of GFs to increase their bioactivity for bone regeneration.

Nanoscaffolds have considerable drug loading abilities, high mobility of drug loaded particles, and efficient in vivo reactivity toward nearby tissues [167]. They can be used for labelling cells, in order to enable monitoring and continuous cell tracking [168], as well as enhancing osteoinduction, osteoconduction, and osseointegration [169].

Tissue engineering and regenerative medicine (TERM) aims to create functional substitutes for diseased and damaged tissues. The strategy behind TERM combines three essential elements, namely, scaffolds, GFs, and stem cells, as can be seen in Figure 6. Scaffolds provide support for tissue formation and are seeded with stem cells. GFs are also included as they regulate the differentiation and proliferation processes [169].

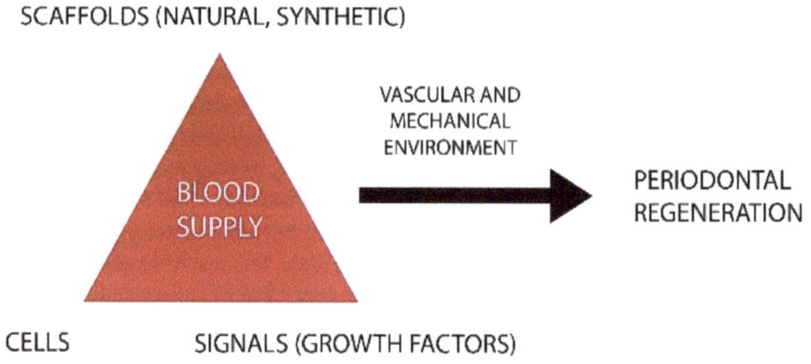

Figure 6. Tissue engineering triad.

5.8. Hydrogels

Microengineering technologies can be successfully used in order to make hydrogel scaffolds mimic in vivo extracellular matrix (ECM). These techniques include lithography, micromolding, biopaterning, and microfluidics. According to Geckil et al. (2010) [170], hydrogels are three-dimensional, insoluble, cross-linked hydrophilic polymeric networks that are capable of providing scaffold to facilitate cell growth, infiltration, and differentiation [171], and can be used to deliver cells with regenerative functions, as pictured in Figure 7. Polymers in hydrogel can absorb a large amount of biological fluid or water with the help of interconnected microscopic pores. In order to increase the biological (hydrophilicity, cell-adhesiveness), mechanical (stiffness, viscoelasticity), and biophysical properties like porosity, combinations of either synthetic or natural hydrogels can be utilized. Such biomaterial composition makes them amenable to surface modification and biomimetic coatings. It is a type of polymer scaffold that has several potential advantages in bone repair. Hydrogels are materials that are able to mimic natural ECM of the bone, allowing to encapsulate bioactive molecules or cells. Network structure of the aforementioned materials allows proteins that are entrapped inside to be confined in the meshes of gel and released as required [172]. Moreover, those materials are absorbable and demonstrate magnificent integration with surrounding tissues, which removes the necessity of its surgical removal and additional trauma [173].

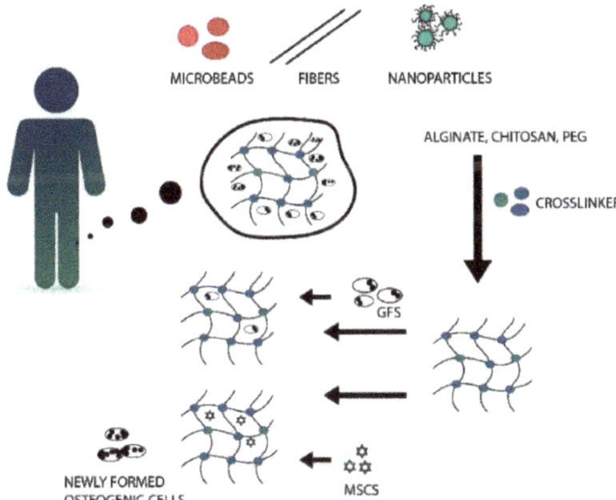

Figure 7. Schematic illustration of hydrogel-assisted bone regeneration. MSC, mesenchymal stem cell; GF, growth factor.

There are still some challenges that require further investigation. A controlled release of encapsulated drugs is one of them. Both burst and delayed release of the drug can affect actual therapeutic effect, and the use of inappropriate polymers can also cause toxic reactions [174]. Mimicking such a 3D-cell microenvironment in vitro with the use of hydrogels is crucial for various applications like constructing tissues for repair.

5.9. Nanostructured Scaffolds for Bone Tissue Engineering

Scaffolds can be utilised in bone tissue engineering in order to deliver biofactors including cells, genes, and proteins to generate bone and assessment of vascularity formation, together with overall tissue maturation [175]. There are three rules that a scaffold needs to comply with in order to be useful in tissue engineering. Firstly, it is required that the scaffold enhances the regenerative capability of the chosen biofactor; secondly, it must provide the correct anatomic geometry in order to maintain space for tissue regeneration; and thirdly, the scaffold needs to provide temporary mechanical load bearing within the specific tissue defect.

Many materials have been proposed as synthetic bone substitutes. Hydroxyapatite is regarded as one of the most bioactive bone substitute materials, mainly because of its superior osteoconductivity. On the other hand, synthetic octacalcium phosphate has been shown to be a good precursor of biological apatite in both teeth and bones, and it also presented better biodegradable and regenerative characteristics when compared with the other calcium phosphate bone substitute materials [176]. Hence, one of the disadvantages of such materials is their inability to achieve close apposition of the material to the neighbouring bony tissue, as well as brittleness of the ceramic materials. This can be overcome by mixing ceramic with, for example, polyesters, in order to form a composite that has good biodegradability, a high affinity for cells of polyesters, as well as osteoconductivity together with mechanical strength of calcium phosphates.

Mechanical properties can be enhanced by cross-linking. Arvidson et al. (2011) [175] give examples of polypropylene fumarate and CaSO$_4$/-TCP materials that are similar to those of cancellous bone substitutes, with compressive strength of 5 MPa and modulus of 50 MPa during degradation [177].

Within the stem cell niche, nanoscale interactions with ECM components form another source of passive mechanical forces that can influence stem cell behaviours [178]. The ECM is built of a broad spectrum of structural polysaccharides and proteins that span over different length scales. Such a connection between stem cells and their nano-environment enables long-term maintenance and control of stem cell behaviour. The possibility to fabricate such small-scale technologies and platforms makes it possible to gain valuable insights into stem cell biomechanics [179].

Currently, scaffolds manufactured from nanotubes, nanoparticles, and nanofibres have emerged as promising candidates for better mimicking the nanostructure of natural ECM. They resemble it, and can be efficiently used to replace damaged tissues [110].

The ideal bone tissue scaffold should be osteogenic, osteoinductive, and osteoconductive [180]. The aforementioned biomimetic efforts include choosing biomaterials that are naturally present in bone, like collagen or hydroxyapatite. Other factors include incorporating growth factors like BMPs and fabricating multiple scale architectures in the scaffold.

It was confirmed by Gong et al. (2015) [178] and Kim et al. (2013) [181] that using nanogrooved matrices mimicking the natural tissues made it possible for the body and nucleus of hMSCs with the sparser nanogrooved pattern to become elongated and orientated more along the direction of nanogrooves than those with the relatively denser nanogroove patterns. The formation of cytoskeleton is crucial for the shape effect on the stem cell differentiation, while a type of synthetic ECM comprised of hierarchically multiscale structures can provide native ECM-like topographical cues for controlling the adhesion and differentiation of hMSCs. Interestingly, the platform that integrates hMSCs into the PLGA scaffold showed potential to regenerate the osseous tissues without the need for further surgical treatments.

Synthetically nanofabricated topography is also a factor that can influence the cell morphology, alignment, adhesion, migration, proliferation, and cytoskeleton organisation [182]. The conclusion is that there is an involvement of cytoskeleton into the stem cells' physiology, suggesting the importance of the force balance along the mechanical axis of the ECM–integrin–cytoskeleton linkage, and their regulation by the mechanical signals in the stem cell niche [183]. Nowadays, there are additional possibilities of printing biocompatible scaffolds using the selective laser melting (SLM) method [184].

6. Tissue Engineering-Stem Cell Application in Bone Augmentation

Stem cells play vital roles in the repair of every tissue of the organism. They are undifferentiated cells, capable of renewing themselves and, by differentiation, they can be induced to develop into many different cell lineages [185]. They have been proved to be promising in tissue regeneration, as well as in the augmentation process. During bone reconstruction procedures, surgeons harvest autologous bone from the patient in order to transplant the graft to the injured site [186]. Moreover, autologous bone grafts still have an unpredictable resorption rate [187]. Nowadays, regeneration of large bony defects is still difficult to manage, despite many advances in bone regeneration treatment [37]. A tissue-engineering model is being promoted as an efficient, "state-of-the-art" technology for major osteogenesis [188,189].

In the view of increasing demands for bone grafting and limitations of "gold standard" procedures, surgeons are looking for a better approach. Tissue engineering allows combining synthetic scaffolds and molecular signals together with mesenchymal or bone marrow stem cells [175] to form hybrid constructs. The classical approach of tissue engineering concerns harvesting stem cells from the bone marrow; then isolating and expanding them; and, at the end, inserting the cells on a suitable synthetic or natural scaffold, before implantation into the same patient [190].

The goal of the modern approach is to reach stem cells present in more accessible sources in the human body, like periodontal ligament or deciduous and permanent teeth. According to Arvidson et al. (2011) [175], it is assumed that the perivascular region in the dental pulp is the niche for pulp-derived stem cells (PDSCs) of mesenchymal origin. In in vitro studies, PDSC are able to regenerate, and they have multilineage potential and plasticity [191] (including chondrocytes and osteoblasts).

Stem cell therapy demonstrates extraordinary value for many severe injuries and diseases. It includes key elements like extracellular matrix scaffolds and stem cells [192]. Furthermore, clinical trials about jaw bone regeneration applied in dental areas have demonstrated positive results [185].

Fortunately, different pre-osteogenic cells types can be used in the practice of bone regeneration. This type of cell is further differentiated into osteogenic cell lineages.

Nowadays, tissue engineering is a process focused on harvesting of multipotent stem cells from an autologous source and their successive in vitro culture. Then, the amount of such cells is increased within the injured tissue [193]. A major disadvantage of the stem cell transplantation is the need for large amounts of cells and accessibility.

There are different kinds of stem cells present in the human body; nevertheless, currently, there are only two main types of them that are used in clinical practice, namely mesenchymal stem cells (MSCs) and hematopoietic stem cells (HSCs).

6.1. Mesenchymal Stem Cells Use

MSCs are multipotent stromal cells that can differentiate into a variety of cells [194], including chondrocytes and osteoblasts. Even though bone marrow was the original source of MSCs, there are alternatives that have been drawn from the other adult's tissues [195,196]. MSCs are nonhematopoietic, which results in their lack of contribution to the formation of blood cells like that of hematopoietic stem cells [197].

MSCs have been isolated from nearly every tissue possible, including brain, spleen, liver, kidney, lung, synovial membrane, or muscles [198–201]. They can be relatively easily expanded and differentiated into multiple tissue lineages, which makes them crucial in present and future tissue engineering [202]. Another key feature of MSCs is their rapid expansion in vitro without loss of their characteristics. Bruder et al. [203] and Cancedda et al. [204] proved that bone marrow stem cells (BMSCs) are even capable of retaining their undifferentiated phenotype for 38 doublings, which eventually results in billion-fold expansion. MSCs transplanted systemically are able to migrate to specific site of injury in animals, which proves their migratory capacity.

They can be identified by the expression of several molecules, including CD105 (SH2) and CD73 (SH3/4). Additionally, they are negative for hematopoietic markers CD34, CD45, and CD14 [205].

Ashton et al. [206] have carried out an interesting experiment. They cultured freshly isolated rabbit marrow cells both in vitro and in diffusion chambers in vivo. The differentiation of osteogenic tissues in the diffusion chambers had to be divided into two categories: the first was the formation of bone in a fibrous layer surrounding cartilage, and the second was an intramembranous bone, formed directly within fibrous tissue that was not associated with cartilage. The authors suggested the presence of osteogenic precursors that had the potential to control the differentiation process via either of two major paths of skeletal development in embryo.

In the past, the MSC differentiation process in vitro involved incubating a confluent monolayer of MSCs together with β-glycerophosphate, ascorbic acid, and dexamethasone for 2–3 weeks [207]. The problem with the use of such factors in order to influence the cells was that it implausibly reflected physiological signals that MSCs received during osteogenesis in vivo. On the other hand, currently, the role of BMPs was investigated, which resulted in the promotion of bone growth in both humans and animals models [208].

6.2. Hematopoietic Stem Cells Use (BMMSCs)

Bone marrow-derived mesenchymal stem cells (BMMSCs) remain the most widely used osteogenic cells in bone tissue engineering research. They are present in adult bone marrow. They can be successfully used as an alternative for bone grafting, because of their immense replicative and differentiation capacity to form numerous connective tissue cells. BMMSCs can be isolated from the iliac crest [196]. Additionally, they can be obtained from orofacial bones, such as mandible and maxilla bone marrow suctioned during dental treatments (dental implantation), orthodontic osteotomy, cyst extirpation, or third molar extraction [209]. It has to be emphasised that, in both human and animal studies, the bone grafted from the craniofacial area was characterised with greater results and higher bone volume than when it was extracted from rib or iliac crest [210,211].

Mashimo et al. [212] positively evaluated alveolar ridge regeneration in the extraction sockets of mice. Stem cells were implanted immediately into the injured area in femur and tibia. Histological analysis proved that, after 3 and 6 weeks, the experimental group contained a greater quantity of bone marrow than the control group. BMMSCs have a number of advantages strictly connected to bone regeneration, including dynamic proliferation, relatively easy isolation, and expansion in vitro, as well as a lack of ethical controversy related to their medical use [213,214].

7. Conclusions

This review mainly focused on summing up the utility of nanomaterials and stem cells in maxillofacial bone tissue regeneration. Many recent scientific works in the fields of tissue engineering were investigated to demonstrate that both nanomaterials and stem cells can be successfully used as materials for guiding cell differentiation, proliferation, and organisation. It is important to underline that all the previously mentioned studies indicate the fact that nanomaterials alone may not be a complete answer to generate successful bone regenerative scaffolds. One of the primary challenges is to use innovative processing technologies in combination with nanomaterials. The construction of an ideal biomimetic nanocomposite would also require incorporation of a hierarchical design. Eventually, it could be possible to obtain an optimal scaffold in combination with several materials and techniques.

Recent advances in the fields of nanotechnology and tissue engineering have established great promise for finding treatments in bone defects and have led to considerable progress in designing and fabricating bone graft substitutes. For instance, impressive progress in the synthesis and functionalisation of graphene materials has opened up new possibilities for exploring their applications in tissue engineering. The primary function of an optimal biomaterial scaffold is to support the area undergoing reconstruction, providing adequate initial mechanical strength. It should also trigger a new bone formation, and later on gradually degrade, without causing an inflammatory response. The selection of the most appropriate scaffolding material is crucial in a tissue-engineered construct.

Scientific works focusing on stem cells have confirmed that mesenchymal stem cells derived from bone marrow are multipotential. They are attractive candidates for cell-based therapy, owing to self-renewal and immunosuppressive properties. Depending on culture conditions, they may differentiate into a plethora of cell types, including osteoblasts and chondrocytes. Thus, MSC comprise a readily available and abundant source of cells for tissue engineering applications. The lack of immunogenicity of MSC has opened up the potential of using those cells in tissue repair. The idea of such strategies is to take advantage of the body's natural ability to repair injured bony tissue with new bone cells, and to remodel the newly formed tissue. Regardless of cell source, live cell-based implants appear to be superior to cell-free alternatives for bone tissue regeneration. Further research should be focused on developing techniques that combine both nanomaterials and stem cells therapies in order to allow even better clinical outcomes in the future.

Nanomaterials **2020**, *10*, 1216

Author Contributions: Conceptualization: M.D. and R.J.W.; Methodology: W.Z., M.D., and R.J.W.; Software: W.Z., M.D., and R.J.W.; Validation: M.D. and R.J.W.; Formal analysis: W.Z., M.D., and R.J.W.; Investigation: W.Z., M.D., Z.R., M.S., and R.J.W.; Data curation: W.Z., M.D., Z.R., M.S., and R.J.W.; Writing—original draft preparation: W.Z., M.D., and. R.J.W.; Writing—review and editing: W.Z., M.D., and R.J.W.; Supervision: M.D. and R.J.W.; Project administration: R.J.W.; Funding acquisition: R.J.W. All authors read and approved the final manuscript.

Funding: This research was funded by National Science Centre (NCN) grant number UMO-2015/19/B/ST5/01330 entitled '*Preparation and characterisation of biocomposites based on nanoapatites for theranostics*' and grant number UMO-2019/33/B/ST5/02247 entitled '*Preparation and modulation of spectroscopic properties of $YXZO_4$, where X and Z-P^{5+}, V^{5+}, As^{5+}, doped with "s^2-like" ions and co-doped with rare earth ions*'.

Conflicts of Interest: The authors declare no conflicts of interest.

References

1. Lekholm, U.; Ericsson, I.; Adell, R.; Slots, J. The condition of the soft tissues at tooth and fixture abutments supporting fixed bridges. A microbiological and histological study. *J. Clin. Periodontol.* **1986**, *13*, 558–562. [CrossRef]

2. Sakkas, A.; Ioannis, K.; Winter, K.; Schramm, A.; Wilde, F. Clinical results of autologous bone augmentation harvested from the mandibular ramus prior to implant placement. An analysis of 104 cases. *GMS Interdiscip. Plast. Reconstr. Surg. DGPW* **2016**, *5*, Doc21. [CrossRef]

3. Demetriades, N.; Park, J.I.; Laskarides, C. Alternative Bone Expansion Technique for Implant Placement in Atrophic Edentulous Maxilla and Mandible. *J. Oral Implantol.* **2011**, *37*, 463–471. [CrossRef]

4. Basa, S.; Varol, A.; Turker, N. Alternative bone expansion technique for immediate placement of implants in the edentulous posterior mandibular ridge: A clinical report. *Int. J. Oral Maxillofac. Implant.* **2004**, *19*, 554–558.

5. Pistilli, R.; Felice, P.; Piatelli, M.; Nisii, A.; Barausse, C.; Esposito, M. Blocks of autogenous bone versus xenografts for the rehabilitation of atrophic jaws with dental implants: Preliminary data from a pilot randomised controlled trial. *Eur. J. Oral Implantol.* **2014**, *7*, 153–171. [PubMed]

6. Goyal, S.; Iyer, S. Bone Manipulation Techniques. *Int. J. Clin. Implant. Dent.* **2009**, *1*, 22–23. [CrossRef]

7. Tonetti, M.S.; Hämmerle, C.H.F. European Workshop on Periodontology Group C Advances in bone augmentation to enable dental implant placement: Consensus Report of the Sixth European Workshop on Periodontology. *J. Clin. Periodontol.* **2008**, *35*, 168–172. [CrossRef] [PubMed]

8. Pagni, G.; Pellegrini, G.; Giannobile, W.V.; Rasperini, G. Postextraction Alveolar Ridge Preservation: Biological Basis and Treatments. *Int. J. Dent.* **2012**, *2012*, 151030. [CrossRef]

9. Javed, A.; Chen, H.; Ghori, F.Y. Genetic and Transcriptional Control of Bone Formation. *Oral Maxillofac. Surg. Clin. N. Am.* **2010**, *22*, 283–293. [CrossRef]

10. Vieira, A.E.; Repeke, C.E.; Junior, S.D.; Colavite, P.M.; Biguetti, C.C.; Oliveira, R.C.; Assis, G.F.; Taga, R.; Trombone, A.P.F.; Garlet, G.P. Intramembranous bone healing process subsequent to tooth extraction in mice: Micro-computed tomography, histomorphometric and molecular characterization. *PLoS ONE* **2015**, *10*, e0128021. [CrossRef]

11. Bodic, F.; Hamel, L.; Lerouxel, E.; Baslé, M.F.; Chappard, D. Bone loss and teeth. *Jt. Bone Spine* **2005**, *72*, 215–221. [CrossRef] [PubMed]

12. Schropp, L.; Wenzel, A.; Kostopoulos, L.; Karring, T. Bone healing and soft tissue contour changes following single-tooth extraction: A clinical and radiographic 12-month prospective study. *Int. J. Periodontics Restor. Dent.* **2003**, *23*, 313–323

13. Devlin, H.; Ferguson, M.W. Alveolar ridge resorption and mandibular atrophy. A review of the role of local and systemic factors. *Br. Dent. J.* **1991**, *170*, 101–104. [CrossRef] [PubMed]

14. Araujo, M.G.; Lindhe, J. Dimensional ridge alterations following tooth extraction. An experimental study in the dog. *J. Clin. Periodontol.* **2005**, *32*, 212–218. [CrossRef] [PubMed]

15. Mittal, Y.; Jindal, G.; Garg, S. Bone manipulation procedures in dental implants. *Indian J. Dent.* **2016**, *7*, 86–94. [CrossRef]

16. Kanyama, M.; Kuboki, T.; Akiyama, K.; Nawachi, K.; Miyauchi, F.M.; Yatani, H.; Kubota, S.; Nakanishi, T.; Takigawa, M. Connective tissue growth factor expressed in rat alveolar bone regeneration sites after tooth extraction. *Arch. Oral Biol.* **2003**, *48*, 723–730. [CrossRef]

17. Thomas, M.V.; Puleo, D.A. Infection, Inflammation, and Bone Regeneration: A Paradoxical Relationship. *J. Dent. Res.* **2011**, *90*, 1052–1061. [CrossRef]

18. DeVlin, H.; Hoyland, J.; Newall, J.F.; Ayad, S. Trabecular Bone Formation in the Healing of the Rodent Molar Tooth Extraction Socket. *J. Bone Miner. Res.* **1997**, *12*, 2061–2067. [CrossRef]

19. Könnecke, I.; Serra, A.; El Khassawna, T.; Schlundt, C.; Schell, H.; Hauser, A.; Ellinghaus, A.; Volk, H.-D.; Radbruch, A.; Duda, G.N.; et al. T and B cells participate in bone repair by infiltrating the fracture callus in a two-wave fashion. *Bone* **2014**, *64*, 155–165. [CrossRef]

20. Einhorn, T.A.; Gerstenfeld, L.C. Fracture healing: Mechanisms and interventions. *Nat. Rev. Rheumatol.* **2015**, *11*, 45–54. [CrossRef]

21. Knight, M.N.; Hankenson, K.D. Mesenchymal Stem Cells in Bone Regeneration. *Adv. Wound Care* **2013**, *2*, 306–316. [CrossRef] [PubMed]

22. Cardaropoli, G.; Araújo, M.; Lindhe, J. Dynamics of bone tissue formation in tooth extraction sites. An experimental study in dogs. *J. Clin. Periodontol.* **2003**, *30*, 809–818. [CrossRef] [PubMed]

23. Scala, A.; Lang, N.P.; Schweikert, M.T.; de Oliveira, J.A.; Rangel-Garcia, I.; Botticelli, D. Sequential healing of open extraction sockets. An experimental study in monkeys. *Clin. Oral Implant. Res.* **2014**, *25*, 288–295. [CrossRef] [PubMed]

24. Frost, H.M. A 2003 update of bone physiology and Wolff's Law for clinicians. *Angle Orthod.* **2004**, *74*, 3–15. [CrossRef]

25. Reich, K.M.; Huber, C.D.; Lippnig, W.R.; Ulm, C.; Watzek, G.; Tangl, S. Atrophy of the residual alveolar ridge following tooth loss in an historical population. *Oral Dis.* **2011**, *17*, 33–44. [CrossRef]

26. Atwood, D.A. Postextraction changes in the adult mandible as illustrated by microradiographs of midsagittal sections and serial cephalometric roentgenograms. *J. Prosthet. Dent.* **1963**, *13*, 810–824. [CrossRef]

27. Hillmann, G.; Geurtsen, W. Pathohistology of undecalcified primary teeth in vitamin D-resistant rickets: Review and report of two cases. *Oral Surg. Oral Med. Oral Pathol. Oral Radiol. Endod.* **1996**, *82*, 218–224. [CrossRef]

28. Bender, I.B.; Naidorf, I.J. Dental observations in vitamin D-resistant rickets with special reference to periapical lesions. *J. Endod.* **1985**, *11*, 514–520. [CrossRef]

29. Amler, M.H.; Johnson, P.L.; Salman, I. Histological and histochemical investigation of human alveolar socket healing in undisturbed extraction wounds. *J. Am. Dent. Assoc.* **1960**, *61*, 32–44. [CrossRef]

30. Liu, J.; Kerns, D.G. Mechanisms of guided bone regeneration: A review. *Open Dent. J.* **2014**, *8*, 56–65. [CrossRef] [PubMed]

31. Albrektsson, T.; Johansson, C. Osteoinduction, osteoconduction and osseointegration. *Eur. Spine J.* **2001**, *10*, S96–S101. [CrossRef]

32. Donaruma, L.G. Definitions in biomaterials, D.F. Williams, Ed., Elsevier, Amsterdam, 1987, 72 pp. *J. Polym. Sci. Polym. Lett. Ed.* **1988**, *26*, 414. [CrossRef]

33. Sykaras, N.; Opperman, L.A. Bone morphogenetic proteins (BMPs): How do they function and what can they offer the clinician? *J. Oral Sci.* **2003**, *45*, 57–73. [CrossRef] [PubMed]

34. Burchardt, H. The biology of bone graft repair. *Clin. Orthop. Relat. Res.* **1983**, *174*, 28–42. [CrossRef]

35. Tonelli, P.; Duvina, M.; Barbato, L.; Biondi, E.; Nuti, N.; Brancato, L.; Rose, G.D. Bone regeneration in dentistry. *Clin. Cases Miner. Bone Metab.* **2011**, *8*, 24–28. [PubMed]

36. Aukhil, I.; Simpson, D.M.; Suggs, C.; Pettersson, E. In vivo differentiation of progenitor cells of the periodontal ligament. *J. Clin. Periodontol.* **1986**, *13*, 862–868. [CrossRef]

37. Lewandrowski, K.U.; Bondre, S.P.; Gresser, J.D.; Wise, D.L.; Tomford, W.W.; Trantolo, D.J. Improved osteoconduction of cortical bone grafts by biodegradable foam coating. *Biomed. Mater. Eng.* **1999**, *9*, 265–275.

38. Bengazi, F.; Wennström, J.L.; Lekholm, U. Recession of the soft tissue margin at oral implants. A 2-year longitudinal prospective study. *Clin. Oral Implant. Res.* **1996**, *7*, 303–310. [CrossRef]

39. Mandracchia, V.J.; Nelson, S.C.; Barp, E.A. Current concepts of bone healing. *Clin. Podiatr. Med. Surg.* **2001**, *18*, 55–77.

40. Street, J.; Bao, M.; deGuzman, L.; Bunting, S.; Peale, F.V.; Ferrara, N.; Steinmetz, H.; Hoeffel, J.; Cleland, J.L.; Daugherty, A.; et al. Vascular endothelial growth factor stimulates bone repair by promoting angiogenesis and bone turnover. *Proc. Natl. Acad. Sci. USA* **2002**, *99*, 9656–9661. [CrossRef]

41. Kanczler, J.M.; Oreffo, R.O.C. Osteogenesis and angiogenesis: The potential for engineering bone. *Eur. Cell. Mater.* **2008**, *15*, 100–114. [CrossRef]

42. Peng, H.; Usas, A.; Olshanski, A.; Ho, A.M.; Gearhart, B.; Cooper, G.M.; Huard, J. VEGF Improves, Whereas sFlt1 Inhibits, BMP2-Induced Bone Formation and Bone Healing Through Modulation of Angiogenesis. *J. Bone Miner. Res.* **2005**, *20*, 2017–2027. [CrossRef]

43. Zhang, W.; Wang, X.; Wang, S.; Zhao, J.; Xu, L.; Zhu, C.; Zeng, D.; Chen, J.; Zhang, Z.; Kaplan, D.L.; et al. The use of injectable sonication-induced silk hydrogel for VEGF165 and BMP-2 delivery for elevation of the maxillary sinus floor. *Biomaterials* **2011**, *32*, 9415–9424. [CrossRef] [PubMed]

44. King, G.N.; Cochrant, D.L. Factors That Modulate the Effects of Bone Morphogenetic Protein-Induced Periodontal Regeneration: A Critical Review. *J. Periodontol.* **2002**, *73*, 925–936. [CrossRef] [PubMed]

45. Even, J.; Eskander, M.; Kang, J. Bone Morphogenetic Protein in Spine Surgery: Current and Future Uses. *J. Am. Acad. Orthop. Surg.* **2012**, *20*, 547–552. [CrossRef]

46. Hughes, F.J.; Turner, W.; Belibasakis, G.; Martuscelli, G. Effects of growth factors and cytokines on osteoblast differentiation. *Periodontol. 2000* **2006**, *41*, 48–72. [CrossRef] [PubMed]

47. Sasikumar, K.P.; Elavarasu, S.; Gadagi, J.S. The application of bone morphogenetic proteins to periodontal and peri-implant tissue regeneration: A literature review. *J. Pharm. Bioallied Sci.* **2012**, *4*, S427–S430. [CrossRef] [PubMed]

48. Saito, A.; Saito, E.; Handa, R.; Honma, Y.; Kawanami, M. Influence of Residual Bone on Recombinant Human Bone Morphogenetic Protein-2–Induced Periodontal Regeneration in Experimental Periodontitis in Dogs. *J. Periodontol.* **2009**, *80*, 961–968. [CrossRef]

49. Rutherford, R.B.; Sampath, T.K.; Rueger, D.C.; Taylor, T.D. Use of bovine osteogenic protein to promote rapid osseointegration of endosseous dental implants. *Int. J. Oral Maxillofac. Implant.* **1992**, *7*, 297–301.

50. Behr, B.; Tang, C.; Germann, G.; Longaker, M.T.; Quarto, N. Locally Applied Vascular Endothelial Growth Factor A Increases the Osteogenic Healing Capacity of Human Adipose-Derived Stem Cells by Promoting Osteogenic and Endothelial Differentiation. *Stem Cells* **2011**, *29*, 286–296. [CrossRef]

51. Kent Leach, J.; Kaigler, D.; Wang, Z.; Krebsbach, P.H.; Mooney, D.J. Coating of VEGF-releasing scaffolds with bioactive glass for angiogenesis and bone regeneration. *Biomaterials* **2006**, *27*, 3249–3255. [CrossRef] [PubMed]

52. Cooper, M.E.; Vranes, D.; Youssef, S.; Stacker, S.A.; Cox, A.J.; Rizkalla, B.; Casley, D.J.; Bach, L.A.; Kelly, D.J.; Gilbert, R.E. Increased renal expression of vascular endothelial growth factor (VEGF) and its receptor VEGFR-2 in experimental diabetes. *Diabetes* **1999**, *48*, 2229–2239. [CrossRef]

53. Fragiskos, F.D.; Alexandridis, C. Osseointegrated Implants. In *Oral Surgery*; Springer: Berlin/Heidelberg, Germany, 2018; pp. 337–347.

54. McAllister, B.S.; Haghighat, K. Bone Augmentation Techniques. *J. Periodontol.* **2007**, *78*, 377–396. [CrossRef] [PubMed]

55. Becker, W.; Becker, B.E.; Handlesman, M.; Celletti, R.; Ochsenbein, C.; Hardwick, R.; Langer, B. Bone formation at dehisced dental implant sites treated with implant augmentation material: A pilot study in dogs. *Int. J. Periodontics Restor. Dent.* **1990**, *10*, 92–101.

56. Dahlin, C.; Linde, A.; Gottlow, J.; Nyman, S. Healing of Bone Defects by Guided Tissue Regeneration. *Plast. Reconstr. Surg.* **1988**, *81*, 672–676. [CrossRef]

57. Gher, M.E.; Quintero, G.; Assad, D.; Monaco, E.; Richardson, A.C. Bone Grafting and Guided Bone Regeneration for Immediate Dental Implants in Humans. *J. Periodontol.* **1994**, *65*, 881–891. [CrossRef]

58. Wang, H.-L.; Boyapati, L. "PASS" Principles for Predictable Bone Regeneration. *Implant Dent.* **2006**, *15*, 8–17. [CrossRef]

59. Kostopoulos, L.; Karring, T. Guided bone regeneration in mandibular defects in rats using a bioresorbable polymer. *Clin. Oral Implant. Res.* **1994**, *5*, 66–74. [CrossRef]

60. Tarnow, D.P.; Eskow, R.N.; Zamzok, J. Aesthetics and implant dentistry. *Periodontol. 2000* **1996**, *11*, 85–94. [CrossRef]

61. Alghamdi, H. Methods to Improve Osseointegration of Dental Implants in Low Quality (Type-IV) Bone: An Overview. *J. Funct. Biomater.* **2018**, *9*, 7. [CrossRef]

62. Rana, R.; Ramachandra, S.S.; Lahori, M.; Singhal, R.; Jithendra, K.D. Combined soft and hard tissue augmentation for a localized alveolar ridge defect. *Contemp. Clin. Dent.* **2013**, *4*, 556–558. [CrossRef] [PubMed]

63. Thoma, D.S.; Hämmerle, C.H.F.; Cochran, D.L.; Jones, A.A.; Görlach, C.; Uebersax, L.; Mathes, S.; Graf-Hausner, U.; Jung, R.E. Soft tissue volume augmentation by the use of collagen-based matrices in the dog mandible—A histological analysis. *J. Clin. Periodontol.* **2011**, *38*, 1063–1070. [CrossRef] [PubMed]

64. Bagavad Gita, V.; Chandrasekaran, S. Hard and soft tissue augmentation to enhance implant predictability and esthetics: 'The perio-esthetic approach'. *J. Indian Soc. Periodontol.* **2011**, *15*, 59. [CrossRef] [PubMed]

65. Seibert, J.S. Reconstruction of deformed, partially edentulous ridges, using full thickness onlay grafts. Part II. Prosthetic/periodontal interrelationships. *Compend. Contin. Educ. Dent.* **1983**, *4*, 549–562. [PubMed]

66. Kan, J.Y.K.; Rungcharassaeng, K.; Umezu, K.; Kois, J.C. Dimensions of Peri-Implant Mucosa: An Evaluation of Maxillary Anterior Single Implants in Humans. *J. Periodontol.* **2003**, *74*, 557–562. [CrossRef]

67. Tolstunov, L. *Horizontal Alveolar Ridge Augmentation in Implant Dentistry: A Surgical Manual*; John Wiley & Sons: San Francisco, CA, USA, 2015; ISBN 9781119019886.

68. Shalabi, M.M.; Wolke, J.G.C.; de Ruijter, A.J.E.; Jansen, J.A. A Mechanical Evaluation of Implants Placed With Different Surgical Techniques Into the Trabecular Bone of Goats. *J. Oral Implantol.* **2007**, *33*, 51–58. [CrossRef]

69. Tabassum, A.; Meijer, G.J.; Wolke, J.G.C.; Jansen, J.A. Influence of the surgical technique and surface roughness on the primary stability of an implant in artificial bone with a density equivalent to maxillary bone: A laboratory study. *Clin. Oral Implant. Res.* **2009**, *20*, 327–332. [CrossRef]

70. Tarnow, D.P.; Cho, S.C.; Wallace, S.S. The Effect of Inter-Implant Distance on the Height of Inter-Implant Bone Crest. *J. Periodontol.* **2000**, *71*, 546–549. [CrossRef]

71. Miloro, M.; Peterson, L.J. *Peterson's Principles of Oral and Maxillofacial Surgery*; People's Medical Pub. House: Shelton, CT, USA, 2012; ISBN 1607951118.

72. Albrektsson, T.; Zarb, G.; Worthington, P.; Eriksson, A.R. The long-term efficacy of currently used dental implants: A review and proposed criteria of success. *Int. J. Oral Maxillofac. Implant.* **1986**, *1*, 11–25.

73. Chan, M.F.; Närhi, T.O.; de Baat, C.; Kalk, W. Treatment of the atrophic edentulous maxilla with implant-supported overdentures: A review of the literature. *Int. J. Prosthodont.* **1998**, *11*, 7–15.

74. Abou Neel, E.A.; Young, A.M. Antibacterial adhesives for bone and tooth repair. In *Joining and Assembly of Medical Materials and Devices*; Elsevier: Philadelphia, PA, USA, 2013; pp. 491–513.

75. Croisier, F.; Jérôme, C. Chitosan-based biomaterials for tissue engineering. *Eur. Polym. J.* **2013**, *49*, 780–792. [CrossRef]

76. Zhou, H.; Lee, J. Nanoscale hydroxyapatite particles for bone tissue engineering. *Acta Biomater.* **2011**, *7*, 2769–2781. [CrossRef] [PubMed]

77. Danhier, F.; Ansorena, E.; Silva, J.M.; Coco, R.; le Breton, A.; Préat, V. PLGA-based nanoparticles: An overview of biomedical applications. *J. Control. Release* **2012**, *161*, 505–522. [CrossRef] [PubMed]

78. Bauer, T.W.; Muschler, G.F. Bone graft materials. An overview of the basic science. *Clin. Orthop. Relat. Res.* **2000**, *371*, 10–27. [CrossRef]

79. Liu, Y.; Lu, Y.; Tian, X.; Cui, G.; Zhao, Y.; Yang, Q.; Yu, S.; Xing, G.; Zhang, B. Segmental bone regeneration using an rhBMP-2-loaded gelatin/nanohydroxyapatite/fibrin scaffold in a rabbit model. *Biomaterials* **2009**, *30*, 6276–6285. [CrossRef] [PubMed]

80. Li, W.; Zara, J.N.; Siu, R.K.; Lee, M.; Aghaloo, T.; Zhang, X.; Wu, B.M.; Gertzman, A.A.; Ting, K.; Soo, C. Nell-1 Enhances Bone Regeneration in a Rat Critical-Sized Femoral Segmental Defect Model. *Plast. Reconstr. Surg.* **2011**, *127*, 580–587. [CrossRef] [PubMed]

81. Nemcovsky, C.E.; Artzi, Z.; Moses, O. Rotated palatal flap in immediate implant procedures. *Clin. Oral Implant. Res.* **2000**, *11*, 83–90. [CrossRef]

82. Lieberman, J.R.; Daluiski, A.; Einhorn, T.A. The role of growth factors in the repair of bone. *J. Bone Jt. Surg.* **2002**, *84*, 1032–1044. [CrossRef]

83. Florjanski, W.; Orzeszek, S.; Olchowy, A.; Grychowska, N.; Wieckiewicz, W.; Malysa, A.; Smardz, J.; Wieckiewicz, M. Modifications of Polymeric Membranes Used in Guided Tissue and Bone Regeneration. *Polymers* **2019**, *11*, 782. [CrossRef]

84. Neiva, W.V.R. Giannobile Barrier Membrane—An Overview|Science Direct Topics. Available online: https://www.sciencedirect.com/topics/medicine-and-dentistry/barrier-membrane (accessed on 9 February 2020).

85. Alman, B.A.; de Bari, A.; Krajbich, J.I. Massive allografts in the treatment of osteosarcoma and Ewing sarcoma in children and adolescents. *J. Bone Jt. Surg.* **1995**, *77*, 54–64. [CrossRef]

86. Kapoor, S.K.; Thiyam, R. Management of infection following reconstruction in bone tumors. *J. Clin. Orthop. Trauma* **2015**, *6*, 244–251. [CrossRef] [PubMed]

87. Lee, K.; Chan, C.K.; Patil, N.; Goodman, S.B. Cell therapy for bone regeneration-Bench to bedside. *J. Biomed. Mater. Res. Part B Appl. Biomater.* **2009**, *89B*, 252–263. [CrossRef] [PubMed]

88. Quarto, R.; Mastrogiacomo, M.; Cancedda, R.; Kutepov, S.M.; Mukhachev, V.; Lavroukov, A.; Kon, E.; Marcacci, M. Repair of Large Bone Defects with the Use of Autologous Bone Marrow Stromal Cells. *N. Engl. J. Med.* **2001**, *344*, 385–386. [CrossRef]

89. Nkenke, E.; Schultze-Mosgau, S.; Radespiel-Tröger, M.; Kloss, F.; Neukam, F.W. Morbidity of harvesting of chin grafts: A prospective study. *Clin. Oral Implant. Res.* **2001**, *12*, 495–502. [CrossRef]

90. Misch, C.M.; Misch, C.E. The repair of localized severe ridge defects for implant placement using mandibular bone grafts. *Implant Dent.* **1995**, *4*, 261–267. [CrossRef] [PubMed]

91. Misch, C.M.; Misch, C.E.; Resnik, R.R.; Ismail, Y.H. Reconstruction of maxillary alveolar defects with mandibular symphysis grafts for dental implants: A preliminary procedural report. *Int. J. Oral Maxillofac. Implant.* **1992**, *7*, 360–366.

92. Clavero, J.; Lundgren, S. Ramus or chin grafts for maxillary sinus inlay and local onlay augmentation: Comparison of donor site morbidity and complications. *Clin. Implant Dent. Relat. Res.* **2003**, *5*, 154–160. [CrossRef]

93. Blanchaert, R.H.; Harris, C.M. Microvascular Free Bone Flaps. *Atlas Oral Maxillofac. Surg. Clin.* **2005**, *13*, 151–171. [CrossRef]

94. Torroni, A. Engineered Bone Grafts and Bone Flaps for Maxillofacial Defects: State of the Art. *J. Oral Maxillofac. Surg.* **2009**, *67*, 1121–1127. [CrossRef]

95. Harii, K. Clinical application of free omental flap transfer. *Clin. Plast. Surg.* **1978**, *5*, 273–281.

96. Schliephake, H. Revascularized tissue transfer for the repair of complex midfacial defects in oncologic patients. *J. Oral Maxillofac. Surg.* **2000**, *58*, 1212–1218. [CrossRef]

97. Kumar, P.; Vinitha, B.; Fathima, G. Bone grafts in dentistry. *J. Pharm. Bioallied Sci.* **2013**, *5*, S125–S127. [CrossRef]

98. Wei, P.-C.; Laurell, L.; Geivelis, M.; Lingen, M.W.; Maddalozzo, D. Acellular Dermal Matrix Allografts to Achieve Increased Attached Gingiva. Part 1. A Clinical Study. *J. Periodontol.* **2000**, *71*, 1297–1305. [CrossRef] [PubMed]

99. Wagshall, E.; Lewis, Z.; Babich, S.B.; Sinensky, M.C.; Hochberg, M. Acellular dermal matrix allograft in the treatment of mucogingival defects in children: Illustrative case report. *J. Dent. Child.* **2002**, *69*, 39–43.

100. Kimelman Bleich, N.; Kallai, I.; Lieberman, J.R.; Schwarz, E.M.; Pelled, G.; Gazit, D. Gene therapy approaches to regenerating bone. *Adv. Drug Deliv. Rev.* **2012**, *64*, 1320–1330. [CrossRef] [PubMed]

101. Govender, S.; Csimma, C.; Genant, H.K.; Valentin-Opran, A.; Amit, Y.; Arbel, R.; Aro, H.; Atar, D.; Bishay, M.; Börner, M.G.; et al. Recombinant human bone morphogenetic protein-2 for treatment of open tibial fractures. *J. Bone Jt. Surg.* **2002**, *84*, 2123–2134. [CrossRef]

102. Hernigou, P. Bone transplantation and tissue engineering. Part II: Bone graft and osteogenesis in the seventeenth, eighteenth and nineteenth centuries (Duhamel, Haller, Ollier and MacEwen). *Int. Orthop.* **2015**, *39*, 193–204. [CrossRef]

103. Gjerde, C.; Mustafa, K.; Hellem, S.; Rojewski, M.; Gjengedal, H.; Yassin, M.A.; Feng, X.; Skaale, S.; Berge, T.; Rosen, A.; et al. Cell therapy induced regeneration of severely atrophied mandibular bone in a clinical trial. *Stem Cell Res. Ther.* **2018**, *9*, 213. [CrossRef]

104. Zimmermann, G.; Moghaddam, A. Allograft bone matrix versus synthetic bone graft substitutes. *Injury* **2011**, *42*, S16–S21. [CrossRef]

105. Calori, G.M.; Mazza, E.; Colombo, M.; Ripamonti, C. The use of bone-graft substitutes in large bone defects: Any specific needs? *Injury* **2011**, *42*, S56–S63. [CrossRef]

106. Walmsley, G.G.; McArdle, A.; Tevlin, R.; Momeni, A.; Atashroo, D.; Hu, M.S.; Feroze, A.H.; Wong, V.W.; Lorenz, P.H.; Longaker, M.T.; et al. Nanotechnology in bone tissue engineering. *Nanomed. Nanotechnol. Biol. Med.* **2015**, *11*, 1253–1263. [CrossRef]

107. Funda, G.; Taschieri, S.; Bruno, G.A.; Grecchi, E.; Paolo, S.; Girolamo, D.; del Fabbro, M. Nanotechnology Scaffolds for Alveolar Bone Regeneration. *Materials* **2020**, *13*, 201. [CrossRef] [PubMed]

108. Mistry, A.S.; Mikos, A.G. Tissue Engineering Strategies for Bone Regeneration. In *Advances in Biochemical Engineering/Biotechnology*; Springer: Berlin/Heidelberg, Germany, 2005; Volume 94, pp. 1–22.

109. Thorpe, A.A.; Creasey, S.; Sammon, C.; Le Maitre, C.L. Hydroxyapatite nanoparticle injectable hydrogel scaffold to support osteogenic differentiation of human mesenchymal stem cells. *Eur. Cell. Mater.* **2016**, *32*, 1–23. [CrossRef] [PubMed]
110. McMahon, R.E.; Wang, L.; Skoracki, R.; Mathur, A.B. Development of nanomaterials for bone repair and regeneration. *J. Biomed. Mater. Res. Part B Appl. Biomater.* **2013**, *101B*, 387–397. [CrossRef] [PubMed]
111. Tang, D.; Tare, R.S.; Yang, L.-Y.; Williams, D.F.; Ou, K.-L.; Oreffo, R.O.C. Biofabrication of bone tissue: Approaches, challenges and translation for bone regeneration. *Biomaterials* **2016**, *83*, 363–382. [CrossRef] [PubMed]
112. Li, Y.; Liu, C. Nanomaterial-based bone regeneration. *Nanoscale* **2017**, *9*, 4862–4874. [CrossRef] [PubMed]
113. Thian, E.S.; Huang, J.; Barber, Z.H.; Best, S.M.; Bonfield, W. Surface modification of magnetron-sputtered hydroxyapatite thin films via silicon substitution for orthopaedic and dental applications. *Surf. Coat. Technol.* **2011**, *205*, 3472–3477. [CrossRef]
114. Notodihardjo, F.Z.; Kakudo, N.; Kushida, S.; Suzuki, K.; Kusumoto, K. Bone regeneration with BMP-2 and hydroxyapatite in critical-size calvarial defects in rats. *J. Cranio-Maxillofac. Surg.* **2012**, *40*, 287–291. [CrossRef]
115. Ahmed, R.; Faisal, N.H.; Paradowska, A.M.; Fitzpatrick, M.E.; Khor, K.A. Neutron diffraction residual strain measurements in nanostructured hydroxyapatite coatings for orthopaedic implants. *J. Mech. Behav. Biomed. Mater.* **2011**, *4*, 2043–2054. [CrossRef]
116. Habibah, T.U.; Salisbury, H.G. *Hydroxyapatite Dental Material*; StatPearls Publishing: Petersburg, FL, USA, 2020.
117. Pan, S.; Yu, H.; Yang, X.; Yang, X.; Wang, Y.; Liu, Q.; Jin, L.; Yang, Y. Application of Nanomaterials in Stem Cell Regenerative Medicine of Orthopedic Surgery. *J. Nanomater.* **2017**, *2017*, 1985942. [CrossRef]
118. Song, W.; Ge, S. Application of Antimicrobial Nanoparticles in Dentistry. *Molecules* **2019**, *24*, 1033. [CrossRef]
119. Pepla, E.; Besharat, L.K.; Palaia, G.; Tenore, G.; Migliau, G. Nano-hydroxyapatite and its applications in preventive, restorative and regenerative dentistry: A review of literature. *Ann. Stomatol.* **2014**, *5*, 108–114. [CrossRef]
120. Al-Hazmi, F.; Alnowaiser, F.; Al-Ghamdi, A.A.; Al-Ghamdi, A.A.; Aly, M.M.; Al-Tuwirqi, R.M.; El-Tantawy, F. A new large—Scale synthesis of magnesium oxide nanowires: Structural and antibacterial properties. *Superlattices Microstruct.* **2012**, *52*, 200–209. [CrossRef] [PubMed]
121. Nauth, A.; Lane, J.; Watson, J.T.; Giannoudis, P. Bone Graft Substitution and Augmentation. *J. Orthop. Trauma* **2015**, *29*, S34–S38. [CrossRef]
122. Kim, S.J.; Shin, Y.W.; Yang, K.H.; Kim, S.B.; Yoo, M.J.; Han, S.K.; Im, S.A.; Won, Y.D.; Sung, Y.B.; Jeon, T.S.; et al. A multi-center, randomized, clinical study to compare the effect and safety of autologous cultured osteoblast(Ossron) injection to treat fractures. *BMC Musculoskelet Disord.* **2009**, *10*, 20. [CrossRef] [PubMed]
123. Leupold, J.A.; Barfield, W.R.; An, Y.H.; Hartsock, L.A. A comparison of ProOsteon, DBX, and collagraft in a rabbit model. *J. Biomed. Mater. Res. Part B Appl. Biomater.* **2006**, *79B*, 292–297. [CrossRef] [PubMed]
124. Gupta, A.; Kukkar, N.; Sharif, K.; Main, B.J.; Albers, C.E., III. Bone graft substites for spine fusion: A brief review. *World J. Orthop.* **2015**, *6*, 449. [CrossRef]
125. Van Vugt, T.A.G.; Geurts, J.; Arts, J.J. Clinical Application of Antimicrobial Bone Graft Substitute in Osteomyelitis Treatment: A Systematic Review of Different Bone Graft Substitutes Available in Clinical Treatment of Osteomyelitis. *Biomed Res. Int.* **2016**, *2016*, 6984656. [CrossRef]
126. Basha, R.Y.; Sampath Kumar, T.S.; Doble, M. Design of biocomposite materials for bone tissue regeneration. *Mater. Sci. Eng. C* **2015**, *57*, 452–463. [CrossRef]
127. Lee, S.-H.; Shin, H. Matrices and scaffolds for delivery of bioactive molecules in bone and cartilage tissue engineering. *Adv. Drug Deliv. Rev.* **2007**, *59*, 339–359. [CrossRef]
128. Dahabreh, Z.; Panteli, M.; Pountos, I.; Howard, M.; Campbell, P.; Giannoudis, P.V. Ability of bone graft substitutes to support the osteoprogenitor cells: An *in-vitro* study. *World J. Stem Cells* **2014**, *6*, 497. [CrossRef]
129. Bojar, W.; Ciach, T.; Kucharska, M.; Maurin, J.; Gruber, B.; Krzysztoń-Russjan, J.; Bubko, I.; Anuszewska, E. Cytotoxicity Evaluation and Crystallochemical Analysis of a Novel and Commercially Available Bone Substitute Material. *Adv. Clin. Exp. Med.* **2015**, *24*, 511–516. [CrossRef] [PubMed]
130. Laurencin, C.; Khan, Y.; El-Amin, S.F. Bone graft substitutes. *Expert Rev. Med. Devices* **2006**, *3*, 49–57. [CrossRef]

131. Oryan, A.; Alidadi, S.; Moshiri, A.; Maffulli, N. Bone regenerative medicine: Classic options, novel strategies, and future directions. *J. Orthop. Surg. Res.* **2014**, *9*, 18. [CrossRef] [PubMed]

132. Kattimani, V.; Lingamaneni, K.P.; Chakravarthi, P.S.; Kumar, T.S.S.; Siddharthan, A. Eggshell-Derived Hydroxyapatite. *J. Craniofac. Surg.* **2016**, *27*, 112–117. [CrossRef]

133. Webster, T.J.; Siegel, R.W.; Bizios, R. Osteoblast adhesion on nanophase ceramics. *Biomaterials* **1999**, *20*, 1221–1227. [CrossRef]

134. Luo, X.; Zhang, L.; Morsi, Y.; Zou, Q.; Wang, Y.; Gao, S.; Li, Y. Hydroxyapatite/polyamide66 porous scaffold with an ethylene vinyl acetate surface layer used for simultaneous substitute and repair of articular cartilage and underlying bone. *Appl. Surf. Sci.* **2011**, *257*, 9888–9894. [CrossRef]

135. Tavakol, S.; Nikpour, M.R.; Amani, A.; Soltani, M.; Rabiee, S.M.; Rezayat, S.M.; Chen, P.; Jahanshahi, M. Bone regeneration based on nano-hydroxyapatite and hydroxyapatite/chitosan nanocomposites: An in vitro and in vivo comparative study. *J. Nanoparticle Res.* **2013**, *15*, 1373. [CrossRef]

136. Fang, L.; Leng, Y.; Gao, P. Processing of hydroxyapatite reinforced ultrahigh molecular weight polyethylene for biomedical applications. *Biomaterials* **2005**, *26*, 3471–3478. [CrossRef]

137. Huang, Z.; Chen, Y.; Feng, Q.-L.; Zhao, W.; Yu, B.; Tian, J.; Li, S.-J.; Lin, B.-M. In vivo bone regeneration with injectable chitosan/hydroxyapatite/collagen composites and mesenchymal stem cells. *Front. Mater. Sci.* **2011**, *5*, 301–310. [CrossRef]

138. Wang, F.; Su, X.-X.; Guo, Y.-C.; Li, A.; Zhang, Y.-C.; Zhou, H.; Qiao, H.; Guan, L.-M.; Zou, M.; Si, X.-Q. Bone Regeneration by Nanohydroxyapatite/Chitosan/Poly(lactide-co-glycolide) Scaffolds Seeded with Human Umbilical Cord Mesenchymal Stem Cells in the Calvarial Defects of the Nude Mice. *Biomed. Res. Int.* **2015**, *2015*, 261938. [CrossRef] [PubMed]

139. Bhuiyan, D.B.; Middleton, J.C.; Tannenbaum, R.; Wick, T.M. Mechanical properties and osteogenic potential of hydroxyapatite-PLGA-collagen biomaterial for bone regeneration. *J. Biomater. Sci. Polym. Ed.* **2016**, *27*, 1139–1154. [CrossRef] [PubMed]

140. Kruger, T.E.; Miller, A.H.; Wang, J. Collagen Scaffolds in Bone Sialoprotein-Mediated Bone Regeneration. *Sci. World J.* **2013**, *2013*, 812718. [CrossRef] [PubMed]

141. Tsai, K.-S.; Kao, S.-Y.; Wang, C.-Y.; Wang, Y.-J.; Wang, J.-P.; Hung, S.-C. Type I collagen promotes proliferation and osteogenesis of human mesenchymal stem cells via activation of ERK and Akt pathways. *J. Biomed. Mater. Res. Part A* **2010**, *94*, 6732013682. [CrossRef] [PubMed]

142. Somaiah, C.; Kumar, A.; Mawrie, D.; Sharma, A.; Patil, S.D.; Bhattacharyya, J.; Swaminathan, R.; Jaganathan, B.G. Collagen Promotes Higher Adhesion, Survival and Proliferation of Mesenchymal Stem Cells. *PLoS ONE* **2015**, *10*, e0145068. [CrossRef]

143. Kalita, S.J.; Bhatt, H.A. Nanocrystalline hydroxyapatite doped with magnesium and zinc: Synthesis and characterization. *Mater. Sci. Eng. C* **2007**, *27*, 837–848. [CrossRef]

144. Best, S.M.; Porter, A.E.; Thian, E.S.; Huang, J. Bioceramics: Past, present and for the future. *J. Eur. Ceram. Soc.* **2008**, *28*, 1319–1327. [CrossRef]

145. Qin, H.; Zhu, C.; An, Z.; Jiang, Y.; Zhao, Y.; Wang, J.; Liu, X.; Hui, B.; Zhang, X.; Wang, Y. Silver nanoparticles promote osteogenic differentiation of human urine-derived stem cells at noncytotoxic concentrations. *Int. J. Nanomed.* **2014**, *9*, 2469. [CrossRef]

146. Mahmood, M.; Li, Z.; Casciano, D.; Khodakovskaya, M.V.; Chen, T.; Karmakar, A.; Dervishi, E.; Xu, Y.; Mustafa, T.; Watanabe, F.; et al. Nanostructural materials increase mineralization in bone cells and affect gene expression through miRNA regulation. *J. Cell. Mol. Med.* **2011**, *15*, 2297–2306. [CrossRef]

147. Hsu, S.; Yen, H.-J.; Tsai, C.-L. The Response of Articular Chondrocytes to Type II Collagen–Au Nanocomposites. *Artif. Organs* **2007**, *31*, 854–868. [CrossRef]

148. Kumar, V.B.; Khajuria, D.K.; Karasik, D.; Gedanken, A. Silver and gold doped hydroxyapatite nanocomposites for enhanced bone regeneration. *Biomed. Mater.* **2019**, *14*, 055002. [CrossRef] [PubMed]

149. Kattimani, V.S.; Kondaka, S.; Lingamaneni, K.P. Hydroxyapatite—Past, Present, and Future in Bone Regeneration. *Bone Tissue Regen. Insights* **2016**, *7*, BTRI.S36138. [CrossRef]

150. Qing, T.; Mahmood, M.; Zheng, Y.; Biris, A.S.; Shi, L.; Casciano, D.A. A genomic characterization of the influence of silver nanoparticles on bone differentiation in MC3T3-E1 cells. *J. Appl. Toxicol.* **2018**, *38*, 172–179. [CrossRef] [PubMed]

151. Tanaka, M.; Sato, Y.; Haniu, H.; Nomura, H.; Kobayashi, S.; Takanashi, S.; Okamoto, M.; Takizawa, T.; Aoki, K.; Usui, Y.; et al. A three-dimensional block structure consisting exclusively of carbon nanotubes serving as bone regeneration scaffold and as bone defect filler. *PLoS ONE* **2017**, *12*, e0172601. [CrossRef] [PubMed]

152. Saito, N.; Haniu, H.; Usui, Y.; Aoki, K.; Hara, K.; Takanashi, S.; Shimizu, M.; Narita, N.; Okamoto, M.; Kobayashi, S.; et al. Safe Clinical Use of Carbon Nanotubes as Innovative Biomaterials. *Chem. Rev.* **2014**, *114*, 6040–6079. [CrossRef]

153. Lin, C.; Wang, Y.; Lai, Y.; Yang, W.; Jiao, F.; Zhang, H.; Ye, S.; Zhang, Q. Incorporation of carboxylation multiwalled carbon nanotubes into biodegradable poly(lactic-co-glycolic acid) for bone tissue engineering. *Colloids Surf. B Biointerfaces* **2011**, *83*, 367–375. [CrossRef]

154. Saito, N.; Usui, Y.; Aoki, K.; Narita, N.; Shimizu, M.; Ogiwara, N.; Nakamura, K.; Ishigaki, N.; Kato, H.; Taruta, S.; et al. Carbon Nanotubes for Biomaterials in Contact with Bone. *Curr. Med. Chem.* **2008**, *15*, 523–527. [CrossRef]

155. Venkatesan, J.; Pallela, R.; Kim, S.-K. Applications of Carbon Nanomaterials in Bone Tissue Engineering. *J. Biomed. Nanotechnol.* **2014**, *10*, 3105–3123. [CrossRef]

156. Shimizu, M.; Kobayashi, Y.; Mizoguchi, T.; Nakamura, H.; Kawahara, I.; Narita, N.; Usui, Y.; Aoki, K.; Hara, K.; Haniu, H.; et al. Carbon Nanotubes Induce Bone Calcification by Bidirectional Interaction with Osteoblasts. *Adv. Mater.* **2012**, *24*, 2176–2185. [CrossRef]

157. Keselowsky, B.G.; Lewis, J.S. Dendritic cells in the host response to implanted materials. *Semin. Immunol.* **2017**, *29*, 33–40. [CrossRef]

158. Bottoni, C.R.; Smith, E.L.; Shaha, J.; Shaha, S.S.; Raybin, S.G.; Tokish, J.M.; Rowles, D.J. Autograft Versus Allograft Anterior Cruciate Ligament Reconstruction. *Am. J. Sports Med.* **2015**, *43*, 2501–2509. [CrossRef] [PubMed]

159. Bose, S.; Vahabzadeh, S.; Bandyopadhyay, A. Bone tissue engineering using 3D printing. *Mater. Today* **2013**, *16*, 496–504. [CrossRef]

160. Sher, F.; Rößler, R.; Brouwer, N.; Balasubramaniyan, V.; Boddeke, E.; Copray, S. Differentiation of Neural Stem Cells into Oligodendrocytes: Involvement of the Polycomb Group Protein Ezh2. *Stem Cells* **2008**, *26*, 2875–2883. [CrossRef]

161. Biris, A.R.; Mahmood, M.; Lazar, M.D.; Dervishi, E.; Watanabe, F.; Mustafa, T.; Baciut, G.; Baciut, M.; Bran, S.; Ali, S.; et al. Novel Multicomponent and Biocompatible Nanocomposite Materials Based on Few-Layer Graphenes Synthesized on a Gold/Hydroxyapatite Catalytic System with Applications in Bone Regeneration. *J. Phys. Chem. C* **2011**, *115*, 18967–18976. [CrossRef]

162. Crisan, L.; Crisan, B.V.; Bran, S.; Onisor, F.; Armencea, G.; Vacaras, S.; Lucaciu, O.P.; Mitre, I.; Baciut, M.; Baciut, G.; et al. Carbon-based nanomaterials as scaffolds in bone regeneration. *Part. Sci. Technol.* **2019**, 1–10. [CrossRef]

163. Cheng, X.; Wan, Q.; Pei, X. Graphene Family Materials in Bone Tissue Regeneration: Perspectives and Challenges. *Nanoscale Res. Lett.* **2018**, *13*, 289. [CrossRef] [PubMed]

164. Darbandi, A.; Gottardo, E.; Huff, J.; Stroscio, M.; Shokuhfar, T. A Review of the Cell to Graphene-Based Nanomaterial Interface. *JOM* **2018**, *70*, 566–574. [CrossRef]

165. Xu, H.H.K.; Weir, M.D.; Simon, C.G. Injectable and strong nano-apatite scaffolds for cell/growth factor delivery and bone regeneration. *Dent. Mater.* **2008**, *24*, 1212–1222. [CrossRef]

166. Engler, A.J.; Sen, S.; Sweeney, H.L.; Discher, D.E. Matrix Elasticity Directs Stem Cell Lineage Specification. *Cell* **2006**, *126*, 677–689. [CrossRef]

167. Wang, H.; Leeuwenburgh, S.C.G.; Li, Y.; Jansen, J.A. The Use of Micro- and Nanospheres as Functional Components for Bone Tissue Regeneration. *Tissue Eng. Part B Rev.* **2012**, *18*, 24–39. [CrossRef]

168. Carbone, E.J.; Jiang, T.; Nelson, C.; Henry, N.; Lo, K.W.-H. Small molecule delivery through nanofibrous scaffolds for musculoskeletal regenerative engineering. *Nanomed. Nanotechnol. Biol. Med.* **2014**, *10*, 1691–1699. [CrossRef] [PubMed]

169. Vieira, S.; Vial, S.; Reis, R.L.; Oliveira, J.M. Nanoparticles for bone tissue engineering. *Biotechnol. Prog.* **2017**, *33*, 590–611. [CrossRef] [PubMed]

170. Geckil, H.; Xu, F.; Zhang, X.; Moon, S.; Demirci, U. Engineering hydrogels as extracellular matrix mimics. *Nanomedicine* **2010**, *5*, 469–484. [CrossRef]

171. Park, H.; Temenoff, J.S.; Tabata, Y.; Caplan, A.I.; Mikos, A.G. Injectable biodegradable hydrogel composites for rabbit marrow mesenchymal stem cell and growth factor delivery for cartilage tissue engineering. *Biomaterials* **2007**, *28*, 3217–3227. [CrossRef]

172. Wu, G.; Feng, C.; Quan, J.; Wang, Z.; Wei, W.; Zang, S.; Kang, S.; Hui, G.; Chen, X.; Wang, Q. In situ controlled release of stromal cell-derived factor-1α and antimiR-138 for on-demand cranial bone regeneration. *Carbohydr. Polym.* **2018**, *182*, 215–224. [CrossRef]

173. Silva, R.; Fabry, B.; Boccaccini, A.R. Fibrous protein-based hydrogels for cell encapsulation. *Biomaterials* **2014**, *35*, 6727–6738. [CrossRef] [PubMed]

174. Bai, X.; Gao, M.; Syed, S.; Zhuang, J.; Xu, X.; Zhang, X.-Q. Bioactive hydrogels for bone regeneration. *Bioact. Mater.* **2018**, *3*, 401–417. [CrossRef] [PubMed]

175. Arvidson, K.; Abdallah, B.M.; Applegate, L.A.; Baldini, N.; Cenni, E.; Gomez-Barrena, E.; Granchi, D.; Kassem, M.; Konttinen, Y.T.; Mustafa, K.; et al. Bone regeneration and stem cells. *J. Cell. Mol. Med.* **2011**, *15*, 718–746. [CrossRef] [PubMed]

176. Suzuki, O.; Imaizumi, H.; Kamakura, S.; Katagiri, T. Bone Regeneration by Synthetic Octacalcium Phosphate and its Role in Biological Mineralization. *Curr. Med. Chem.* **2008**, *15*, 305–313. [CrossRef] [PubMed]

177. Cai, Z.-Y.; Yang, D.-A.; Zhang, N.; Ji, C.-G.; Zhu, L.; Zhang, T. Poly(propylene fumarate)/(calcium sulphate/β-tricalcium phosphate) composites: Preparation, characterization and in vitro degradation. *Acta Biomater.* **2009**, *5*, 628–635. [CrossRef] [PubMed]

178. Gong, T.; Xie, J.; Liao, J.; Zhang, T.; Lin, S.; Lin, Y. Nanomaterials and bone regeneration. *Bone Res.* **2015**, *3*, 15029. [CrossRef] [PubMed]

179. Stevens, M.M. Biomaterials for bone tissue engineering. *Mater. Today* **2008**, *11*, 18–25. [CrossRef]

180. Vo, T.N.; Kasper, F.K.; Mikos, A.G. Strategies for controlled delivery of growth factors and cells for bone regeneration. *Adv. Drug Deliv. Rev.* **2012**, *64*, 1292–1309. [CrossRef] [PubMed]

181. Kim, J.; Kim, H.N.; Lim, K.-T.; Kim, Y.; Seonwoo, H.; Park, S.H.; Lim, H.J.; Kim, D.-H.; Suh, K.-Y.; Choung, P.-H.; et al. Designing nanotopographical density of extracellular matrix for controlled morphology and function of human mesenchymal stem cells. *Sci. Rep.* **2013**, *3*, 3552. [CrossRef]

182. Li, X.; van Blitterswijk, C.A.; Feng, Q.; Cui, F.; Watari, F. The effect of calcium phosphate microstructure on bone-related cells in vitro. *Biomaterials* **2008**, *29*, 3306–3316. [CrossRef]

183. Yao, X.; Peng, R.; Ding, J. Effects of aspect ratios of stem cells on lineage commitments with and without induction media. *Biomaterials* **2013**, *34*, 930–939. [CrossRef]

184. Janeczek, M.; Szymczyk, P.; Dobrzynski, M.; Parulska, O.; Szymonowicz, M.; Kuropka, P.; Rybak, Z.; Zywicka, B.; Ziolkowski, G.; Marycz, K.; et al. Influence of surface modifications of a nanostructured implant on osseointegration capacity—Preliminary in vivo study. *RSC Adv.* **2018**, *8*, 15533–15546. [CrossRef]

185. Paz, A.G.; Maghaireh, H.; Mangano, F.G. Stem Cells in Dentistry: Types of Intra- and Extraoral Tissue-Derived Stem Cells and Clinical Applications. *Stem Cells Int.* **2018**, *2018*, 4313610. [CrossRef]

186. Sakkas, A.; Wilde, F.; Heufelder, M.; Winter, K.; Schramm, A. Autogenous bone grafts in oral implantology—Is it still a "gold standard"? A consecutive review of 279 patients with 456 clinical procedures. *Int. J. Implant Dent.* **2017**, *3*, 23. [CrossRef]

187. Felice, P.; Pistilli, R.; Lizio, G.; Pellegrino, G.; Nisii, A.; Marchetti, C. Inlay versus Onlay Iliac Bone Grafting in Atrophic Posterior Mandible: A Prospective Controlled Clinical Trial for the Comparison of Two Techniques. *Clin. Implant Dent. Relat. Res.* **2009**, *11*, e69–e82. [CrossRef]

188. Khojasteh, A.; Behnia, H.; Dashti, S.G.; Stevens, M. Current Trends in Mesenchymal Stem Cell Application in Bone Augmentation: A Review of the Literature. *J. Oral Maxillofac. Surg.* **2012**, *70*, 972–982. [CrossRef] [PubMed]

189. Slater, B.J.; Kwan, M.D.; Gupta, D.M.; Panetta, N.J.; Longaker, M.T. Mesenchymal cells for skeletal tissue engineering. *Expert Opin. Biol. Ther.* **2008**, *8*, 885–893. [CrossRef] [PubMed]

190. Petite, H.; Viateau, V.; Bensaïd, W.; Meunier, A.; de Pollak, C.; Bourguignon, M.; Oudina, K.; Sedel, L.; Guillemin, G. Tissue-engineered bone regeneration. *Nat. Biotechnol.* **2000**, *18*, 959–963. [CrossRef] [PubMed]

191. Gronthos, S.; Brahim, J.; Li, W.; Fisher, L.W.; Cherman, N.; Boyde, A.; DenBesten, P.; Robey, P.G.; Shi, S. Stem Cell Properties of Human Dental Pulp Stem Cells. *J. Dent. Res.* **2002**, *81*, 531–535. [CrossRef]

192. Kubo, T.; Doi, K.; Hayashi, K.; Morita, K.; Matsuura, A.; Teixeira, E.R.; Akagawa, Y. Comparative evaluation of bone regeneration using spherical and irregularly shaped granules of interconnected porous hydroxylapatite. A beagle dog study. *J. Prosthodont. Res.* **2011**, *55*, 104–109. [CrossRef]

193. Egusa, H.; Sonoyama, W.; Nishimura, M.; Atsuta, I.; Akiyama, K. Stem cells in dentistry—Part II: Clinical applications. *J. Prosthodont. Res.* **2012**, *56*, 229–248. [CrossRef]

194. Zakrzewski, W.; Dobrzyński, M.; Szymonowicz, M.; Rybak, Z. Stem cells: Past, present, and future. *Stem Cell Res. Ther.* **2019**, *10*, 68. [CrossRef]

195. Ahn, H.H.; Kim, K.S.; Lee, J.H.; Lee, J.Y.; Kim, B.S.; Lee, I.W.; Chun, H.J.; Kim, J.H.; Lee, H.B.; Kim, M.S. *In Vivo* Osteogenic Differentiation of Human Adipose-Derived Stem Cells in an Injectable *In Situ* –Forming Gel Scaffold. *Tissue Eng. Part A* **2009**, *15*, 1821–1832. [CrossRef]

196. Meijer, G.J.; de Bruijn, J.D.; Koole, R.; van Blitterswijk, C.A. Cell based bone tissue engineering in jaw defects. *Biomaterials* **2008**, *29*, 3053–3061. [CrossRef]

197. Gregory, C.A.; Prockop, D.J.; Spees, J.L. Non-hematopoietic bone marrow stem cells: Molecular control of expansion and differentiation. *Exp. Cell Res.* **2005**, *306*, 330–335. [CrossRef]

198. Williams, A.R.; Hare, J.M. Mesenchymal stem cells: Biology, pathophysiology, translational findings, and therapeutic implications for cardiac disease. *Circ. Res.* **2011**, *109*, 923–940. [CrossRef] [PubMed]

199. da Silva Meirelles, L.; Chagastelles, P.C.; Nardi, N.B. Mesenchymal stem cells reside in virtually all post-natal organs and tissues. *J. Cell Sci.* **2006**, *119*, 2204–2213. [CrossRef]

200. De Bari, C.; dell'Accio, F.; Tylzanowski, P.; Luyten, F.P. Multipotent mesenchymal stem cells from adult human synovial membrane. *Arthritis Rheum.* **2001**, *44*, 1928–1942. [CrossRef]

201. Egusa, H.; Sonoyama, W.; Nishimura, M.; Atsuta, I.; Akiyama, K. Stem cells in dentistry—Part I: Stem cell sources. *J. Prosthodont. Res.* **2012**, *56*, 151–165. [CrossRef] [PubMed]

202. Rawadi, G.; Vayssière, B.; Dunn, F.; Baron, R.; Roman-Roman, S. BMP-2 Controls Alkaline Phosphatase Expression and Osteoblast Mineralization by a Wnt Autocrine Loop. *J. Bone Miner. Res.* **2003**, *18*, 1842–1853. [CrossRef]

203. Bruder, S.P.; Jaiswal, N.; Haynesworth, S.E. Growth kinetics, self-renewal, and the osteogenic potential of purified human mesenchymal stem cells during extensive subcultivation and following cryopreservation. *J. Cell. Biochem.* **1997**, *64*, 278–294. [CrossRef]

204. Cancedda, R.; Mastrogiacomo, M.; Bianchi, G.; Derubeis, A.; Muraglia, A.; Quarto, R. Bone marrow stromal cells and their use in regenerating bone. *Novartis Found. Symp.* **2003**, *249*, 133–143, discussion 143-7, 170–174, 239–41.

205. Chamberlain, G.; Fox, J.; Ashton, B.; Middleton, J. Concise Review: Mesenchymal Stem Cells: Their Phenotype, Differentiation Capacity, Immunological Features, and Potential for Homing. *Stem Cells* **2007**, *25*, 2739–2749. [CrossRef]

206. Ashton, B.A.; Allen, T.D.; Howlett, C.R.; Eaglesom, C.C.; Hattori, A.; Owen, M. Formation of bone and cartilage by marrow stromal cells in diffusion chambers in vivo. *Clin. Orthop. Relat. Res.* **1980**, *151*, 294–307. [CrossRef]

207. Pereira, R.F.; Halford, K.W.; O'Hara, M.D.; Leeper, D.B.; Sokolov, B.P.; Pollard, M.D.; Bagasra, O.; Prockop, D.J. Cultured adherent cells from marrow can serve as long-lasting precursor cells for bone, cartilage, and lung in irradiated mice. *Proc. Natl. Acad. Sci. USA* **1995**, *92*, 4857–4861. [CrossRef]

208. Diefenderfer, D.L.; Osyczka, A.M.; Reilly, G.C.; Leboy, P.S. BMP responsiveness in human mesenchymal stem cells. *Connect. Tissue Res.* **2003**, *44*, 305–311. [CrossRef] [PubMed]

209. Mueller, S.M.; Glowacki, J. Age-related decline in the osteogenic potential of human bone marrow cells cultured in three-dimensional collagen sponges. *J. Cell. Biochem.* **2001**, *82*, 583–590. [CrossRef] [PubMed]

210. Zins, J.E.; Whitaker, L.A. Membranous vs endochondral bone autografts: Implications for craniofacial reconstruction. *Surg. Forum* **1979**, *30*, 521–523.

211. Koole, R.; Bosker, H.; van der Dussen, F.N. Late secondary autogenous bone grafting in cleft patients comparing mandibular (ectomesenchymal) and iliac crest (mesenchymal) grafts. *J. Craniomaxillofac. Surg.* **1989**, *17*, 28–30. [CrossRef]

212. Mashimo, T.; Sato, Y.; Akita, D.; Toriumi, T.; Namaki, S.; Matsuzaki, Y.; Yonehara, Y.; Honda, M. Bone marrow-derived mesenchymal stem cells enhance bone marrow regeneration in dental extraction sockets. *J. Oral Sci.* **2019**, *61*, 284–293. [CrossRef]

213. Toma, C.; Pittenger, M.F.; Cahill, K.S.; Byrne, B.J.; Kessler, P.D. Human Mesenchymal Stem Cells Differentiate to a Cardiomyocyte Phenotype in the Adult Murine Heart. *Circulation* **2002**, *105*, 93–98. [CrossRef]
214. Sasaki, M.; Abe, R.; Fujita, Y.; Ando, S.; Inokuma, D.; Shimizu, H. Mesenchymal Stem Cells Are Recruited into Wounded Skin and Contribute to Wound Repair by Transdifferentiation into Multiple Skin Cell Type. *J. Immunol.* **2008**, *180*, 2581–2587. [CrossRef]

nanomaterials

Review

Application of Selected Nanomaterials and Ozone in Modern Clinical Dentistry

Adam Lubojanski [1], Maciej Dobrzynski [2], Nicole Nowak [3], Justyna Rewak-Soroczynska [3], Klaudia Sztyler [4], Wojciech Zakrzewski [1], Wojciech Dobrzynski [5], Maria Szymonowicz [1], Zbigniew Rybak [1], Katarzyna Wiglusz [6] and Rafal J. Wiglusz [3,*]

[1] Department of Experimental Surgery and Biomaterial Research, Wroclaw Medical University, Bujwida 44, 50-345 Wrocław, Poland; adam.lubojanski@student.umed.wroc.pl (A.L.); wojciech.zakrzewski@student.umed.wroc.pl (W.Z.); maria.szymonowicz@umed.wroc.pl (M.S.); zbigniew.rybak@umed.wroc.pl (Z.R.)

[2] Department of Pediatric Dentistry and Preclinical Dentistry, Wroclaw Medical University, Krakowska 26, 50-425 Wrocław, Poland; maciej.dobrzynski@umed.wroc.pl

[3] Institute of Low Temperature and Structure Research, Polish Academy of Sciences, Okólna 2, 50-422 Wrocław, Poland; n.nowak@intibs.pl (N.N.); j.rewak@intibs.pl (J.R.-S.)

[4] Department of Pediatric Dentistry, Academic Dental Polyclinics, Krakowska 26, 50-435 Wrocław, Poland; klaudia.sztyler1@gmail.com

[5] Student Scientific Circle at the Department of Dental Materials, School of Medicine with the Division of Dentistry in Zabrze, Medical University of Silesia in Katowice, Akademicki Sq. 17, 41-902 Bytom, Poland; wojt.dobrzynski@wp.pl

[6] Department of Analytical Chemistry, Wroclaw Medical University, Borowska 211 A, 50-566 Wroclaw, Poland; katarzyna.wiglusz@umed.wroc.pl

* Correspondence: r.wiglusz@intibs.pl; Tel.: +48-71-3954-159; Fax: +48-71-344-10-29

Citation: Lubojanski, A.; Dobrzynski, M.; Nowak, N.; Rewak-Soroczynska, J.; Sztyler, K.; Zakrzewski, W.; Dobrzynski, W.; Szymonowicz, M.; Rybak, Z.; Wiglusz, K.; et al. Application of Selected Nanomaterials and Ozone in Modern Clinical Dentistry. *Nanomaterials* **2021**, *11*, 259. https://doi.org/10.3390/nano11020259

Received: 21 December 2020
Accepted: 11 January 2021
Published: 20 January 2021

Publisher's Note: MDPI stays neutral with regard to jurisdictional claims in published maps and institutional affiliations.

Abstract: This review is an attempt to summarize current research on ozone, titanium dioxide (TiO_2), silver (Ag), copper oxide CuO and platinum (Pt) nanoparticles (NPs). These agents can be used in various fields of dentistry such as conservative dentistry, endodontic, prosthetic or dental surgery. Nanotechnology and ozone can facilitate the dentist's work by providing antimicrobial properties to dental materials or ensuring a decontaminated work area. However, the high potential of these agents for use in medicine should be confirmed in further research due to possible side effects, especially in long duration of observation so that the best way to apply them can be obtained.

Keywords: dental nanomaterials; dental implants; endodontics; prosthetic; ozone; antimicrobial activity; microorganisms

1. Introduction

Ozone is a gas composed of three oxygen atoms: the molecular weight is 47.98, it is an unstable substance, which quickly releases single oxygen atoms, and the half-life is 40 min at 20 °C and nearly 140 min at 0 °C. Ozone is colorless and has a characteristic smell. It occurs in nature as the ozone layer which protects organisms from ultraviolet rays. Ozone is heavier than air and it moves down where gas combines with pollutants, which is known as the self-cleansing phenomenon [1–3]. It is an extremely strong oxidizer, 1.5 times stronger than chloride when comparing their anti-microbial potential [4]. A gas mixture of 0.5 to 5% of ozone and 95 to 99.5% of oxygen is used in medicine. For almost a century, singlet (nascent) oxygen is known for its ability to inactivate bacteria, fungi, and viruses. Progress in medicine opens new applications for ozone therapy [5,6].

Silver particles have been demonstrated to be effective components in adhesives, implants, or prosthetic materials. They have an electron configuration of $[Kr]4d^{10}5s^1$, and it is now possible to produce silver nanoparticles with controlled morphology and size, as well as specific target functions and homogeneity [7,8]. Additionally, they can be successfully used in orthodontics [9] as multifunctional building blocks for dental

37

materials [10], components for tissue conditioner, as well as act synergistically with several types of antibiotics [11,12]. Antifungal effects of silver particles against *Candida albicans* can be especially observed when the material is added to silicone-based liners or resins [13]. With regard to antibacterial activity, particles of silver influence the permeability of bacterial membrane, leading to its disruption. The material is responsible for the stimulation of oxidative stress resulting in thedestruction of cellular structures, such as DNA [14], lipids or proteins, eventually causing the destruction of the entire bacterial cell. Abroad range of aspects of the synthesis, as well as application and toxicology of silver particles, have been covered in recent reviews [15,16].The time needed to fully release Ag^+ from a particle depends on dissolution processes which depend on pH, the concentration of the dissolved O_2, surface coating or ionic strength of the medium [17].

Copper oxide (CuO) plays a significant role as a bactericidal and antifungal agent [18]. It is the simplest element of Cu compounds, that reveal a range of potential physical properties such as high-temperature superconductivity, spin dynamics or electron correlation effects [19], and is cheaper than silver oxide. It has high surface areas with uncommon crystalline structures [20,21] and additionally can improve fluid viscosity and enhance thermal conductivity. These characteristics make CuO a potentially useful energy-saving material. Its nanoparticles are successfullyused as additives in lubricants, metallic coatings or polymers [22]. The extremely high surface areas and atypical morphologies make CuO nanoparticles enhance the shear bond strength of adhesives, additionally influencing their antimicrobial characteristics, which allow them to dose-dependently inhibit, for example, *Escherichia coli* bacilli, but not *Salmonella* Typhimurium [23]. *Streptococcus mutans* cocci, on the other hand, are affected by CuO in a similar way as they are affected by particles of silver [24].According to other studies, CuO can decrease biofilm formation from 70% to up to 80% [25].

Titanium dioxide (TiO_2) is a photocatalyst and is widely used as a self-disinfecting and self-cleaning material for surface coating in a variety of applications [26].Currently, titanium and its alloys are broadly used in dental implantology due to their excellent biocompatibility and good mechanical characteristics. Thanks to its properties, including nontoxicity and super-hydrophobicity, it has been applied in removing bacteria and harmful organic materials from both air and water, acting as a sterilizing agent in places such as medical centers [27]. However, TiO_2 can be activated only in the low-ultraviolet (LUV)range (<400 nm). Apart from its disinfecting characteristics, recent research has proven that TiO_2 allows localized drug delivery with the use of nanotubes [28].

Metal nanoparticles can be obtained by converting metals into fine particles with a diameter smaller than 100 nm. Platinum nanoparticles (PtNPs) have been reported in several studies [29,30] to have antibacterial and anti-inflammatory effects, and because PtNPs were demonstrated to be a potent antioxidant in vitro, their addition to resin-based materials may also improve their biocompatibility [31]. Additionally, platinum particles mixed with a 4-methacryloyloxyethyl trimellitic anhydride (4-META)/methyl methacrylate (MMA) adhesive increase the dentin bond strength twice when compared to the regular material [32]. Contact between Pt particles and bacteria has reportedly led to the decomposition of the latter [33]. Yang et al. [34] confirmed that platinum particles are unlikely to cause allergy, have a potential for clinical application and do not cause genotoxic potential.This review is focused on using nanotechnology and ozone in modern dentistry. Moreover, the antibacterial properties of the agents can allow for more effective methods of treatment with fewer complications.

2. Antimicrobial Properties

2.1. Gaseous Ozone, Ozonated Water and Ozonated Oil

In stomatology, three different forms of ozone are used: gaseous ozone, ozonated water and ozonated oil (sunflower oil, olive oil, groundnut oil) [35].

There are numerous research studies describing the antibacterial activity of ozone against oral pathogens. One of such studies evaluated the antibacterial effect of ozone against the pathogenic *Enterococcus faecalis* culture on a model of human teeth prepared using a specific protocol [36]. Samples were divided into four groups and for each of them, a different treatment was applied (ozone, photo-activated disinfection, saline as a positive control and NaOCl as a negative control). After 7 days of incubation, CFU (colony-forming unit) values were calculated. Both ozonation and photo-activated disinfection were revealed to be effective because the numbers of grown colonies were reduced [36]. Similar research was performed by Camacho-Alonso et al., who compared the activity of ozone with that of NaOCl solution, chlorhexidine, tri-antibiotic mixture, propolis and photodynamic therapy against *E. faecalis*. All the applied procedures were effective [37]. A comparison of a greater group of antibacterial procedures was conducted by Sancakli et al., who analyzed not only ozonation but also chlorhexidine and laser therapy Er:YAG (Erbium-doped Yttrium Aluminium Garnet laser), KTP crystal (PotassiumTitanylPhosphate crystal ($KTiOPO_4$)), alone or in combination, against *Streptococcus mutans*. The obtained results revealed that the most efficient procedures were chlorhexidine, the combination of Er:YAG laser with ozone, Er:YAG with ozone and chlorhexidine (but its effect was weaker than for chlorhexidine alone) and the combination of KTP laser with ozone and ozone and chlorhexidine. Generally, ozonation and laser therapy applied alone were insufficient and required combining with other factors [38]. Besides gaseous ozone, also the antibacterial activity of ozonated olive oil was investigated using the direct contact agar diffusion test (the measurement of inhibition zones) on *Aggregatibacter actinomycetemcomitans*, *S. mutans* and *Prevotella intermedia* isolates. Ozonated olive oil inhibited the growth of all the tested strains but antibacterial effect was weaker than for chlorhexidine, which was used as a reference sample. The authors also evaluated minimal inhibitory concentrations (MIC) and minimal bactericidal concentrations (MBC) using the standard broth dilution method. Unfortunately, only a 10% solution of ozonated olive was able to inhibit the growth of *A. actinomycetencomitans*, and for *S. mutans* and *P. intermedia*, no MIC values could be determined using 0.01–10% solutions of the oil. Consequently, MBC values were also not determined. This result is not satisfactory, especially when compared with the formulation of chlorhexidine, where a 0.01% solution was already bactericidal [35]. There are also numerous research studies describing clinical trials with ozone. In the work of Krunić et al., ozone treatment was applied to patients with deep carious lesions. Chlorhexidine was applied in other patients as a reference. The experiments were performed on isolated *Lactobacillus* spp. as well as on the total population of bacteria. The data demonstrate that both applied treatmentprocedures caused a reduction in the number of bacteria, however, chlorhexidine was more effective [39]. In vivo experiments were also performed by Ajeti et al. The antibacterial efficacy of ozone combined with 0.9% NaCl, 2% chlorhexidine or 2.5% NaOCl was compared. The most effective of all of the tested combinations was the combination of ozone with NaOCl [40]. Apart from gaseous ozone and ozonated olive oil, ozonated water is also applied to reduce the risk of a microbial infection. In a clinical study by Anumula et al., patients were divided into two groups. One group received chlorhexidine, and the other ozonated water as an oral rinse. The saliva samples after 0, 7 and 14 days were diluted and cultivated to calculate CFU/mL values. The obtained results were very promising because the colony number of *S. mutans* decreased, and for ozonated water, the effect was even more visible than for chlorhexidine [41]. The findings of the research studies listed above confirm the antibacterial potential of different forms of ozone but, if possible, it should be combined with other techniques to obtain better results.

2.2. Titanium Dioxide Nanoparticles

Titanium oxide can reduce bacterial growth by damaging bacterial cell membranes, which leads to enhanced permeability and, as a consequence ofthe loss of vital cellular components, to death [42]. Antibacterial and antibiofilm effect of pure TiO_2 coated on the surface of metal washers was investigated using the common strains inhabiting human oral cavity: *Streptococcus sanguinis*, *S. mutans* and *L. acidophilus*. In dentistry, biofilm contributes to formation of manyoral diseases (e.g. caries, periodontis, periimplantitis) therefore it is needed to ensure that necessary prevention and rapid reaction capabilities are in place to deal with any such problems. The simplified process of biofilm formation is presented in Figure 1.

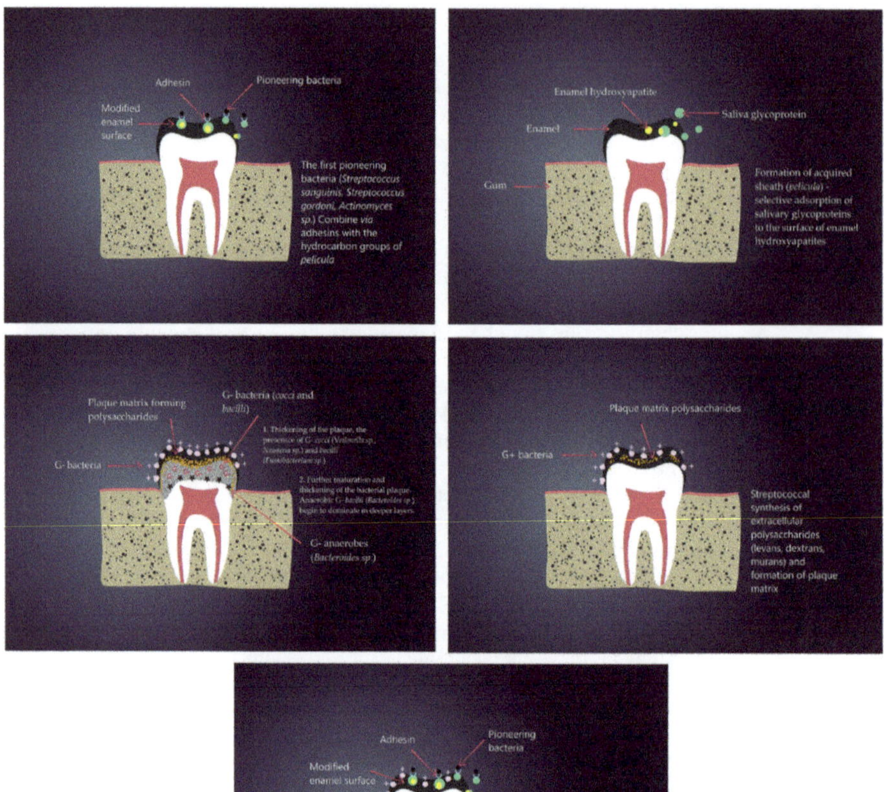

Figure 1. A scheme showing the stages of biofilm formation.

The smallest reduction in biofilm formation was observed for *L. acidophilus*, but for *S. sanguinis* and *S. mutans*, the antibiofilm effect was more significant and dose-dependent—it increased with dopant enhancement [42]. A combination of TiO_2 with metals is frequently described. In a work by Chen et al., the antibacterial activity against *S. mutans* was also investigated. TiO_2 as well as Ag/TiO_2 were mixed with polymethyl methacrylate (PMMA), silanized aluminum borate whiskers (ABWs) and nano-ZrO_2 to obtain a composite. An additive of pure titanium oxide and its silver-modified derivative to the resin decreased bacterial growth without increasing the cytotoxicity of the material. Nevertheless, it should be noted that the sample with incorporated silver was much more effective [43]. Titanium dioxide, alone and combined with silver, was also combined with polyacrylate resin polymer to obtain nanohybrid coatings on titanium discs. These materials have been tested against *Streptococcus salivarius* and the obtained results are promising [44]. The antibacterial effect of TiO_2 combined with silver was also investigated on *S. sanguinis* along with evaluating bacterial adhesion to the material in the form of titanium discs coated with TiO_2 or Ag (alone or combined with hydroxyapatite). The results indicate that the discs coated with silver combined with hydroxyapatite were the most effective of all the tested samples [45]. In the work of Lavaee et al., *S. mutans* and *S. sanguinis* were isolated from saliva samples of patients and antimicrobial activity of TiO_2, alone and combined with nano-Ag or nano-Fe_3O_4, was tested. The combination of these three nanoparticles (nano-TiO_2 + nano-Ag + nano-Fe_3O_4) turned out to be the most effective of all the tested materials and this effect was observed for both tested strains. Moreover, the antibiofilm activity of this mixture was also observed [46]. Apart from silver, copper is another material that can be combined with titanium oxide to enhance its antimicrobial effect [47]. Titanium dioxide is also a promising material due to its photocatalytic properties: after UV exposure, it produces ROS, which enhances its antibacterial properties [48].

2.3. Silver Nanoparticles and Ions

Silver, especially in its ionic form, has the highest antimicrobial activity among metals. Its mode of action is complex and involves protein impairment by forming Ag-S bonds, which leads to respiratory chain dysfunction and damage to membrane pumps. Ag^+ ionsare also genotoxic—they interact with DNA which causes its condensation and inhibition of the replication process. Silver also stimulates ROS production, which is characteristic of metals [49]. Besides releasing ions, silver nanoparticles can also affect bacterial cells by themselves. One mode of action is based on attaching to the surface of a cell and leading to the denaturation of cell membrane as a result ofthe accumulation of particles. Due to their nano-size, they can also penetrate through the cell wall, and disturb metabolic processes and bacterial signal transduction [50]. There is an extensive body of research describing antibacterial activity of silver-based nanomaterials against oral pathogens. In the work of Liu et al., polyetheretherketone (PEEK) was coated with 3–12 nm of nano-silver coatings. The antimicrobial potential of the obtained materials was evaluated usingan *S. mutans* strain. The obtained results indicate that efficacy increases with the thickness of the silver layer, however even a 3 nm-thick sample effectively prevented bacterial growth. The number of bacterial colonies attached to the surface of silver-coated materials was also reduced [51]. *S. mutans* was also used in the research of Kim et al., who evaluated the antimicrobial potential of feldspathic porcelain combined with nano-sized silver at concentrations of 0%, 5%, 10%, 20% and 30%. All the tested samples, even the non-doped one, inhibited bacterial growth, but the most significant reduction was observed for the highest concentration of silver (30%) [52]. In order to evaluate antibacterial activity of silver nanoparticles, a model of human teeth (after extraction) was prepared and, after cleansing, contaminated with *E. faecalis*. After 21 days of incubation, the samples were divided into three groups treated with different intracanal dressings: $Ca(OH)_2$ paste, $Ca(OH)_2$ paste mixed with chlorhexidine and $Ca(OH)_2$ paste supplemented with a suspension of silver nanoparticles. Samples prepared in such a way were incubated for 1 week and 1 month and then the colonies were counted. Moreover, SEM (Scannicelectron Microscopy) images were measured. The obtained data revealed that all of the applied dressings effectively reduced bacterial growth and adhesion, and the silver additive improved this effect [53]. Interesting results were also obtained by Yin et al., who prepared sodium fluoride solution mixed with different concentrations of polyethylene glycol-coated silver nanoparticles (PEG-AgNPs). Antibacterial activity (half maximal inhibitory concentration, IC_{50}) of these materials was examined using an *S. mutans* strain. The obtained data indicate that PEG-AgNPs effectively kill bacteria at a concentration of 21.16 ppm ofsilver and this concentration was non-cytotoxic for a human gingival fibroblast cell line (HGF-1) [54]. In the work by Bacali et al., antibacterial activity of polymethyl methacrylate (PMMA) denture resin combined with graphene and Ag nanoparticles was tested against, among others, *S. mutans*. Interestingly, pure PMMA also inhibited *S. mutans* growth but the addition of graphene and silver nanoparticles intensified this effect [55]. Three oral microbes (*E. faecalis*, *S. mutans* and *Streptococcus oralis*) were used to evaluate the antimicrobial activity of bioactive glass combined with silver nanoparticles and tetracycline. The prepared materials effectively released the loaded drug, which led to the inhibition of bacterial growth. However, materials without the addition of tetracycline were not so effective. Silver-combined bio-glass slightly inhibited the growth of *E. faecalis* but no such effect was observed for *S. mutans* [56]. Chitosan was also used as a carrier of drug (gentamicin) and silver nanoparticles. Pure chitosan slightly reduced bacterial growth (*S. mutans*) but adding gentamicin or the combination of Ag/chitosan or Ag/chitosan/gentamicin enhanced the antibacterial effect. Surprisingly, no differences between the action of these three materials were detected [57]. Epigallocatechin gallate was another material used as a carrier for nanoparticles. Antimicrobial efficacy was evaluated for silver nanoparticles and compared with ionic silver in the form of $AgNO_3$. The results indicate that the effectiveness of silver nanoparticles against *S. mutans* is higher than that of the $AgNO_3$ solution. The red-to-green ratio calculated from confocal laser microscopy (dead/live cells proportion) was also higher for AgNPs. Moreover, in microscopic images, it can be seen that bacteria did not form a biofilm structure in the AgNPs sample, whilst for $AgNO_3$ and control (water), a fully grown biofilm can be noted. Additionally, AgNPs significantly reduced the production of lactic acid and polysaccharides by *S. mutans* biofilms [58].

2.4. Platinum Nanoparticles

The antibacterial effect of platinum is based on the ability to decompose bacterial cell structure. The exact mechanism is uncertain and, as for metals in general, probably connected with the increase of reactive oxygen species (ROS) production, which results in cellular damage via oxidative stress [33]. Platinum nanoparticles (PtNPs) have an antimicrobial effect against oral pathogens. In the work of Itohiya et al., PtNPs were obtained via direct irradiation of platinum with an infrared pulsed laser and tested on common dental bacteria: *S.mutans*, *E. faecalis* and *Porphyromonasgingivalis*. The range of applied concentrations of PtNPs was 1–20 ppm and the concentrations above 5 ppm completely inhibited the growth of all the tested strains [33]. Platinum nanoparticles were also combined with polymeric PMMA and the antibacterial activity of such a combination was evaluated using *S. mutans* and *Streptococcus sobrinus* reference strains. The results indicate that platinum nanoparticles exhibit antimicrobial activity against the planktonic form of both tested strains (growth reduction). Moreover, platinum-combined polymer turned out to be less prone to bacterial adhesion than pure PMMA [59].

2.5. Copper Oxide

Copper oxide nanoparticles (CuONPs) can be obtainedwith the use of various methods of synthesis. By using different preparation methods, particles with different sizes can be produced; for example, through colloidal-thermal synthesis, precipitation synthesis, microwave irradiation or sol-gel techniques, very small particles can be obtained which are estimated from 3 to 10 nm. Larger particles from 10 to even 30 nm can be received via sonochemical synthesis, spinning disk reactor, solid-state reaction, microemulsion system or thermal decomposition [60].

CuONPs, due to their small size, are well known for their optical and magnetic properties and electrical conductivity. However, copper oxide nanoparticles are also known for their biological properties, in particular for their antimicrobial activity; nevertheless, cytotoxic properties were well estimated in both in vitro and in vivo models [60].

Skin and the respiratory system are parts of the human body which are the most exposed to CuOPNs. The cytotoxicity of copper oxide nanoparticles against normal cell lines such as human keratinocytes (HaCaT)and mouse embryonic fibroblasts (MEF) was estimated by Luo et al. Copper oxide nanoparticles, in a dose-dependent manner (from a concentration of 20 µg/mL), distinctly induced c-Jun N-terminal kinase (JNK) and extracellular signal-regulated kinase (Erk); moreover, level of p53 was significantly decreased in HaCaT and MEF cell lines, and a reduction of the amounts of viable cells was also observed in a concentration-dependent way, which clearly indicates the toxic effect on the above-mentioned cell types [61]. Another study confirmed the cytotoxic effect of CuONPs on human lung epithelial cells (A549). Concentration from 10mg/mL to 100µg/mL drastically decreased cell viability. CuONPs particles were compared with CuO bulk particles and it was clearly demonstrated that CuO bulk particles exhibitedaweaker toxic effect (concentration 58 mg/mL) than CuO nanoparticles (concentration of 15 mg/mL). TEM images (Transmission Electron Microscopy) show the entry of CuOPNs into the intracellular environment, but also into the nucleus, mitochondria and lysosomes. Mitochondrial influx of CuO nanoparticles may lead to inducing its depolarization and probably caused the generation of reactive oxygen species (ROS) as well, which elevated oxidative stress [62]. The cytotoxic effect of CuONPson the A549 cell line was also confirmed by an independent research of Akhtar et al. [63]. The toxic effect of CuO nanoparticles against human blood lymphocytes was also evaluated by estimating cell viability, ROS generation, peroxidation of lipids, but also by estimating lysosomal and mitochondrial disruption. The data clearly indicated the disintegration of mitochondrial membrane and an elevated generation of reactive oxygen species (ROS), and lysosomal damage was also visible. The viability of human blood lymphocytes drastically dropped and was estimated at IC_{50} after the treatment with CuO nanoparticles at a concentration of 385 µM [64]. Sun et al. compared the cytotoxic properties of several different metal nanoparticles, such as Fe_3O_4, Fe_2O_3, SiO_2, TiO_2 and CuO, and exposed to them a human type II alveolar epithelial cell line (A549), human non-small cell lung cancer (H1650) cells and human nasopharyngeal carcinoma (CNE-2Z). Copper nanoparticles were determined to be the most toxic among all the tested nanoparticles and on all the tested cell lines, and the toxic effect was estimated at a concentration of 30 µg/mL [65]. Additionally, Beltrán-Partida et al. estimated the potential cytotoxic effect of CuNPs, $CuCO_3$ and the antifungal drug triclosan on primary human gingival fibroblasts (HGF) isolated from a clinically healthy young (15 years old) male patient as a suitable model for this type of test. Among all the tested compounds, $CuCO_3$ was the most toxic on HGF when compared to CuNPs which exhibited IC_{50} at a concentration of 137.4 µg/mL [66].

The antimicrobial potential of copper has been known for ages but, as a result ofthe development of nanotechnology, nano-sized copper oxide also became an issue of interest. The antibacterial effect of CuO was described by Ramazanzadeh et al., who prepared brackets coated with CuO, ZnO or a 1:1 combination of the materials which was tested on *S. mutans* after 0, 2, 4, 6 and 24 h. No effect was observed for ZnO but CuO completely inhibited bacterial growth after 4 h of incubation. The ZnO/CuO mixture was even more effective because the bacterial growth was affected after 2 h of treatment [67]. The susceptibility of *S. mutans* was also tested by Toodehzaeim et al., who prepared nanocomposites doped with 0.01%, 0.5% and 1% of CuO and the inhibition zones were measured around discs placed on the surface of agar plates with bacteria [68]. MIC_{50} values were determined for CuO by Amiri et al. on the *S. mutans*, *Lactobacillus acidophilus* and *Lactobacillus casei* strains. The obtained values are as follows: <1 µg/mL for *L. acidophilus*, 1–10 µg/mL for *S. mutans* and 10 µg/mL for *L. casei*. At the concentration of 1000 µg/mL, a complete reduction of growth was observed for all the tested bacteria [21].

3. The Use of Ozone and Nanoparticles in Dentistry

3.1. Restorative Dentistry

Tooth decay is a very common disease: The Global Burden of Disease Study 2017 reported that there are 2.4 billion people suffering from this illness, withcaries of primary teeth as a problem concerning 530 million children [69]. Ozonation is an efficient treatment that can help to reduce the amounts of microorganisms in oral cavity preventing to dental caries—the most common disease worldwide. In Figure 2 has been shown the number of oral diseases worldwide indicateing a serious problem for many people.

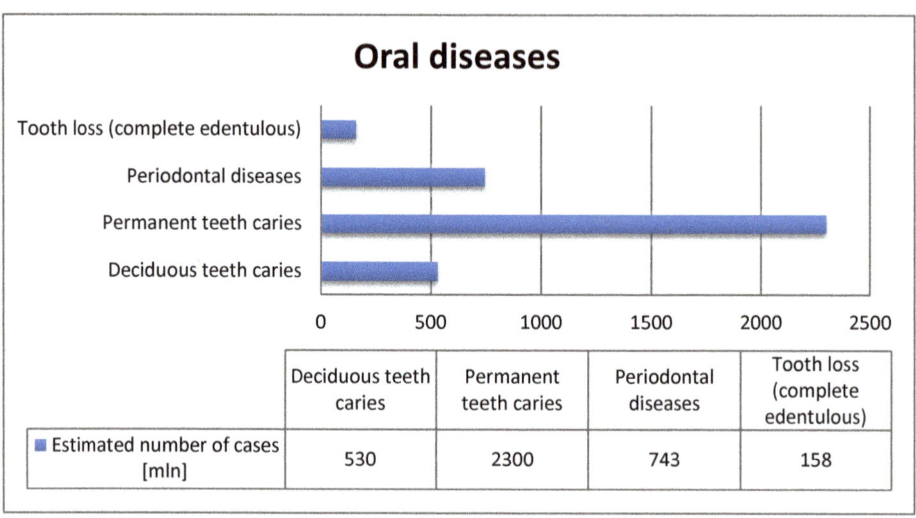

Figure 2. A graph showing the number of patients affected by oral diseases. Teeth caries is one of the most common illnesses in the world.

The consequence of primary tooth caries is a high risk of infection in permanent teeth; for this reason, it is important to prevent early childhood caries (ECC). Ozone therapy seems to be an appreciable method to supervise carious lesions. Ximenes, Cardoso, and Astorga et al. examined the potential of gaseous and aqueous ozone in the reduction of *S. mutans*, *L. acidophilus* and *E. Faecalis*: scientists took note of the last-mentioned bacteria, which is not a typical pathogen for ECC, but it is the species with the highest resistance to antimicrobial factors. Researchers point out good effectiveness of both methods in the evaluation of ECC [70]. Other studies show theefficacy of gaseous ozone in the reduction of *Lactobacillus* sp. in deep carious lesions to be similar to the efficacy of chlorhexidine [39]. On the other hand, Hauser-Gerspach et al. point out in their research that there is no significant difference in the reduction of viable bacteria in cavitated carious lesions in children in vivo, while aqueous ozone seems to be more effective in some cases [71]. Duangthip et al. also indicate a lack of efficacy of ozone in controlling dentin caries [72]. Another study points to ozone therapy as a good alternative for patients experiencing anxiety [6]. Kalnina [73] found no significant difference in the efficacy of ozone and fluoride varnish and sealants in reducing the probability of caries in permanent tooth. Similar study results were observed in the remineralization of initial caries in enamel. Due to these properties, ozone can be a powerful alternative for the prevention and treatment of caries. The author drew attention to the high cost of ozone generator compared to fluoride varnish and sealants [73]. Microleakage is a common problem in dentistry, withsecondary caries as its frequently observed effect. According to the study, ozone has no negative influence on pit and fissure sealants which protect the surface in more than 85% of caries in children. Ozone can be used to disinfect a previously prepared surface without affecting the adhesion of pit and fissure sealants [74]. A similar observation on the subject of initial micro-tensile bond strength of a self-etch adhesive to dentin was provided by Dalkilic et al., whoindicated a safe use of ozone as a disinfectant [75]. Another advantage of ozone is the absence of pain during the therapy, which makes it easier to work with anxious patients [76]. Almaz et al. noted the wide range of applications of ozone and effective properties in caries treatment, prevention, and remineralization. However, some researchers indicate inefficacy of ozone [77].

Mineral Trioxide Aggregate (MTA) is often used in the case of operator errors, such as root perforation or exposure of the pulp. This material can be modified with TiO_2 and other nanoparticles to improve the properties. TiO_2 provides an antimicrobial effect, self-cleaning and photo-elastic properties [78]. Another material commonly used in dentistry is glass ionomer (GIO), whose main components are calcium or strontium alumino-fluoro-silicate glass powder, which reacts with water-soluble acidic polymer. The resultant material is valued for the ability to release fluoride ions to the surface which stop the cariogenic process. Addition of TiO_2 nanoparticles increases the durability of GIO (especially to compressive strength), which is a significant disadvantage of the material. Garcia-Contreras et al. [79] mentioned a better antibacterial effect of GIO modified with TiO_2 nanoparticles. However, the sensitivity of GIO to water has increased, which caused the loss of material. Ferrando-Magraner et al. concluded in their scientific research that the most beneficial application of titania nanoparticles is for dental bonding material [80].

Cavities filled with dental material after inaccurate preparation can cause secondary caries. Theaddition of nanoparticles to materials can prevent the need of replacement. Composite resin modified with Ag doped with ZnO nanoparticles demonstrates antibacterial properties. Studies indicate that compressive strength does not change significantly in comparison to unenriched composite resin. Another property of nanoparticles is the reduction of bacterial biofilm, which is more resistant than planktonic bacterial cells [81], [82]. Koohpeima et al. used Ag nanoparticles as a coating before the application of etch-and-rinse and self-etch adhesive systems. The strength of bond did not change negatively, Ag NPs did not affect the color, which could have been the case under the influence of metal NPs. Together with Ag NPs, it increases the chances of proper treatment [83]. Vazquez-Garcia et al. indicates that calcium silicate cements, white MTA and Portland cement with 30% of ZrO_2, after manipulation with AgNPs solution, have good effectiveness against *E. faecalis* in both forms of biofilm and planktonic cells. Both materials have similar application, and addition of NPs does not negatively affect their properties [84]. Elgamily et al., in studies concerning a cavity disinfectant containing AgNPs and AuNPs in comparison to chlorhexidine (CHX), examined their impact on *S. mutans*. Both agents reduced the amounts of bacteria, but CHX was more effective [85]. Silver diamine fluoride (SDF) is an effective anticariotic agent which currently includes AgNPsas a result ofthe development of nanotechnology. Scientists recommend it, especially for children, as the agent has good antimicrobial properties [86]. On the other hand, Fakhruddin et al. [87] point out that even though in vitro results are positive, there is no evidence of good effectiveness in in vivo studies.

Amiri et al. examined the influence of CuO Np on oral bacteria and *Candida* species. According to research, CuO NPs have good antibacterial properties but they have a weak impact on *Candida* species. Due to antibiotic resistance of bacteria, this nanoparticle may be a good control measure for preventing caries [21].

3.2. Endodontics

Endodontic treatment consists of proper canal preparation followed by hermetic filling. It can be achieved with the use of chemo-mechanical procedures which include abundant irrigation and effective canal shaping. Root canal instruments widen the main root canal space and remove most of the canal content, at the same time, mechanical interference produces smear layer which forecloses proper disinfection and prevents tight sealing of the root canal system. Chemical cleansing removes the smear layer and plays a major role in the overall success of endodontic treatment. The main reason why chemical interference and disinfection protocols are so crucial is the complexity of the root canal system and rich network of channels which are unreachable to instruments. The most important characteristics of irrigators are their tissue-dissolving ability and low toxicity, combined with antimicrobial effect [88]. The goal of modern endodontic therapy and modern endodontic materials is to achieve maximal disinfection of the root canal system, smallest possible area of preparation which will provide long-term success and no reinfections.Ozone is considered as a beneficial choice of canal antiseptic. Its main advantages are high antimicrobial activity, low toxicity, and the fact that they do not generate drug resistance [89]. In root canal treatment, ozone is useful to eliminate *Enteroccocus fecalis*, *Peptostreptococcus micros*,

Pseudomonas aeruginosa and *Candida albicans*. Ozone is used in many forms such as gas, water, and oil. The water and gas forms can be used in the rinsing protocol. The oil form can be used as a medical insert in cases of dental pulp necrosis. The gas form provides high penetrability to lateral channels and root deltas which increases the chance of maximal disinfection. The concept of the device is simple, ozone is generated by the machine, then it is channeled through the handpiece to the affected root canals [90]. In Figure 3 has been shown an application of ozone therapy unit. Ozone is generated in the unit and then delivered to the root canal with an endodontics cannula. Silicone cap prevents ozone leakage. It has been shown deep penetration of ozone providing maximum disinfection rate.

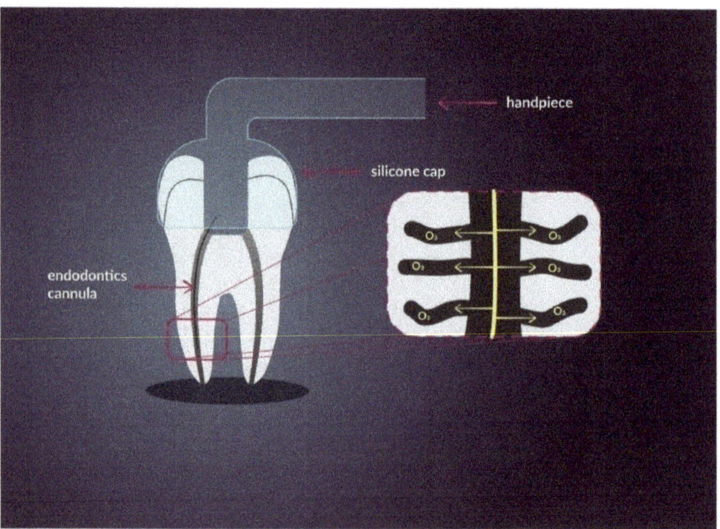

Figure 3. Aschemeof application of ozone treatment.

After treatment, the gas is converted back into oxygen by the ozone neutralizer. According to the studies, the suggested method of ozone treatment in the root canal system is at the end of chemo-mechanical preparation. To improve its effect, the amount of organic debris left inside the root canal should be reduced to a minimum [91]. Many studies were conducted to verify assumptions connected with disinfection strength of ozone. Some of them noted that ozone water efficiency was similar tothat of 5.25% NaOCl [91–96], some of them obtained the same results with ozone gas, and in their case, the efficiency was similar to that of 3% NaOCl (Sodium hypochlorite) [97]. Results to the contrary were reported in some studies, in which after using ozone water irrigation, the canal content contained bacteria [98,99]. Ozone was investigated in many disinfection protocol combinations and according to a study of Hubbezoglu et al., the best action was accomplished when ozone was used at a concentration of 16 ppm with ultrasonic agitation [94]. Noites et al. achieved the best bacteria reduction effect with 2% chlorhexidine followed by 24 s of ozone [100]. This combination can be considered in the treatment of a resorbed apex, wide open foramen. Nevertheless, these indications should be estimated with caution [101]. Recaizan et al. compared the reduction of bacteria with the use of ozone to their reduction using other antiseptics. Bacteria reduction using ozone was 90.4% and it turned out to be more efficient than using Er.YAG and KTP lasers [99]. Scanning electron microscope examination was used in order tocomparedifferent treatment protocol results including a control group, NaOCl + EDTA (Ethylenediaminetetraacetic acid), NaOCl + citric acid, and NaOCl + ozonated water groups. Indentation distance increase turned out to be comparable in every protocol. Tubule densities, tubule diameters, total surface free energy and the Lewis base polar force were higher using NaOCl + EDTA treatment compared to the control group, NaOCl + citric acid and NaOCl + ozonated water treatments. The comparison of contact angle and wettability of three sealer materials shows that the use of NaOCl + ozonated water groups significantly increases the wettability of Acroseal on the root canal dentine and decreases contact angle compared to the control group. The results for Apexite and Endomethasone material yield distinctive results, in which the usage of NaOCl + ozonated water increases the contact angle compared to the control group [102]. The Study of Bojar et al. represented results in which AH- 26 and EX fill root canal sealers shows an improved shear bond after root canal ozone treatment [103]. The conclusion is that ozone has a certain antibacterial effect, its major advantage is non-toxicity and penetrability, however ozone cannot be used alone as acanal antiseptic, although studies show that it might improve root canal therapy and should be included in the treatment protocol.

Nanotechnology can also be applied in endodontic treatment. It is considered to be a stimulating and up-to-date topic that needs to be elaborated. Nanoparticles are minute solid particles with a diameter of 1–100 nm [104]. Their advantages are ultra-small sizes, high surface area to mass ratio and high chemical reactivity. Their antimicrobial effect is the result of large surface area and high charge density. AgNPs is one of the nanomaterials that can be considered in endodontic treatment. It has antibacterial effect on both Gram-positive and Gram-negative bacteria. It also shows antiviral and antifungal action. The mechanism responsible for the antimicrobial effect is still not fully understood. The most plausible mechanism is the interruption of ATP particle and preventing DNA replication by free silver ions, production of reactive oxygen or direct damage of cell membrane by Ag^+. It is certain that free silver ions play a crucial role in antibacterial action. Scientific evidence shows that release of Ag^+ from smaller particles (<10 nm) is higher than from largerones. One of the disadvantages of the AgNPs is possible toxicity. AgNPs above 80 µg/mL could be considered cytotoxic [24].Preclinical studies on rats show effectssuch as accumulation in organs, discoloration, cytotoxicity or even sperm cell production disturbance [105,106].

AgNPs can be used for coating gutta-percha. Shantiaee et al. compared nanosilver-coated gutta-percha to an uncoated one and it resulted in slightly less leakage [107]. Lotfi et al. compared AgNPs solution to NaOCl. The size of silver particle used in this examination was 35 nm. Their results report a similar antibacterial effect of AgNPs and 5.25% NaOCl [108]. González-Luna et al. obtained the same results but with comparison to 2.25% NaOCl [109]. A study of Abbaszadegan et al. shows satisfactory annihilation of *E. faecalis* [110]. Nevertheless, some studies show that AgNPs were less effective against *E. faecalis* than NaOCl, therefore, at the moment, AgNPs can not be considered as a substitute for NaOCl [104]. In Figure 4 has been shown, in simplification, the use of Ag^+ ions in irrigation solution for endodontic application. Their advantages are ultra-small size, high chemical reactivity, and antimicrobial properties. The solution enables good disinfectant penetration and good antibacterial effect.

Another interesting application of AgNPs is adding them to the mineral trioxide aggregate (MTA). The idea was to improve activity against strict anaerobes. The outcome showed efficiency against different bacteria and *C. albicans* higher than in the case of unmodified MTA [111]. Another tested combination was $Ca(OH)_2$ and AgNPs. The result was that $Ca(OH)_2$ showed better antibacterial effect in comparison to AgNPs alone or AgNPs + $Ca(OH)_2$ [109]. The addition of 0.15% AgNPs and 2.5% dimethylaminohexadecyl methacrylate (DMAHDM) to the AH Plus paste significantly increased antibacterial activity against *E. Faecalis* [112]. Chlorhexidine (CHX)-AgNPs containing lyotropic liquid crystals showed excellent sterilization and inhibitory effect on *E. faecalis* lasting for more than one month with a bacterial annihilation rate of 98.5%.

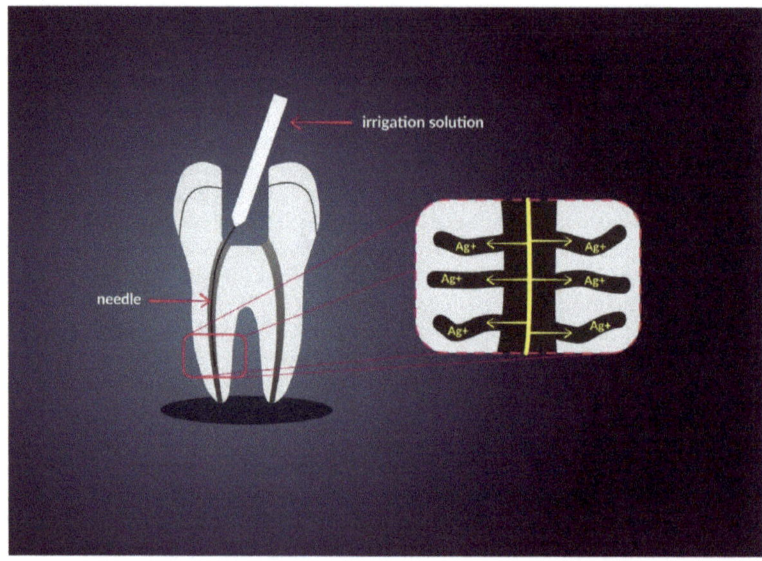

Figure 4. A scheme of irrigation with AgNPs solution. AgNPsreleaseAg$^+$ions.

Nano-MgO can also be included in endodontic treatment. Its advantages are antimicrobial activity and non-toxicity. In one of the studies, nano-MgO was compared to different disinfectants such as CHX and NaOCl. There was no substantial difference between nano-MgO solutions, 5.25% NaOCl and 2% CHX gluconate regarding the time required to inhibit *E. faecalis* and *S. aureus* growth. On the other hand, 5 mg/L nano-MgO presented better overall antibacterial efficacy in the elimination of *E. faecalis* than 5.25% NaOCl [113]. Copper and copper oxide nanoparticles present both antibacterial and antifungal effect. They produce superoxide ions, reactive oxygen and cause leakage of intracellular components that causes cell death. Antibacterial effect may be provided by hydroxyl radicals, damage of DNA and important proteins [114]. A crucial stage to obtain desirable antibacterial effect is the synthesis of NPs followed by stability in the medium, which is usually a polymeric matrix. A study by Hajipour et al. shows that the antibacterial strength of copper nanoparticles depends on both nanoparticle and bacterial concentration, temperature, pH and aeration [115]. Except for *S. aureus*, CuO NPs show higher strength compared to other nanoparticles such as Sb_2O_3, ZnO or NiO. The role of CuO nanoparticles is to create an enriched surface of the inorganic antibacterial agents. Copper NPs show bactericidal effect similar to triclosan and decrease the level of many bacteria populations, such as *S. aureus*, *E. coli* and *B. subtilis*. Cu NPs show potential as a component in endodontic materials, however further studies are needed [116]. A layer of nanometric TiO_2 can be used as a coating on super-elastic rotary NiTi instruments. The idea is to improve mechanical and chemical properties without losing cutting effectiveness. The study showed that TiO_2 coating improved cutting capacity, corrosion behavior and fatigue resistance [117].

3.3. Dental Surgery

Ozone has a wide range of applications in treatments such as implant placement, hemisections, extractions, bone regeneration and tissue healing. Often, it depends on a patient's condition, in these therapies' antibiotics are used prophylactically. It isavery important factor in dentistry to ensure that the presence of microbes in the field of work is reduced to a minimum. Researchshow significantly faster regeneration, reduction of complications and pain. Even in difficult cases of patients with diabetes and microangiopathy, ozone therapy after topical application improves treatment. Animal research show better osseointegration in cases of reduced immunity [118]. Peri-implantitis is a frequently occurring inflammatory process caused by biofilm, it leads to bone loss and other complications [119]. In some research [120], scientists reported positive influence on the clinical attachment loss (CAL) and defect bone fill, which points to ozone decontamination as an effective method in bone healing in reconstructive surgery. Third molar surgery can cause pain, swelling and trismus.Kazancioglu et al. examined the influence of ozone. Due to its antimicrobial and analgetic properties, reduction of pain was observed when the gas was used one week before and one week after the extraction. However, ozone had no impact on trismus and swelling [121]. Other studies compared ozone and laser therapies: ozone was effective in pain treatment, as opposed to low-level laser therapies (LLLT), and it was also effective in reducing swelling and trismus. These methods can be alternative to corticosteroids and nonsteroidal anti-inflammatory drugs, which can cause side effects [122].

Titanium is commonly used in implantology owing to its adequate biocompatibility, however daily hygiene activities cause harm to the surface of implants. TiN coating enhances hardness, hydrothermal and ozone treatment boosts osteoconductivity and provides decontamination. This activity increases the chance of successful treatment outcome [123]. Efficacy of ozone treatment at implant surfaces also depends on material roughness, since higher roughness increases bacterial adhesion.

In the case of titanium and zirconia, both of which were polished, and zirconia was additionally acid etched, research showed high efficiency of gaseous ozone treatment. The study demonstrated reduction of *P. gingivalis* by more than 99.94% and *S. sanguinis* bymore than 90%, ozone was applied for 24 s at 140 ppm [124]. Isler et al. mention peri-implantitis as a serious problem in dentistry, research demonstrates ozone therapy as an effective adjunction to surgical regenerative treatment. Better results of PI, GI, PD and the amount of bacteria in individual quadrants were observed by scientists. Additionally, no infraction of surfaces, in this case made of titanium and zirconia, was observed [125].

Antimicrobial properties of metal nanoparticles such as titanium dioxide, silver or copper dioxide make it possible to use them as implant modification. Amount of plaque biofilm is an important factor of implantation failure [104]. Titanium alloy is commonly used in dentistry; however, it does not show antimicrobial properties by itself [126].These properties can be improved by applying a surface coating, e.g., visible light active TiO_2. This material can also be doped with metals and non-metal particles which improve its properties. However, further research is needed to design toxic-free nanoparticles [127]. The TiO_2 nanoparticle has very beneficial antimicrobial properties which can prevent dental plaque formation [128]. Wettability of surface is important for correct osseointegration but, on the other hand, there are studies that point out better antimicrobial properties at the hydrophobic surface which characterizes TiO_2 nanotubes. They have good biocompatibility which makes them suitable material for dental implants, they are also suitable for drugdelivery. These properties increase chances of successful treatment [129]. Nano-titania coating also has a positive effect on the peri-implant tissue whose osseointegration process is accelerated [130]. There are many microbes in the oral cavity which is why antibiotics are not effective enough for every pathogen; in addition, bacterial resistance reduces the efficacy of chemotherapy. In TiO_2 nanotubes can be suspended in various substances, which further enhance its well known microbicidal properties. For better biocompability of implant nanohydroxyapatite, a coating can be added which does not affect the desired properties of titania nanotubes coated with Ag nanoparticles [126,131]. *S. aureus, S. epidermidis, E. coli* and *P. aeruginosa* are common causes of dental implant complications, such as peri-implantitis. Bacterial infection takes place even 30 min after dental treatment and nanoparticles can help in ensuring correct treatment [132]. Some research demonstrate that nanoporous titania has better antimicrobial properties than nanotube titania material, also the former ensures the release of the loaded drug for 7 days [133]. In Figure 5 has been shown the advantages of using nanoporous TiO_2 loaded with LL37 which gives positive effect on osseointegration.

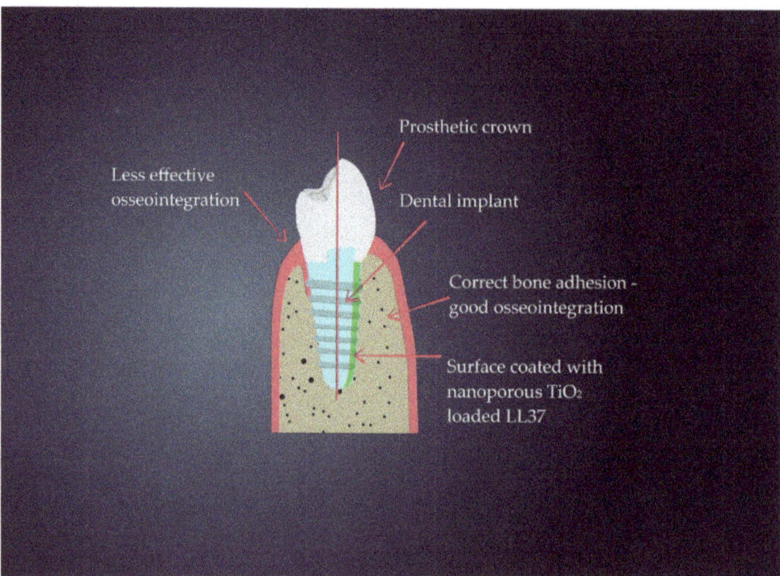

Figure 5. Ascheme showing difference in osseointegration between implant surface with and without coating.

Silver NP has a high antimicrobial and osseointegration value which makes this material good for implant coating. It reduces bacterial adhesion up to 50%. The previously mentioned titanium implant has good biocompatibility, which with the addition of silver NP coating, providesan outstanding antimicrobial effect. Nanoporous or nanotube titania coating in combination with silver NP ensure a wide range of antimicrobial properties [133,134]. In addition to the antimicrobial effect which depends on the size of nanoparticle (the smaller it is, the stronger its effect), silver NP coating applied to a titanium implant increases the density of bone [50]. Platinum nanoparticles have a wide range of applications in medicine, such as cancer treatment, bone allograft, bone loss and others. Due to its properties (antimicrobial, anti-inflammatory), PtNP may be suitable for the dental surgeries of guided bone regeneration and guided tissue regeneration [134].

3.4. Prosthetic

For removable partial dentures affect, i.e., periodontal tissues, maintaining their hygiene is a very important aspect that affects the long-term use of the prosthetic in the oral cavity by the patient. Therefore, new sterilization methods are needed that help maintain the hygiene of the prosthesis, and thus eliminate the formation of plaque (control its formation). Recently, the use of ozone, which has antibacterial properties, seems to be very promising. The plaque on the prosthesis that causes periodontitis consists mainly of *Lactobacillus*, *Propionibacterium* and *Arcalinia* bacteria. The *Propionibacterium* species were the dominant part of the flora. An equally important factor influencing inflammation of the oral mucosa from the prosthesis are yeasts of the *Candida albicans* genus. *C. albicans* produces large amounts of acetic and pyruvic acid, which causes inflammation. The yeast adheres well to the denture base materials in vitro, although the adhesion depends on the strains and environmental conditions. Denture-induced stomatitis is routinely encountered in clinical practice as a symptom of the plaque build-up on the dental surface (dentures), therefore effective plaque control should be initiated to prevent these consequences. An effective method is the use of ozone as a denture cleaning agent. The advantages of ozone in the water phase are its potency, ease of application, lack of mutagenicity, quick bactericidal action and suitability for use as a solution for soaking medical and dental instruments [134,135]. The use of ozone as a denture cleaner is effective against the methicillin-resistant *S. aureus* and viruses [136].Ozone can be applied (used) to clean the surface of alloys in removable partial dentures with little effect on alloy quality in terms of reflectance, surface roughness and weight. Direct exposure to ozone gas was a more effective germicide compared to ozonated water. Therefore, ozone gas can be clinically useful for the disinfection of removable dentures [137]. In dentistry, ulcers can occur as a result of using dentures. They are usually located in the vestibular furrows of the maxilla and mandible, causing pain. The treatment of ulcers requires the discontinuation of the use of the prosthesis, its correction, laser therapy and oral and prosthesis hygiene. Treatment also includes the use of topical medications such as chlorhexidine mouthwashes, topical hydrogels, hydrogel dressings, and sometimes topical cortisone. The use of ozone in the treatment of ulcers also seems to be an interesting method due to its properties. Research by Bader et al. on a group of women and men wearing prostheses showed that ozone reduces the size and pain of traumatic ulcer, and accelerates the regeneration of diseased tissue, and thus shortens the treatment [138].

The field of prosthetics is another branch of dentistry in which nanoparticles can be implemented. Their main role is to provide better antimicrobial properties of materials used for making removable dentures. Without proper hygiene, these restorative alternatives can lead to denture stomatitis and cause both reversible and irreversible consequences. One of the studiescompared antifungal activity between a standard denture base (polymethylmethacrylate—PMMA) and one modified with AgNPs. The conclusion was that the addition of AgNPs to the denture base provides 10^5 less *C. albicans* adhesion than the control group after 24 and 48 h incubation [139]. Another study examined incorporation of silver-sulfadiazine-loaded MSNs into PMMA at up to 5%. Researchers examined mechanical properties and microbial effect against *Candida albicans* and *Streptococcus oralis* on removable and provisional dental restoratives. Results showed that the addition of Ag-MSNs decreased the adhesion of microbes and at the same time improvedmechanical properties of PMMA. The inorganic part of NPs provided better mechanical properties, at the same time, silver ions provided a microbial anti-adhesive effect. Li et al. also examined the addition of silver NPs and their effect on *Candida albicans* adhesion and biofilm formation. Their results demonstrated that the minimal concentration of silver NPs which provides the anti-adhesion effect on *C. albicans* is 5%. Lower concentrations do not ensure this effect [140]. Another study examined the addition of AgNPs to irreversible hydrocolloid impression materials such as Zelgan or Tropicalgin. Results showed better antimicrobial activity compared to standard materials. Adding silver NPs can also provide an increase of gel strength, permanent deformation or the flow of the material, although many of mechanical properties show different variations depending on the wt% of AgNPs [141].AgNPs can also be added to alginate impression powder to increase antimicrobial activity and reduce the risk of cross-contamination by bacteria, viruses or fungi [142]. Köroğlu et al. investigated the effect of adding solution of AgNPs to acryl liquid used for mixing with the powder part of acrylic material. In their study, 0.3, 0.8 and 1.6 wt% of AgNPs were used. Results showed that the addition of 0.8 and 1.6 wt% AgNPs decreased the flexural strength and elastic modulus of microwave-polymerized acrylic resin [143]. In the study of Oei et al., silver ions improved both antimicrobial and mechanical properties of PMMA and the antibacterial activity increased to 28. The study of Matsuura et al. showed that AgNPs implemented in tissue conditioners support antifungal activity against *C. albicans* and antibacterial activity against *S. aureus*, *Pseudomonas aeruginosa* for 4 weeks [144]. Many studies indicate that the addition of AgNPs increases antimicrobial properties of the material and at the same time, doesnotaffect its mechanical properties. It can be considered as a promising component of denture base, dental impression material or tissue conditioner, although further studies are needed. AgNPs can also be considered as a component of permanent prosthetic restoration materials such as porcelain. The idea is to improve fracture toughness and prevention of crack propagation. One of the studies revealed that AgNPs increase the fatigue parameter, shorten the time required for slow crack growth and reduce crack growth rate [145]. Fujieda et al. proved that fracture toughness and Young's modulus increased with addition of AgNPs and Pt-NPs [146]. Another study showed that the addition of silver ions improved mechanical properties of Computer Aided Design/Computer Aided Manufacturing (CAD/CAM) blocks and at the same time, decreased crack length [147]. Mohsen et al. examined the effect of AgNPs on the color of ceramics. Their results showed that the addition of AgNPs affected the color quality of dental ceramic. According to numerous studies, AgNPs should be considered as an addition to porcelain, due to improvement of mechanical properties. Another nanoparticle tested in the field of prosthetics is Pt nanoparticles. One of the studies examined modified PMMA denture acrylic containing platinum nanoparticles. The size of the processed Pt NPs was less than 5 nm. Results showed an anti-adherent effect provided by reducing the area available for bacterial adhesion, at the same time, no Pt ion leaching was observed, orit was observed at extremely small amounts [59]. The addition of TiO_2 nanoparticles in prosthetic treatment has been investigated in many studies. Researchers examined the physical, mechanical and biological behavior of a $PMMA/TiO_2$. In one of studies, $PMMA/TiO_2$ nanocomposite specimen containing 3 wt. % TiO_2 NPs was examined. The results revealed that TiO_2 NPs improved stiffness and mechanical behavior of the PMMA matrix. They also provided both antibacterial and anti-adhesive effects [148]. Some studies suggest that TiO_2 antimicrobial effect is based on the photocatalytic effect, that is attributable to deactivation of cellular enzymes that lead to higher permeability and to cell death [149]. The study by Tsuji et al. showed that an increased amount of TiO_2 can weaken the material and cause internal decomposition, therefore adequate amounts of NPs needs to be used [150].

4. Cytotoxicity

4.1. Ozone

As it was mentioned before, there are three different therapeutic ways in which ozone canbe used in dentistry and other fields of medicine. In dentistry, ozone can be applied in the form of ozonated oil (sunflower oil etc.), ozonated water or simple gaseous ozone.

An in vitro study provided by Borges et al. showed the safety of ozone treatment towards a human skin keratinocyte cell line (HaCaT) and a murine fibroblast cell line (L929). Data demonstrated increased relative cell number of murine fibroblasts and human keratinocytes after treatment with ozone at a concentration of 8 μg/mL.With the use ofa wound healing assay and a scratch assay, it can be clearlyseen that the rate of the wound healing process is the same as in the control group for both cell lines [151]. Another study conducted by Huth et al. evaluated the cytotoxicity of gaseous and aqueous ozone (in PBS) and other antimicrobial agents such as sodium hypochlorite (NaOCl), chlorhexidine digluconate (CHX), hydrogen peroxide 3% (H_2O_2) for 1min exposure (a clinically relevant time period for ozonated water and gaseous ozone application) and metronidazole for 24 h exposure to gingival fibroblasts (HGF-1) and human oral epithelial (BHY) cells. Results reveal that less toxic features were observed in HGF-1 and BHY cells treated with ozonated water (1.25 μg/mL), which showed the most biocompatible properties, when compared to gaseous ozone, NaOCl, CHX or H_2O_2, which occurred to be the most toxic. An apoptotic assay, used to measure caspase-3 and -7 activity, revealed no alteration after treatment with ozonated water (ozonated PBS) when compared to control in both cell lines. Additionally, metronidazole demonstrated no significant cytotoxic effect towards gingival fibroblasts nor human oral epithelial cells [152]. Similar data was provided later by Colombo et al., who used ozonated olive oil and the CHX agents Corsodyl Dental Gel® and Plak Gel®. They evaluated their biocompatibility on immortalized human gingival fibroblasts (HGFs). Each agent was diluted 1:10 four times in Dulbecco's Modified Eagle's Medium (DMEM), which resulted in receiving different concentrations of the tested compounds (1:10, $1:10^2$, $1:10^3$, $1:10^4$). After 2 and 24 h of treatment with the aforementioned agents, cells showed 100% viability when treated with ozonated olive oil in all tested concentrations when compared to the control. CHX compounds showed severe cytotoxicity in the highest concentration (1:10) for Corsodyl Dental Gel® and Plak Gel® and moderate cytotoxicity in lower concentrations (1:10², 1:10³, 1:10⁴) [153]. Kashiwazaki et al. tested ozonated water and hand disinfectants which contained 83% ethanol, 1% chlorhexidine, 1% chlorhexidine ethanol, 0.2% benzalkonium chloride and 0.5% povidone-iodine on a human keratinocyte cell line in a three-dimensional cultured human epidermis model. Cells were divided into two groups, the first group, 1-week cultured, which developed an immature stratum corneum (SC), and the second group, 2-week cultured, which developed a mature SC. Histological changes, cell viability and release of interleukin 1α were evaluated after treatment with ozonated water and hand disinfectant. There was no histological alteration after treatment in 1-week cultured cells and 2-week cultured cells, and also, no vacuolar cell formation was observed. A viability assay showed no cytotoxic effect of ozonated water and interleukin 8 activation was not observed, which indicatesthe high biocompatibility and immuno-compatibility features of the tested ozonated water [154]. Another study evaluated non-toxic and anti-tumor properties of ozonated water in anin vivo mouse model with the useof tumor-bearing mice and the control group. No injurious effect on normal tissue, such as spleen, small intestine, liver, kidney and muscle, was observed after 24 h direct application of ozonated water, even at a relatively high concentration (208 mM). Nonetheless, tumor tissue exhibited a decreased proliferation rate, inhibited growth and showed features of necrosis, which indicatesan increase of ROS levels [155].

4.2. Titanium Dioxide

Titanium dioxide nanoparticles (TiO_2 NPs) can be synthesized via numerous methods such as the sol-gel method, the hydrothermal method, the co-precipitation method, sluggish precipitation, hydrolysis, and simple precipitation [156], but also with the use of the solvothermal or direct oxidation method [157]. With the use of different synthesis methods, we can obtain various sizes of nanoparticles which oscillate from ≤ 10 to ≥ 100 nm and even from 200 to 300 nm, and different polymorphic phases such as rutile, brookite or anatase [156,158–160].

TiO_2 NPs are widely used in physics and chemistry due to their photochemical activity and physicochemical stability [158]. Titanium dioxide nanoparticles find their application in gas sensors, solar energy converters, pigments, ceramic supports, but can also be used to purify wastewater [159,161]. Moreover, TiO_2 NPs are extensively used in UV filters in sunscreens and other everyday cosmeticssuch as lip balm, creams, foundations and also toothpaste [162]. In the food industry, TiO_2NPs are used as a white pigment in food or food supplements [160]. In dentistry, titanium dioxide nanoparticles can be found in tooth-bleaching gels or dental composites used in orthodontics, in dental acryl resins [163,164].

Due to the many applications of TiO_2 NPs, it is crucial to evaluate their potential cytotoxic effect on cells and tissues. Skin, gastrointestinal tract and respiratory system are the most exposed to TiO_2 NPs, especially when applied in cosmetics such as sunscreens, food, or when used in the field of dentistry.

The cytotoxic and genotoxic effect of titanium dioxide nanoparticles was evaluated by Meena et al. The results obtained by them pointed out thedose-dependent (50, 100 and 200 μg/mL) and time-dependent (24, 48 and 72 h) toxic effect of TiO_2 NPs on human embryonic kidney cells (HEK-293). TiO_2 nanoparticles induced elevated lactate dehydrogenase releasedfrom the cell, and damage of cell membrane which led to cell death. Moreover, cells treated with 100 and 200 μg/mL of TiO_2 nanoparticles exhibited an increased level of reactive oxygen species and elevated level of proapoptotic proteins such as Bax, caspase-3 and upregulation of the p53 protein in response to DNA breakage [165]. Another study conducted by Shukla et al. confirmed the cytotoxic and genotoxic effect of nano-TiO_2 on a human liver cell line (HepG2). Relatively low concentrations of nanoparticles (20, 40, 80 μg/mL) increased the level of ROS, which led to DNA breakage via the oxidative stress-dependent pathway. Additionally, upregulation of p53, Bax, caspase-9 and caspase-3 was observed, however Bcl-2 expression was reduced, which points to apoptosis via the mitochondrial, and thus by the caspase-dependent pathway [166]. This indicates that titanium dioxide nanoparticles should be used with caution and in very low concentrations. Due to the fact that the respiratory system is also exposed to the inhalation of TiO_2 NPs, it may lead to an increased risk for lung health. TiO_2 NPs were tested towardsa human lung cancer cell line (A549) to evaluate its potential genotoxic and cytotoxic characteristics. A relatively low concentration of the tested titanium dioxide NPs (10 and 50 µg/mL) after 6to 24 h of incubation led to an increase in the level of ROS, p53 and p21. Similar to the above-mentioned research, Bcl-2 level was downregulated at the mRNA and protein level; moreover, the cleavage of caspase-3 was observed. This data suggests that TiO_2 NPs cause alteration in gene expression which leads to apoptotic changes and, as a consequence, to human lung cancer cell death [167]. Another research compared five different TiO_2 NPs: 9 nm rutile (R9), 5 nm rutile (R5) and 14 nm anatase (A14), the commercially available 60 nm anatase (A60) and P25, which contained 80% anatase and 20% rutile, and their potential cytotoxic effect on a human bronchial epithelial cell line (BEAS-2B), a human type II alveolar epithelial cell line (A549) and human bronchial epithelial cells (NHBE). It occurred that the level of ROS was increased especially after 24 h of incubation in all five different TiO_2 NPs; however, the strongest effect was observed in the P25 sample in NHBE and BEAS-2B cells after 2 and 24 h. For A549 cells, smaller particles (5 and 9 nm) of TiO_2induced elevated release of intracellular ROS. Cell viability did not drastically alter allofthe tested cell lines, however, after incubation with a relatively high concentration (400 µg/mL), the viability of NHBE and A549 cells slightly dropped. Interleukin 8 (IL-8) plays a key role as a chemoattractant for neutrophils and other granulocytes and stimulates their migration to the infection site. After treatment with five different types of TiO_2 nanoparticles, the level of the pro-inflammatory mediator IL-8 was substantially elevatedin all the cell lines when compared to the control group. The strongest effect was observed for anatase (A60), however A14 seems not to induce IL-8 upregulation [168]. Titanium dioxide nanoparticles (TiO_2 NPs) were also tested against a human epidermal cell line (A431) to establish their potential risk to human skin cells. At low concentrations of 8 and 80 µg/mL and short time of incubation (6 h), a decreased level of glutathione and strongly increased level of reactive oxygen species was observed, which led to apoptosis. Genotoxic characteristics, evaluated with the use of a commitment assay, showed damage of DNA at the concentrations of 8 and 80 µg/mL after 6 h of treatment. These data clearly indicate that TiO_2 nanoparticles should be used at very low concentrations and with high caution [169].

Independent studies conducted by Xue et al., Gao et al. and Wright et al. [170–172] have proven the cytotoxic effect of titanium dioxide nanoparticles on the human skin keratinocyte cell line. Xue tested different sizes (4, 10, 21, 25 and 60 nm) and different forms (anatase/rutile, rutile and anatase) under UVA radiation. Results clearly pointed to the toxic effect on HaCaT cells and an increased amount of apoptotic cells, which was induced in a dose-dependent manner, inwhen treated with 10 and 25 nm particles at a concentration of 200 µg/mL. However, UVA radiation alone did not affect cell viability. Elevated levels of reactive oxygen species (ROS) and malondialdehyde (MDA) were also observed and, at the same time, there was a decreased amount of superoxide dismutase, which indicates cell damage [170]. TiO_2 nanoparticles (25 nm) when compared to nanosized bismuth oxybromide (BiOBr) demonstrated a stronger toxic effect on ahuman keratinocyte cell line. Enhanced ROS was also confirmed at a concentration of 25 µg/mL of TiO_2NPs. Titanium dioxide nanoparticles also induced early and late apoptosis of HaCaT cells and the cell cycle was found to be arrested after treatment with TiO_2NPs [171]. Wright analyzed three different sizes of particles (a 1 mm particle composed of 100% rutile, a 21 nm particle composed 80% of anatase and 20% of rutile and a 12 nm particle which contained only anatase). Data demonstrated dose-dependent elevated caspase-8 and caspase-9 activity and increased apoptosis of the tested cells. It is worth noting that cells did not exhibit malignant transformation when treated with TiO_2 NPs and exposed to UVC radiation [172]. In vivo studies conducted on mice and rat models also confirmed toxic characteristics and increased inflammatory response, which indicates oxidative tissue damage in rats [173].

Mouse model showed in vivo toxicity and DNA disruption of liver tissue and a decrease of glutathione peroxidase, which clearly indicate liver tissue damage [174].

4.3. Silver

Many methods are known for the synthesis of silver nanoparticles (AgNPs). With the use of physical, chemical and biological methods, various AgNPs with desirable morphology can be obtained in order to achieve the best properties [175]. Therefore, silver nanoparticles can be used in different fields of science; for example, for single molecule detection [176] or by enhancing electrical conductivity, AgNPs can be intended for high-frequency electronic applications [177]. However, in the biological approach, silver nanoparticles have been known mostly for their antibacterial properties. Due to these characteristics, they can be used as an efficient disinfectant to sterilize surfaces or medical equipment, and devices in the industry during the food packaging process or in environmental usage as air and water disinfectant [178]. AgNPs can also be used to impregnate and functionalize textiles [179], but they can also be used as coating for wood flooring [180].

Despite their various properties, silver nanoparticles exhibit a toxic effect in vitro and in vivo. Many studies revealed that silver nanoparticles increase the level of ROS in cells and therefore increase oxidative stress and cause defects in DNA structure [181,182]. AshaRani et al. [183] evaluated the cytotoxicity effect on human cell lines using human glioblastoma cells (U251) and normal human lung fibroblast cells (IMR-90). The size of the tested Ag nanoparticles oscillated from 6 to 20 nm, however cell viability and ATP level drastically dropped in a time- and concentration-dependent manner, especially after 48 and 72 h incubation. Further investigation revealed mitochondrial damage and increased level of ROS, as well as arrest of the cell cycle in the G2/M phase and disruption of DNA structure [183]. Another study confirmed a dose-dependent and time-dependent toxic effect of silver nanoparticles and silver ions on human osteoblasts (OB) and primary human mesenchymal stem cells (MSC). A severe decrease of cell viability as well as increased oxidative stress was observed already in 10 µg/g of the tested AgNPs after 21 days of incubation with the tested compounds. The human osteoblast (OB) cell line proved to be more sensitive to exposition to silver nanoparticles [184]. Another research confirmed toxic properties of both silver ions and silver nanoparticles towards the human lymphoblastoid TK6 cell line. It occurred that both forms of AgNPs induced cytotoxicity and genotoxicity at comparable concentrations, which oscillated from 1.00 to 1.75 µg/mL of the tested compounds. Additionally, genotoxicity was confirmed with the use of the micronucleus assay and the results clearly showed an increased level of reactive oxygen species (ROS) at relatively low concentrations of both forms of silver after 24 h treatment. The expression of genes involved in oxidative stress, such as glutathione peroxidase 7, thyroid peroxidase and heme oxygenase, and expression of genes in response to cell cycle arrest caused by DNA damage, such as cyclin-dependent kinase inhibitor 1A, were substantially elevated. This study clearly indicates that both silver ions and silver nanoparticles are toxic by leading to an increase in cellular stress and DNA structure damage [185].

The main problem is that in vitro toxicity of silver nanoparticles and silver ions occurs at relatively low concentrations and, on the other hand, antimicrobial properties demand substantially elevated concentrations of AgNPs. Therefore, it is of immense importance to estimate the most favorable conditions, ensuring both antibacterial properties and safe use with regard to cell lines and, even further, for in vivo applications. Hence, Albers et al. [184] estimated in vitro cytotoxicity of silver ions and nanoparticles in osteoclast (OCs) and osteoblast (OBs) cells at antibacterial concentrations against *S. epidermidis*. The study showed a size-dependent and concentration-dependent toxic effect of AgNPs. The most cytotoxic effect was observed after treating OBs and OCs cells with silver nanoparticles sized 50 nm, and the larger the particles were, the weaker cytotoxic effect was observed. This suggests that the smaller the size of the nanoparticle surface is, the more silver is released to the environment. Moreover, OBs cells were more sensitive to the used nanoparticles. As compared with antibacterial activity, which was obtained at a concentration of 8 mg/mL for

AgNPs towards *S. epidermidis*, 50% of cell viability was maintained at a concentration of 0.048 mM for osteoclast cells, and for the same concentration, the viability of osteoblast cells was highly below 50%. Thus, at the present moment, there are no possibilities to combine antimicrobial and cyto-safety properties [184]. Pérez-Díaz et al. showed similar data but compared the viability of dermal human fibroblasts at an antimicrobial concentration of sliver nanoparticles against *Streptococcus mutans*. The data clearly indicated that AgNPs concentrations higher than 10 ppm represented acytotoxic level, while reduction of biofilm growth was observed at the concentration of 100 ppm [186].

On the other hand, combined materials, such as Ag NP-coated titanium dental implants with hydroxyapatite (HA) applied to the surface, seem to be promising solutions, because after 7 days, primary human osteoblasts showed biocompatibility with the tested material. Osteoblasts also demonstrated adhesive properties toward the tested material and there were no alterations in cell morphology, and elevated levels of alkaline phosphatase and lactate dehydrogenase were not observed either [187]. This suggests that complex materials such as coating titanium implants with nanoparticles may be the best solution to obtain cyto-safety properties in vivo and in vitro.

4.4. Platinum

Platinum, due to its high biocompatibility, its corrosion resistance and the possibility to visualize it under X-rays, is widely used in the field of biomedical sciencesinsurgical instruments, implantable electronic devices and implantssuch as cardioverter-defibrillators, stents, knee or hip implants as well as dental implants [188].

However, platinum nanoparticles (PtNPs) demonstrate different features when compared to platinum solid implants. PtNPs can be obtained via a variety of methods, usually through the reduction of Pt ions in the liquid phase with a stabilizing or capping agent to produce nanoparticles in the form of colloids, eventually via the microemulsion method or via reduction and impregnation of Pt ions into micro-porous base [189]. Due to their method of synthesis, different shapes can be obtained, such as isolated nanoparticles, dendrites or crystalline nanowires with various sizes, which can be used in the industrial application, in optical fields, as a catalyst in fuel cells and biosensors [190–193].

Some researchers pointed out a size-dependent toxic effect on macrophage cell Raw 264.7 viability. An increased cytotoxic effect was observed when cells were treated with platinum nanoparticles sized 5 nm as compared to PtNPs sized 30 nm, asa smaller size of nanoparticles can be easily taken upby cells; nevertheless, both samples of nanoparticles exhibit a toxic effect. A dose-dependent toxic effect was also observed, cell viability decreased drastically at a concentration of 5 ppm of the tested nanoparticles. Moreover, data indicated that density and cell morphology were altered after treatment with PtNPs. Additionally, platinum nanoparticles induced the activation of caspase-3 and caspase-7, which led to nucleus fragmentation and apoptosis, which also indicates that an elevated level of caspases arrested the DNA repair process [194]. Another study conducted by Konieczny et al. [195] confirmed the cytotoxic effect of platinum nanoparticles on human skin cells. The authors have used PtNPs with an estimated size of around 5.8 and 57 nm, and both were protected with polyvinylpyrrolidone. Nanoparticles were used with three different concentrations of 6.25, 12.5 and 25 µg/mL and, after 24 and 48 h treatment, the viability of normal human epidermal keratinocytes (NheKs) slightly decreased, especially when treated with smaller particles. Nonetheless, cell viability seems to be unaffected, and genotoxicity and DNA damage via activation of caspase-3 and caspase-7 were observed, primarily caused by smaller PtNPs. This study indicated dose-dependent and concentration-dependent toxic properties of platinum nanoparticles [195]. In Labrador-Rached et al.'s research, it was demonstrated that 70 nm PtNPs in a concentration of 100 µg/mL exhibit a relatively high cytotoxic effect, however a toxic effect observed at lower concentrations of 5 and 25 µg/mL was indiscernible. They also obtained elevated ROS levels in response to an upregulated release of pro-inflammatory factors such as IL-1β, IL-8 and TNF-α in a HepG2 liver model [196].

An in vivo model provided by Lin et al. showed that small platinum nanoparticles of 5 nm and larger, up to 70 nm, by acting on ion channels by the extracellular site of ventricular cardiomyocytes of neonatal mice, disrupted cardiac electrophysiology, which can lead to the threat of cardiac conduction block. However, a relatively high dose of 5 and 70 nm PtNPs did not meaningfully elevate ROS generation and lactate dehydrogenase level [197].

To compare toxic nanoparticles, green synthesis of platinum nanoparticles could be a good solution to reduce their toxic properties [198]. Some studies still demonstrate toxicity and genotoxic effect on various cell lines, such as human embryonic kidney (HEK293) cells [199] or human breast cancer (MCF-7) cells [200]. However, novel studies reveal a selective cytotoxic effect on cancer cells (MCF-7) when compared to normal human embryonic kidney cells (HEK293) [201], which could be a good direction for the usage of this particles.

5. Discussion

Ozone has a wide range of applications. In dentistry, its properties are not only antimicrobial, but immuno-stimulating, analgesic and anti-inflammatory. The use of ozone in dental surgery seems promising, especially because of its wound-healing properties. There are many caries prevention agents on the market: ozone seems to be effective against microbes after cavity preparation, but the opinions of scientists are divided [71,72]. The problem in the case of ozone is its efficacy against biofilm. Ozone therapy can help maximize disinfection effect, at the same time providing low toxicity and low possibility of drug resistance. Although ozone effectiveness shows a wide range in many studies, it can be considered as an additional disinfection protocol step. New technologies in dentistry are most often associated with a high cost of their introduction to the dental office. What also poses a problem in the case of ozone generators are substitutes that do not require investment and are at the same time effective, examples are fluoride varnish and sealants for caries prevention or NaOCl used in endodontics [73]. Apart from the discussed ozone applications in dentistry, there are many other fields in medicine where ozone can have a significant impact. Acceleration of wound-healing of oral mucosa is an example of positive properties, which offers a wide range of application possibilities [202].

Ag NPs are the most frequently discussed NPs in the dentistry literature. Due to their properties, they are widely used in medicine, for example in the form of coating in dental implants or as a modifier of dental materials. Also, many studies suggest that AgNPs can be considered in endodontic protocols as an additional irrigation or amplifier of the antimicrobial effect of other agents, such as MTA and CHX. However, the problem is to find the most favorable ratio between cytotoxicity and antimicrobial properties, which depends on the size of NPs. TiO_2 are used especially as coatings of implants or rotary NiTi instruments. They can be combined with other NPs such as Ag or ZrO_2 NPs to obtain better properties of materials. Nanoporous and nanotube titania coating may contain various substances, such as antibiotics, silver and zirconia, which improve their properties [126]. TiO_2 NPs appear to have a promising reduction effect on biofilm which is a serious problem in dentistry. However, the problem of cytotoxicity arises again. Silver and titania NPs appear more often in literature and their antimicrobial properties seem to be well proven in contrast to CuO and Pt NPs, which do not appear frequently in articles. Pt NPs have promising properties, especially in guided bone and tissue regeneration. Also, in the field of prosthetics, they are used in modified PMMA denture acrylic. However, again the problem is their cytotoxicity and the possibility of using other NPs [203]. CuO nanoparticles are used in conservative dentistry as a caries preventing agent and they seem to have a potential for application in endodontic materials. Unfortunately, cytotoxicity is a significant problem.

Different antimicrobial agents are reported to be used as the components of orthodontic adhesives. Among them are metal nanoparticles which, regarding the growing bacterial resistance to the commonly used drugs, may be considered as an alternative treatment applied as a peri-implant infection prevention measure [68]. Generally, nanoparticles, due to their small size as well as large surface area, possess unique features and, especially when combined with metal ions, can help prevent bacterial growth in the vicinity of the material [68]. The efficacy of antimicrobial treatment with inorganic nanoparticles is highly correlated with their physicochemical properties. In general, the smaller size of grains the material has, the more effective it is. Another important factor, apart from the size of particles, is their morphology, and it was established that needle-shaped particles are more likely to damage bacterial cells than spherical ones [204]. Metal nanoparticles may have bactericidal or bacteriostatic effects. The result of the former is bacterial death and the latter leads to the inhibition of growth or multiplication [204]. Most frequently, the antimicrobial effect of applied nanoparticles is based on the release of free metal ions, although other modes of action have also been noted, such as direct mechanical damage of bacterial cell after the internalization of the particles into the cell. Another mechanism is connected with the production of reactive oxygen species (ROS) [204]. Nanoparticles can also adsorb to the cell wall and, as a result of depolarization, increaseits permeability [205]. Moreover, metals lead to protein dysfunction and impair the enzymatic activity, which results in cellular metabolism malfunctioning [206]. Nanotechnology provides a wide range of modifications in dentistry, where new, better materials or implants are constantly being sought. The addition of nanoparticles can improve their properties such as their bactericidal, adhesion, or osseointegration properties. The discussed materials have a variety of applications in many fields of dentistry. The issue which is mainly elaborated in this reviewisantimicrobial properties. However, nanomaterials offer many more possibilities, as they have even more properties which enhance materials and tools used in medicine.

6. Conclusions

In summary, both ozone treatment and nanotechnology seem to have a prosperous future in dentistry, offering a wide range of applications. Ozone can be used in every field of dentistry due to its efficient antibacterial properties. Treatment with ozone may be more appropriate for people with visit anxiety. The role of Ag and TiO_2 NPs in dentistry seems to be promising. In the case of CuO and Pt NPs, the problem may be the cytotoxicity and the presence of alternative materials, e.g., other NPs. It is important to compare the ratio of the effectiveness of the discussed agents to those commonly used. Their price may also pose a problem, especially in the case of prophylactic treatments, such asozonehygienization and povidone iodine impregnation.Both ozone and nanoparticles discussed in this review have antimicrobial properties. For this reason, they can be used as an improvement to treatment methods or materials. Another advantage is the prevention of complications such as peri-implantitis and secondary caries. Nanotechnology gives a huge field for the development of new materials and methods of treatment in modern dentistry. Further research is needed for each of the agents to rediscover or find the most advantageous method of obtaining the material or using it in dentistry.

Author Contributions: Conceptualization: M.D. and R.J.W.; Methodology: A.L., N.N., J.R.-S.,W.D. and R.J.W.; Software: K.S., M.D., W.D. and R.J.W.; Validation: M.D. and R.J.W.; Formal analysis: W.Z., M.D., M.S., Z.R., K.W. and R.J.W.; Investigation: W.Z., W.D., M.D., Z.R., M.S. and R.J.W.; Data curation: A.L., N.N., J.R.-S., W.D., W.Z., M.D., Z.R., M.S., K.W. and R.J.W.; Writing—original draft preparation: A.L., N.N., J.R.-S., W.Z., M.D. and. R.J.W.; Writing—review and editing: K.W., M.D. and R.J.W.; Supervision: M.D. and R.J.W.; Project administration: R.J.W.; Funding acquisition: R.J.W. All authors have read and agreed to the published version of the manuscript.

Funding: This research was funded by the National Science Centre (NCN), grant number UMO-2015/19/B/ST5/01330, entitled 'Preparation and characterisation of biocomposites based on nanoapatites for theranostics'.

Institutional Review Board Statement: Not applicable.

Informed Consent Statement: Not applicable.

Data Availability Statement: Not applicable.

Conflicts of Interest: The authors declare no conflict of interest.

References

1. Naik, S.V.; Rajeshwari, K.; Kohli, S.; Zohabhasan, S.; Bhatia, S. Ozone—A biological therapy in dentistry—Reality or Myth????? *Open Dent. J.* **2016**. [CrossRef] [PubMed]
2. Bhateja, S. The miraculous healing therapy—Ozone therapy in dentistry. *J. Dent.* **2012**, *3*, 150–155. [CrossRef]
3. Elvis, A.; Ekta, J. Ozone therapy: A clinical review. *J. Nat. Sci. Biol. Med.* **2011**. [CrossRef] [PubMed]
4. Kumar, A.; Bhagawati, S.; Tyagi, P.; Kumar, P. Current interpretations and scientific rationale of the ozone usage in dentistry: A systematic review of literature. *Eur. J. Gen. Dent.* **2014**, *3*. [CrossRef]
5. Nagarakanti, S.; Athuluru, D. Ozone: A new revolution in dentistry. *WebmedCentral Dent.* **2011**. [CrossRef]
6. Garg, R. Ozone: A new face of dentistry. *Internet J. Dent. Sci.* **2012**, *7*. [CrossRef]
7. Sun, Y. Controlled synthesis of colloidal silver nanoparticles in organic solutions: Empirical rules for nucleation engineering. *Chem. Soc. Rev.* **2013**, *42*, 2497–2511. [CrossRef]
8. Noronha, V.T.; Paula, A.J.; Duran, G.; Galembeck, A.; Cogo-Mueller, K.; Franz-Montan, M. Silver nanoparticles in dentistry. *Dent. Mater.* **2017**, *33*, 1110–1126. [CrossRef]
9. Ahn, S.J.; Lee, S.J.; Kook, J.K.; Lim, B.S. Experimental antimicrobial orthodontic adhesives using nanofillers and silver nanoparticles. *Dent. Mater.* **2009**, *25*, 206–213. [CrossRef]
10. Tomalia, D.A.; Khanna, S.N. A systematic framework and nanoperiodic concept for unifying nanoscience: Hard/soft nanoelements, superatoms, meta-Atoms, new emerging properties, periodic property patterns, and predictive mendeleev-like nanoperiodic tables. *Chem. Rev.* **2016**, *116*, 2705–2774. [CrossRef]
11. Ruden, S.; Hilpert, K.; Berditsch, M.; Wadhwani, P.; Ulrich, A.S. Synergistic interaction between silver nanoparticles and membrane-permeabilizing antimicrobial peptides. *Antimicrob. Agents Chemother.* **2009**, *53*, 3538–3540. [CrossRef] [PubMed]
12. Huang, L.; Dai, T.; Xuan, Y.; Tegos, G.P.; Hamblin, M.R. Synergistic combination of chitosan acetate with nanoparticle silver as a topical antimicrobial: Efficacy against bacterial burn infections. *Antimicrob. Agents Chemother.* **2011**, *55*, 3432–3438. [CrossRef] [PubMed]
13. Acosta-Torres, L.S.; Mendieta, I.; Nuñez-Anita, R.E.; Cajero-Juárez, M.; Castaño, V.M. Cytocompatible antifungal acrylic resin containing silver nanoparticles for dentures. *Int. J. Nanomed.* **2012**, *7*, 4777–4786. [CrossRef]
14. Feng, Q.L.; Wu, J.; Chen, G.Q.; Cui, F.Z.; Kim, T.N.; Kim, J.O. A mechanistic study of the antibacterial effect of silver ions on Escherichia coli and Staphylococcus aureus. *J. Biomed. Mater. Res.* **2000**, *52*, 662–668. [CrossRef]
15. Antony, J.J.; Sivalingam, P.; Chen, B. Toxicological effects of silver nanoparticles. *Environ. Toxicol. Pharmacol.* **2015**, *40*, 729–732. [CrossRef]
16. Durán, N.; Durán, M.; de Jesus, M.B.; Seabra, A.B.; Fávaro, W.J.; Nakazato, G. Silver nanoparticles: A new view on mechanistic aspects on antimicrobial activity. *Nanomed. Nanotechnol. Biol. Med.* **2016**, *12*, 789–799. [CrossRef]
17. Misra, S.K.; Dybowska, A.; Berhanu, D.; Luoma, S.N.; Valsami-Jones, E. The complexity of nanoparticle dissolution and its importance in nanotoxicological studies. *Sci. Total Environ.* **2012**, *438*, 225–232. [CrossRef]
18. Cioffi, N.; Torsi, L.; Ditaranto, N.; Tantillo, G.; Ghibelli, L.; Sabbatini, L.; Bleve-Zacheo, T.; D'Alessio, M.; Giorgio Zambonin, P.; Traversa, E. Copper nanoparticle/polymer composites with antifungal and bacteriostatic properties. *Chem. Mater.* **2005**, *17*, 5255–5262. [CrossRef]
19. Cava, R.J. Structural chemistry and the local charge picture of copper oxide superconductors. *Science* **1990**, *247*, 656–662. [CrossRef]
20. Chang, Y.N.; Zhang, M.; Xia, L.; Zhang, J.; Xing, G. The toxic effects and mechanisms of CuO and ZnO nanoparticles. *Materials* **2012**, *5*, 2850–2871. [CrossRef]
21. Amiri, M.; Etemadifar, Z.; Daneshkazemi, A.; Nateghi, M. Antimicrobial effect of copper oxide nanoparticles on some oral bacteria and candida species. *J. Dent. Biomater.* **2017**, *4*, 347–352. [PubMed]
22. Battez, A.H.; Viesca, J.L.; González, R.; Blanco, D.; Asedegbega, E.; Osorio, A. Friction reduction properties of a CuO nanolubricant used as lubricant for a NiCrBSi coating. *Wear* **2010**, *268*, 325–328. [CrossRef]
23. Pan, X.; Redding, J.E.; Wiley, P.A.; Wen, L.; McConnell, J.S.; Zhang, B. Mutagenicity evaluation of metal oxide nanoparticles by the bacterial reverse mutation assay. *Chemosphere* **2010**, *79*, 113–116. [CrossRef] [PubMed]
24. Bapat, R.A.; Chaubal, T.V.; Joshi, C.P.; Bapat, P.R.; Choudhury, H.; Pandey, M.; Gorain, B.; Kesharwani, P. An overview of application of silver nanoparticles for biomaterials in dentistry. *Mater. Sci. Eng. C* **2018**, *91*, 881–898. [CrossRef]
25. Eshed, M.; Lellouche, J.; Matalon, S.; Gedanken, A.; Banin, E. Sonochemical coatings of ZnO and CuO nanoparticles inhibit streptococcus mutans biofilm formation on teeth model. *Langmuir* **2012**, *28*, 12288–12295. [CrossRef]
26. Antibacterial Effect of TiO_2 Nanoparticles on Pathogenic Strain of *E. coli*. Available online: https://www.researchgate.net/publication/230683452_Antibacterial_effect_of_TiO2_nanoparticles_on_pathogenic_strain_of_E_coli (accessed on 11 September 2020).
27. Sun, H.; Bai, Y.; Cheng, Y.; Jin, W.; Xu, N. Preparation and characterization of visible-light-driven carbon—Sulfur-codoped TiO_2 photocatalysts. *Ind. Eng. Chem. Res.* **2006**, *45*, 4971–4976. [CrossRef]
28. Popat, K.C.; Eltgroth, M.; LaTempa, T.J.; Grimes, C.A.; Desai, T.A. Titania nanotubes: A novel platform for drug-eluting coatings for medical implants? *Small* **2007**, *3*, 1878–1881. [CrossRef]

29. Rosenberg, B.; van Camp, L.; Krigas, T. Inhibition of cell division in Escherichia coli by electrolysis products from a platinum electrode. *Nature* **1965**, *205*, 698–699. [CrossRef]

30. Chwalibog, A.; Sawosz, E.; Hotowy, A.; Szeliga, J.; Sokolowska, A. Visualization of interaction between inorganic nanoparticles and bacteria or fungi. *Int. J. Nanomed.* **2010**, *5*, 1085–1094. [CrossRef]

31. Watanabe, A.; Kajita, M.; Kim, J.; Kanayama, A.S.; Miyamoto, Y. In vitro free radical scavenging activity of platinum nanoparticles. *Nanotechnology* **2009**, *20*. [CrossRef]

32. Shuhei, H.; Futami, N.; Toru, T.; Takatsumi, I.; Takahiro, W.; Kiyotaka, A. Effect of application time of colloidal platinum nanoparticles on the microtensile bond strength to dentin. *Dent. Mater. J.* **2010**, *29*, 682–689. [CrossRef]

33. Itohiya, H.; Matsushima, Y.; Shirakawa, S.; Kajiyama, S.; Yashima, A.; Nagano, T.; Gomi, K. Organic resolution function and effects of platinum nanoparticles on bacteria and organic matter. *PLoS ONE* **2019**, *14*. [CrossRef]

34. Yang, X.; Yang, J.; Wang, L.; Ran, B.; Jia, Y.; Zhang, L. Pharmaceutical intermediate-modified gold nanoparticles: Against multidrug-resistant bacteria and wound-healing application via an electrospun scaffold. *ACS Nano* **2017**, *11*, 5737–5745. [CrossRef] [PubMed]

35. Pietrocola, G.; Ceci, M.; Preda, F.; Poggio, C.; Colombo, M. Evaluation of the antibacterial activity of a new ozonized olive oil against oral and periodontal pathogens. *J. Clin. Exp. Dent.* **2018**. [CrossRef] [PubMed]

36. Tuncay, O.; Dinçer, A.N.; Kuştarci, A.; Er, O.; Dinç, G.; Demirbuga, S. Effects of ozone and photo-activated disinfection against Enterococcus faecalis biofilms in vitro. *Niger. J. Clin. Pract.* **2015**, *18*, 814–818. [CrossRef] [PubMed]

37. Camacho-Alonso, F.; Salmerón-Lozano, P.; Martínez-Beneyto, Y. Effects of photodynamic therapy, 2% chlorhexidine, triantibiotic mixture, propolis and ozone on root canals experimentally infected with Enterococcus faecalis: An in vitro study. *Odontology* **2017**, *105*, 338–346. [CrossRef]

38. Sancakli, G.Y.; Siso, S.H.; Yildiz, S.O. Antibacterial effect of surface pretreatment techniques against streptococcus mutans. *Niger J. Clin. Pr.* **2018**, *21*, 170–175.

39. Krunić, J.; Nikola, S.; Dukic, L.; Jelena, R.; Branka, P.; Ivana, S. Clinical antibacterial effectiveness and biocompatibility of gaseous ozone after incomplete caries removal. *Clin. Oral Investig.* **2019**, *23*, 785–792. [CrossRef]

40. Ajeti, N.; Pustina-Krasniqi, T.; Apostolska, S.; Xhajanka, E. The effect of gaseous ozone in infected root canal. *Open Access Maced. J. Med. Sci.* **2018**, *6*, 389–396. [CrossRef]

41. Anumula, L.; Kumar, K.V.S.; Krishna, C.H.N.V.M.; Lakshmi, K.S. Antibacterial activity of freshly prepared ozonated water and chlorhexidine on Mutans Streptococcus when used as an oral rinse—A randomised clinical study. *J. Clin. Diagnostic Res.* **2017**, *11*, ZC05–ZC08. [CrossRef]

42. Sodagar, A.; Sadegh, M.S.A.; Bahador, A.; Jalali, Y.F.; Behzadi, Z.; Elhaminejad, F. Mirhashemi, A.H. Effect of TiO$_2$ nanoparticles incorporation on antibacterial properties and shear bond strength of dental composite used in orthodontics. *Dent. Press J. Orthod.* **2017**, *22*, 67–74. [CrossRef]

43. Chen, R.; Han, A.; Huang, Z.; Karki, J.; Wang, C. Antibacterial activity, cytotoxicity and mechanical behavior of nano-enhanced denture base resin with different kinds of inorganic antibacterial agents. *Dent. Mater. J.* **2017**, *36*, 693–699. [CrossRef] [PubMed]

44. Györgyey, Á.; Janovák, L.; Ádám, A.; Kopniczky, J.; Tóth, K.; Deák, Á.; Panayotov, I.; Cuisinier, F.; Dékány, I.; Turzó, K. Investigation of the in vitro photocatalytic antibacterial activity of nanocrystalline TiO$_2$ and coupled TiO$_2$/Ag containing copolymer on the surface of medical grade titanium. *J. Biomater. Appl.* **2016**, *31*, 55–67. [CrossRef] [PubMed]

45. Besinis, A.; Hadi, S.D.; Le, H.R.; Tredwin, C.; Handy, R.D. Antibacterial activity and biofilm inhibition by surface modified titanium alloy medical implants following application of silver, titanium dioxide and hydroxyapatite nanocoatings. *Nanotoxicology* **2017**, *11*, 327–338. [CrossRef] [PubMed]

46. Lavaee, F.; Faez, K.; Faez, K.; Hadi, N.; Modaresi, F. Antimicrobial and antibiofilm activity of silver, titanium dioxide and iron nano particles. *Am. J. Dent.* **2016**, *29*, 315–320. [PubMed]

47. Rosenbaum, J.; Versace, D.L.; Abbad-Andallousi, S.; Pires, R.; Azevedo, C. Antibacterial properties of nanostructured Cu-TiO$_2$ surfaces for dental implants. *Biomater. Sci.* **2017**, *5*, 455–462. [CrossRef]

48. Dizaj, S.M.; Lotfipour, F.; Barzegar-Jalali, M.; Zarrintan, M.H.; Adibkia, K. Antimicrobial activity of the metals and metal oxide nanoparticles. *Mater. Sci. Eng. C* **2014**, *44*, 278–284. [CrossRef]

49. Kolmas, J.; Groszyk, E.; Kwiatkowska-Różycka, D. Substituted hydroxyapatites with antibacterial properties. *Biomed. Res. Int.* **2014**, *2014*, 1–15. [CrossRef]

50. Yin, I.X.; Zhang, J.; Zhao, I.S.; Mei, M.L.; Li, Q.; Chu, C.H. The antibacterial mechanism of silver nanoparticles and its application in dentistry. *Int. J.Nanomed.* **2020**, *15*, 2555–2562. [CrossRef]

51. Liu, X.; Gan, K.; Liu, H.; Song, X.; Chen, T.; Liu, C. Antibacterial properties of nano-silver coated PEEK prepared through magnetron sputtering. *Dent. Mater.* **2017**, *33*, e348–e360. [CrossRef]

52. Kim, J.H.; Park, S.W.; Lim, H.P.; Park, C.; Yun, K.D. Biocompatibility evaluation of feldspathic porcelain with nano-sized silver ion particles. *J. Nanosci. Nanotechnol.* **2018**, *18*, 1237–1240. [CrossRef] [PubMed]

53. Afkhami, F.; Pourhashemi, S.J.; Sadegh, M.; Salehi, Y.; Fard, M.J.K. Antibiofilm efficacy of silver nanoparticles as a vehicle for calcium hydroxide medicament against Enterococcus faecalis. *J. Dent.* **2015**, *43*, 1573–1579. [CrossRef] [PubMed]

54. Yin, I.X.; Zhao, I.S.; Mei, M.L.; Lo, E.C.M.; Tang, J.; Li, Q. Synthesis and characterization of fluoridated silver nanoparticles and their potential as a non-staining anti-caries agent. *Int. J. Nanomed.* **2020**, *15*, 3207–3215. [CrossRef] [PubMed]

55. Bacali, C.; Baldea, I.; Moldovan, M.; Carpa, R.; Badea, F. Flexural strength, biocompatibility, and antimicrobial activity of a polymethyl methacrylate denture resin enhanced with graphene and silver nanoparticles. *Clin. Oral Investig.* **2020**, *24*, 2713–2725. [CrossRef]

56. Lee, J.H.; El-Fiqi, A.; Mandakhbayar, N.; Lee, H.H.; Kim, H.W. Drug/ion co-delivery multi-functional nanocarrier to regenerate infected tissue defect. *Biomaterials* **2017**, *142*, 62–76. [CrossRef] [PubMed]

57. Divakar, D.D.; Jastaniyah, N.T.; Altamimi, H.G.; Alnakhli, Y.O.; Muzaheed Alkheraif, A.A.; Haleemg, S. Enhanced antimicrobial activity of naturally derived bioactive molecule chitosan conjugated silver nanoparticle against dental implant pathogens. *Int. J. Biol. Macromol.* **2018**, *108*, 790–797. [CrossRef]

58. Yin, I.X.; Yu, O.Y.; Zhao, I.S.; Mei, M.L.; Li, Q.-L.; Tang, J.; Chu, C.-H. Developing biocompatible silver nanoparticles using epigallocatechin gallate for dental use. *Arch. Oral Biol.* **2019**, *102*, 106–112. [CrossRef]

59. Nam, K.Y. Characterization and bacterial anti-adherent effect on modified PMMA denture acrylic resin containing platinum nanoparticles. *J. Adv. Prosthodont.* **2014**, *6*, 207–214. [CrossRef]

60. Grigore, M.; Biscu, E.; Holban, A.; Gestal, M.; Grumezescu, A. Methods of synthesis, properties and biomedical applications of CuO nanoparticles. *Pharmaceuticals* **2016**, *9*, 75. [CrossRef]

61. Luo, C.; Li, Y.; Yang, L.; Long, J.; Zheng, Y.; Liu, J.; Xiao, S.; Jia, J. Activation of Erk and p53 regulates copper oxide nanoparticle-induced cytotoxicity in keratinocytes and fibroblasts. *Int. J. Nanomed.* **2014**, *9*, 4763. [CrossRef]

62. Wang, Z.; Li, N.; Zhao, J.; White, J.C.; Qu, P.; Xing, B. CuO nanoparticle interaction with human epithelial cells: Cellular uptake, location, export, and genotoxicity. *Chem. Res. Toxicol.* **2012**, *25*, 1512–1521. [CrossRef]

63. Akhtar, M.J.; Kumar, S.; Alhadlaq, H.A.; Alrokayan, S.A.; Abu-Salah, K.M.; Ahamed, M. Dose-dependent genotoxicity of copper oxide nanoparticles stimulated by reactive oxygen species in human lung epithelial cells. *Toxicol. Ind. Health* **2016**, *32*, 809–821. [CrossRef] [PubMed]

64. Assadian, E.; Zarei, M.H.; Gilani, A.G.; Farshin, M.; Degampanah, H.; Pourahmad, J. Toxicity of copper oxide (CuO) nanoparticles on human blood lymphocytes. *Biol. Trace Elem. Res.* **2018**, *184*, 350–357. [CrossRef] [PubMed]

65. Sun, T.; Yan, Y.; Zhao, Y.; Guo, F.; Jiang, C. Copper oxide nanoparticles induce autophagic cell death in A549 cells. *PLoS ONE* **2012**, *7*, e43442. [CrossRef] [PubMed]

66. Beltrán-Partida, E.; Valdez-Salas, B.; Valdez-Salas, E.; Pérez-Cortéz, G.; Nedev, N. Synthesis, characterization, and in situ antifungal and cytotoxicity evaluation of ascorbic acid-capped copper nanoparticles. *J. Nanomater.* **2019**. [CrossRef]

67. Ramazanzadeh, B.; Jahanbin, A.; Yaghoubi, M.; Shahtahmassbi, N.; Shafaee, H. Comparison of antibacterial effects of ZnO and CuO nanoparticles coated brackets against streptococcus mutans. *J. Dent.* **2015**, *16*, 200–205.

68. Toodehzaeim, M.H.; Zandi, H.; Meshkani, H.; Firouzabadi, A.H. The effect of CuO nanoparticles on antimicrobial effects and shear bond strength of orthodontic adhesives. *J. Dent.* **2018**, *19*, 1–5.

69. Peres, M.A.; Macpherson, L.M.D.; Weyant, R.J.; Blánaid, D.; Renato, V.; Mathur, M.R.; Stefan, L.; Roger Keller, C.; Guarnizo-Herreo, C.C.; Cristin, K. Oral diseases: A global public health challenge. *Lancet* **2019**, *394*, 249–260. [CrossRef]

70. Ximenes, M.; Cardoso, M.; Astorga, F.; Arnold, R.; Pimenta, L.A.; Viera, R.D. Antimicrobial activity of ozone and NaF-chlorhexidine on early childhood caries. *Braz. Oral Res.* **2017**, *31*. [CrossRef]

71. Hauser-Gerspach, I.; Pfäffli-Savtchenko, V.; Dähnhardt, J.E.; Meyer, J.; Lussi, A. Comparison of the immediate effects of gaseous ozone and chlorhexidine gel on bacteria in cavitated carious lesions in children in vivo. *Clin. Oral Investig.* **2009**, *13*, 287–291. [CrossRef]

72. Duangthip, D.; Jiang, M.; Chu, C.H.; Lo, E.C.M. Non-surgical treatment of dentin caries in preschool children—Systematic review. *BMC Oral Health* **2015**, *15*. [CrossRef]

73. Kalnina, J. Ozone Therapy in Prevention and Treatment of in Caries in Permanent Teeth. 2017. Available online: www.rsu.lv (accessed on 2 January 2020).

74. Cehreli, S.B.; Yalcinkaya, Z.; Guven-Polat, G.; Çehreli, Z.C. Effect of ozone pretreatment on the microleakage of pit and fissure sealants. *J. Clin. Pediatr. Dent.* **2010**, *35*, 187–190. [CrossRef] [PubMed]

75. Dalkilic, E.E.; Arisu, H.D.; Kivanc, B.H.; Uctasli, M.B.; Omurlu, H. Effect of different disinfectant methods on the initial microtensile bond strength of a self-etch adhesive to dentin. *Lasers Med. Sci.* **2012**, *27*, 819–825. [CrossRef] [PubMed]

76. Reddy, S.; Reddy, N.; Dinapadu, S.; Reddy, M.; Pasari, S. Role of ozone therapy in minimal intervention dentistry and endodontics—A review. *J. Int. Oral Heal.* **2013**, *5*, 102–108.

77. Almaz, M.E.; Sönmez, I.Ş. Ozone therapy in the management and prevention of caries. *J. Formos. Med. Assoc.* **2015**, *114*, 3–11. [CrossRef] [PubMed]

78. Zanjani, V.A.; Tabari, K.; Sheikh-Al-Eslamian, S.M.; Abrandabadi, A.N. Physiochemical properties of experimental nano-hybrid MTA. *J. Med. Life* **2018**, *11*, 51–56.

79. Garcia-Contreras, R.; Scougall-Vilchis, R.J.; Contreras-Bulnes, R.; Kanda, Y.; Nakajima, H.; Sakagami, H. Effects of TiO$_2$ nano glass ionomer cements against normal and cancer oral cells. *In Vivo* **2014**, *28*, 895–908.

80. Ferrando-Magraner, E.; Bellot-Arcís, C.; Paredes-Gallardo, V.; Almerich-Silla, J.M.; García-Sanz, V.; Fernández-Alonso, M.; Montiel-Company, J.M. Antibacterial properties of nanoparticles in dental restorative materials. A systematic review and meta-analysis. *Medicina* **2020**, *56*, 55. [CrossRef]

81. Dias, H.B.; Bernardi, M.I.B.; Marangoni, V.S.; Bernardi, A.C.d.; Rastelli, A.N.d.; Hernandes, A.C. Synthesis, characterization and application of Ag doped ZnO nanoparticles in a composite resin. *Mater. Sci. Eng. C* **2019**, *96*, 391–401. [CrossRef]

82. Kasraei, S.; Sami, L.; Hendi, S.; AliKhani, M.-Y.; Rezaei-Soufi, L.; Khamverdi, Z. Antibacterial properties of composite resins incorporating silver and zinc oxide nanoparticles on Streptococcus mutans and Lactobacillus. *Restor. Dent. Endod.* **2014**, *39*, 109. [CrossRef]

83. Koohpeima, F.; Mokhtari, M.J.; Khalafi, S. The effect of silver nanoparticles on composite shear bond strength to dentin with different adhesion protocols. *J. Appl. Oral Sci.* **2017**, *25*, 367–373. [CrossRef]

84. Vazquez-Garcia, F.; Tanomaru-Filho, M.; Chávez-Andrade, G.M.; Bosso-Martelo, R.; Basso-Bernardi, M.I.; Guerreiro-Tanomaru, J.M. Effect of silver nanoparticles on physicochemical and antibacterial properties of calcium silicate cements. *Braz. Dent. J.* **2016**, *27*, 508–514. [CrossRef] [PubMed]

85. Elgamily, H.M.; El-Sayed, H.S.; Abdelnabi, A. The antibacterial effect of two cavity disinfectants against one of cariogenic pathogen: An in vitro comparative study. *Contemp. Clin. Dent.* **2018**, *9*, 457–462. [CrossRef] [PubMed]

86. Scarpelli, B.B. In vitro evaluation of the remineralizing potential and antimicrobial activity of a cariostatic agent with silver nanoparticles. *Braz. Dent. J.* **2017**, *28*, 738–743. [CrossRef] [PubMed]

87. Fakhruddin, K.S.; Egusa, H.; Ngo, H.C.; Panduwawala, C.; Pesee, S.; Samaranayake, L.P. Clinical efficacy and the antimicrobial potential of silver formulations in arresting dental caries: A systematic review. *BMC Oral Health* **2020**, *20*, 1–13. [CrossRef]

88. Tuncay, Ö.; Er, Ö.; Demirbuga, S.; Zorba, Y.O.; Topçuoğlu, H.S. Effect of gaseous ozone and light-activated disinfection on the surface hardness of resin-based root canal sealers. *Scanning* **2016**, *38*, 141–147. [CrossRef]

89. César Sumita, T.C.; Junqueira, J.C.; Jorge, A.O.C.; do Rego, M.A. Antimicrobial effects of ozonated water on the sanitization of dental instruments contaminated with *E. coli, S. aureus, C. albicans,* or the spores of *B. Atrophaeus. J. Infect. Public Health* **2012**, *5*, 269–274. [CrossRef]

90. Tandan, M.; Gupta, S.; Tandan, P. Ozone in conservative dentistry & endodontics: A Review. *Int. J. Clin. Prev. Dent.* **2012**, *8*, 29–35.

91. Thote, A.; Ikhar, A.; Chandak, M.; Nikhade, P. Ozone an endodontic bliss: A review. *Int. J. Adv. Res.* **2018**, *6*, 951–956. [CrossRef]

92. Cardoso, M.G.; de Oliveira, L.D.; Koga-Ito, C.Y.; Jorge, A.O.C. Effectiveness of ozonated water on *Candida albicans, Enterococcus faecalis,* and endotoxins in root canals. *Oral Surg. Oral Med. Oral Pathol. Oral Radiol. Endodontol.* **2008**, *105*. [CrossRef]

93. Noguchi, F.; Kitamura, C.; Nagayoshi, M.; Chen, K.K.; Terashita, M.; Nishihara, T. Ozonated water improves lipopolysaccharide-induced responses of an odontoblast-like cell line. *J. Endod.* **2009**, *35*, 668–672. [CrossRef]

94. Hubbezoglu, I.; Zan, R.; Tunc, T.; Sumer, Z. Antibacterial efficacy of aqueous ozone in root canals infected by Enterococcus faecalis. *Jundishapur J. Microbiol.* **2014**, *7*, 11411. [CrossRef] [PubMed]

95. Hems, R.S.; Gulabivala, K.; Ng, Y.L.; Ready, D.; Spratt, D.A. An in vitro evaluation of the ability of ozone to kill a strain of *Enterococcus faecalis. Int. Endod. J.* **2005**, *38*, 22–29. [CrossRef] [PubMed]

96. Nagayoshi, M.; Kitamura, C.; Fukuizumi, T.; Nishihara, T.; Terashita, M. Antimicrobial effect of ozonated water on bacteria invading dentinal tubules. *J. Endod.* **2004**, *30*, 778–781. [CrossRef] [PubMed]

97. Kist, S.; Kollmuss, M.; Jung, J.; Schubert, S.; Hickel, R.; Huth, K.C. Comparison of ozone gas and sodium hypochlorite/chlorhexidine two-visit disinfection protocols in treating apical periodontitis: A randomized controlled clinical trial. *Clin. Oral Investig.* **2017**, *21*, 995–1005. [CrossRef] [PubMed]

98. Estrela, C.; Estrela, C.R.A.; Decurcio, D.A.; Hollanda, A.C.B.; Silva, J.A. Antimicrobial efficacy of ozonated water, gaseous ozone, sodium hypochlorite and chlorhexidine in infected human root canals. *Int. Endod. J.* **2007**, *40*, 85–93. [CrossRef]

99. Zan, R.; Hubbezoglu, I.; Sümer, Z.; Tunç, T.; Tanalp, J. Antibacterial effects of two different types of laser and aqueous ozone against enterococcus faecalis in root canals. *Photomed. Laser Surg.* **2013**, *31*, 150–154. [CrossRef]

100. Noites, R.; Pina-Vaz, C.; Rocha, R.; Carvalho, M.F.; Gonçalves, A.; Pina-Vaz, I. Synergistic antimicrobial action of chlorhexidine and ozone in endodontic treatment. *Biomed. Res. Int.* **2014**, *2014*. [CrossRef]

101. Silva, E.J.N.L.; Prado, M.C.; Soares, D.N.; Hecksher, F.; Martins, J.N.R.; Fidalgo, T.K.S. The effect of ozone therapy in root canal disinfection: A systematic review. *Int. Endod. J.* **2019**. [CrossRef]

102. Antonijevic, D.; Milovanovic, P.; Brajkovic, D.; Ilic, D.; Hahn, M.; Amling, M. Microstructure and wettability of root canal dentine and root canal filling materials after different chemical irrigation. *Appl. Surf. Sci.* **2015**, *355*, 369–378. [CrossRef]

103. Mohammadi, Z.; Yaripour, S.; Shalavi, S.; Palazzi, F.; Asgary, S. Root canal irrigants and dentin bonding: An update. *Iran. Endod. J.* **2017**, *12*, 131. [CrossRef]

104. Song, W.; Ge, S. Application of antimicrobial nanoparticles in dentistry. *Molecules* **2019**, *24*, 1033. [CrossRef] [PubMed]

105. Hussain, S.M.; Hess, K.L.; Gearhart, J.M.; Geiss, K.T.; Schlager, J.J. In vitro toxicity of nanoparticles in BRL 3A rat liver cells. *Toxicol. In Vitro* **2005**, *19*, 975–983. [CrossRef] [PubMed]

106. Jeong, G.N.; Jo, U.B.; Ryu, H.Y.; Kim, Y.S.; Song, K.S.; Yu, I.J. Histochemical study of intestinal mucins after administration of silver nanoparticles in Sprague-Dawley rats. *Arch. Toxicol.* **2010**, *84*, 63–69. [CrossRef] [PubMed]

107. Shantiaee, Y.; Dianat, O.; Mohammadkhani, H.; Baghban, A.A. Cytotoxicity comparison of nanosilver coated gutta-percha with guttaflow and normal Gutta-Percha On L929 fibroblast with MTT assay. *J. Dent. Sch. Shahid Beheshti Univ. Med. Sci.* **2011**, *29*, 63–69.

108. Lotfi, M.; Vosoughhosseini, S.; Ranjkesh, B.; Khani, S. Antimicrobial efficacy of nanosilver, sodium hypochlorite and chlorhexidine gluconate against Enterococcus faecalis. *Afr. J. Biotechnol.* **2011**, *10*, 6799–6803.

109. González-Luna, P.I.; Martínez-Castañon, G.-A.; Zavala-Alonso, A.; Patiño-Marin, N.; Nino, N.; Moran, J.; Ramírez-González, J.-H. Bactericide effect of silver nanoparticles as a final irrigation agent in endodontics on enterococcus faecalis: An ex vivo study. *J. Nanomater.* **2016**, *2016*. [CrossRef]

110. Abbaszadegan, A.; Nabavizadeh, M.; Gholami, A.; Aleyasin, Z.S.; Dorostkar, S.; Saliminasab, M.; Ghasemi, Y.; Hemmateenejad, B.; Sharghi, H. Positively charged imidazolium-based ionic liquid-protected silver nanoparticles: A promising disinfectant in root canal treatment. *Int. Endod. J.* **2015**, *48*, 790–800. [CrossRef]
111. Samiei, M.; Aghazadeh, M.; Lotfi, M.; Shakoei, S.; Aghazadeh, Z.; Pakdel, S.M.V. Antimicrobial efficacy of mineral trioxide aggregate with and without silver nanoparticles. *Iran. Endod. J.* **2013**, *8*, 166–170. [CrossRef]
112. Seung, J.; Weir, M.D.; Melo, M.A.S.; Romberg, E.; Nosrat, A.; Xu, H.H.K.; Tordik, P.A. A modified resin sealer: Physical and antibacterial properties. *J. Endod.* **2018**, *44*, 1553–1557. [CrossRef]
113. Monzavi, A.; Eshraghi, S.; Hashemian, R.; Momen-Heravi, F. In vitro and ex vivo antimicrobial efficacy of nano-MgO in the elimination of endodontic pathogens. *Clin. Oral Investig.* **2015**, *19*, 349–356. [CrossRef]
114. Grumezescu, A.M. Nanobiomaterials in dentistry: Applications of nanobiomaterials. *Nanobiomater. Dent. Appl. Nanobiomater.* **2016**, *11*, 1–467.
115. Hajipour, M.J.; Fromm, K.M.; Ashkarran, A.A.; Aberasturi, D.J.D.; Larramendi, I.R.D.; Rojo, T.; Serpooshan, V.; Parak, W.J.; Mahmoudi, M. Antibacterial properties of nanoparticles. *Trends Biotechnol.* **2012**, *30*, 499–511. [CrossRef] [PubMed]
116. Argueta-Figueroa, L.; Morales-Luckie, R.A.; Scougall-Vilchis, R.J.; Olea-Mejía, O.F. Synthesis, characterization and antibacterial activity of copper, nickel and bimetallic Cu-Ni nanoparticles for potential use in dental materials. *Prog. Nat. Sci. Mater. Int.* **2014**, *24*, 321–328. [CrossRef]
117. Aun, D.P.; Peixoto, I.F.D.C.; Houmard, M.; Buono, V.T.L. Enhancement of NiTi superelastic endodontic instruments by TiO_2 coating. *Mater. Sci. Eng. C* **2016**, *68*, 675–680. [CrossRef] [PubMed]
118. Łazarz Półkoszek, M.J.; Loster, J.E.; Wiśniewska, G. Application of ozone to various fields of dentistry—Review of literature. *Protet. Stomatol.* **2020**, *70*, 90–106. [CrossRef]
119. Madi, M.; Htet, M.; Zakaria, O.; Alagl, A.; Kasugai, S. Re-osseointegration of dental implants after periimplantitis treatments: A systematic review. *Implant. Dent.* **2018**, *27*, 101–110. [CrossRef] [PubMed]
120. Isler, S.C.; Soysal, F.; Akca, G.; Bakirarar, B.; Ozcan, G.; Unsal, B. The effects of decontamination methods of dental implant surface on cytokine expression analysis in the reconstructive surgical treatment of peri-implantitis. *Odontology* **2020**. [CrossRef]
121. Kazancioglu, H.O.; Kurklu, E.; Ezirganli, S. Effects of ozone therapy on pain, swelling, and trismus following third molar surgery. *Int. J. Oral Maxillofac. Surg.* **2014**, *43*, 644–648. [CrossRef]
122. Kazancioglu, H.O.; Ezirganli, S.; Demirtas, N. Comparison of the influence of ozone and laser therapies on pain, swelling, and trismus following impacted third-molar surgery. *Lasers Med. Sci.* **2014**, *29*, 1313–1319. [CrossRef]
123. Shi, X.; Xu, L.; Le, T.B.; Zhou, G.; Zheng, C.; Tsure, K.; Ishikawa, K. Partial oxidation of TiN coating by hydrothermal treatment and ozone treatment to improve its osteoconductivity. *Mater. Sci. Eng. C* **2016**, *59*, 542–548. [CrossRef]
124. Hauser-Gerspach, I.; Vadaszan, J.; Deronjic, I.; Gass, C.; Meyer, J.; Dard, M.; Waltimo, T.; Stübinger, S.; Mauth, C. Influence of gaseous ozone in peri-implantitis: Bactericidal efficacy and cellular response. An in vitro study using titanium and zirconia. *Clin. Oral Investig.* **2012**, *16*, 1049–1059. [CrossRef]
125. Isler, S.C.; Unsal, B.; Soysal, F.; Ozcan, G.; Peker, E.; Karaca, I.R. The effects of ozone therapy as an adjunct to the surgical treatment of peri-implantitis. *J. Periodontal Implant. Sci.* **2018**, *48*, 136–151. [CrossRef] [PubMed]
126. Gunputh, U.F.; Le, H.; Handy, R.D.; Tredwin, C. Anodised TiO_2 nanotubes as a scaffold for antibacterial silver nanoparticles on titanium implants. *Mater. Sci. Eng. C* **2018**, *91*, 638–644. [CrossRef] [PubMed]
127. Liao, C.; Li, Y.; Tjong, S.C. Visible-light active titanium dioxide nanomaterials with bactericidal properties. *Nanomaterials* **2020**, *10*, 124. [CrossRef] [PubMed]
128. Bapat, R.A.; Joshi, C.P.; Bapat, P.; Chaubal, T.V.; Pandurangappa, R.; Jnanendrappa, N.; Gorain, B.; Khurana, S.; Kesharwani, P. The use of nanoparticles as biomaterials in dentistry. *Drug Discovery Today* **2019**, *24*, 85–98. [CrossRef]
129. Kunrath, M.F.; Leal, B.F.; Hubler, R.; de Oliveira, S.D.; Teixeira, E.R. Antibacterial potential associated with drug-delivery built TiO_2 nanotubes in biomedical implants. *AMB Express* **2019**, *9*, 51. [CrossRef]
130. Azzawi, Z.G.M.; Hamad, T.I.; Kadhim, S.A.; Naji, G.A.H. Osseointegration evaluation of laser-deposited titanium dioxide nanoparticles on commercially pure titanium dental implants. *J. Mater. Sci. Mater. Med.* **2018**, *29*. [CrossRef]
131. Gunputh, U.F.; Le, H.; Lawton, K.; Besinis, A.; Tredwin, C.; Handy, R.D. Antibacterial properties of silver nanoparticles grown in situ and anchored to titanium dioxide nanotubes on titanium implant against *Staphylococcus aureus*. *Nanotoxicology* **2020**, *14*, 97–110. [CrossRef]
132. Fretwurst, T.; Nelson, K.; Tarnow, D.P.; Wang, H.L.; Giannobile, W.V. Is metal particle release associated with peri-implant bone destruction? An emerging concept. *J. Dent. Res.* **2018**, *97*, 259–265. [CrossRef]
133. Shen, X.; Al-Baadani, M.A.; He, H.; Cai, L.; Liu, J. Antibacterial and osteogenesis performances of LL37-loaded titania nanopores in vitro and in vivo. *Int. J. Nanomedicine* **2019**, *14*, 3043–3054. [CrossRef]
134. Rai, M.; Ingle, A.P.; Birla, S.; Yadav, A.; Santos, C.A.D. Strategic role of selected noble metal nanoparticles in medicine. *Crit. Rev. Microbiol.* **2016**, *42*, 696–719. [CrossRef]
135. Arita, M. Microbicidal efficacy of ozonated water om candida albicans adhered to avrylic plates. *Oral Microbiol. Immunol.* **2005**, *6*, 206–210. [CrossRef] [PubMed]
136. Murakami, H.; Mizuguchi, M.; Hattori, M.; Ito, Y.; Kawai, T.; Hasegawa, J. Effect of denture cleaner using ozone against methicillin-resistant *Staphylococcus aureus* and *E. coli* T1 phage. *Dent. Mater. J.* **2002**, *21*, 53–60. [CrossRef] [PubMed]

137. Suzuki, T.; Oizumi, M.; Furuya, J.; Okamoto, Y.; Rosenstiel, S.F. Influence of ozone on oxidation of dental alloys. *Int. J. Prosthodont.* **1999**, *12*, 179–183. [PubMed]

138. AlZarea, B.K. Management of denture-related traumatic ulcers using ozone. *J. Prosthet. Dent.* **2019**, *121*, 76–82. [CrossRef]

139. Suganya, S.; Ahila, S.C.; Kumar, M.B.; Kumar, V.M. Evaluation and comparison of anti-Candida effect of heat cure polymethyl-methacrylate resin enforced with silver nanoparticles and conventional heat cure resins: An in vitro study. *J. Dent. Res.* **2014**, *25*, 204–207. [CrossRef] [PubMed]

140. Li, Z.; Sun, J.; Lan, J.; Qi, Q. Effect of a denture base acrylic resin containing silver nanoparticles on *Candida albicans* adhesion and biofilm formation. *Gerodontology* **2016**, *33*, 209–216. [CrossRef]

141. Ginjupalli, K.; Alla, R.K.; Tellapragada, C.; Gupta, L.; Perampalli, N.U. Antimicrobial activity and properties of irreversible hydrocolloid impression materials incorporated with silver nanoparticles. *J. Prosthet. Dent.* **2016**, *115*, 722–728. [CrossRef]

142. Şuhani, M.F.; Băciuțcedil, G.; Băciuțcedil, M.; Şuhani, R.; Bran, S. Current perspectives regarding the application and incorporation of silver nanoparticles into dental biomaterials. *Clujul Med.* **2018**, *91*, 274–279. [CrossRef]

143. Köroğlu, A.; Şahın, O.; Kürkçüoğlu, I.; Dede, D.Ö.; Özdemır, T.; Hazer, B. Silver nanoparticle incorporation effect on mechanical and thermal properties of denture base acrylic resins. *J. Appl. Oral Sci.* **2016**, *24*, 590–596. [CrossRef]

144. Matsuura, T.; Abe, Y.; Sato, Y.; Okamoto, K.; Ueshige, M.; Akagawa, Y. Prolonged antimicrobial effect of tissue conditioners containing silver-zeolite. *J. Dent.* **1997**, *25*, 373–377. [CrossRef]

145. Fujieda, T.; Uno, M.; Ishigami, H.; Kurachi, M.; Wakamatsu, N.; Doi, Y. Effects of dental porcelain containing silver nanoparticles on static fatigue. *Dent. Mater. J.* **2013**, *32*, 405–408. [CrossRef] [PubMed]

146. Fujieda, T.; Uno, M.; Ishigami, H.; Kurachi, M.; Wakamatsu, N.; Doi, Y. Addition of platinum and silver nanoparticles to toughen dental porcelain. *Dent. Mater. J.* **2012**, *31*, 711–716. [CrossRef] [PubMed]

147. Uno, M.; Nonogaki, R.; Fujieda, T.; Ishigami, H.; Kurachi, M.; Kamemizu, H.; Wakamatsu, N.; Doi, Y. Toughening of CAD/CAM all-ceramic crowns by staining slurry. *Dent. Mater. J.* **2012**, *31*, 828–834. [CrossRef] [PubMed]

148. Alrahlah, A.; Fouad, H.; Hashem, M.; Niazy, A.A.; AlBadah, A. Titanium Oxide (TiO_2)/polymethylmethacrylate (PMMA) denture base nanocomposites: Mechanical, viscoelastic and antibacterial behavior. *Materials* **2018**, *11*, 1096. [CrossRef]

149. Totu, E.E.; Nechifor, A.C.; Nechifor, G.; Aboul-Enein, H.Y.; Cristache, C.M. Poly(methyl methacrylate) with TiO_2 nanoparticles inclusion for stereolitographic complete denture manufacturing—The fututre in dental care for elderly edentulous patients? *J. Dent.* **2017**, *59*, 68–77. [CrossRef]

150. Tsuji, H.; Sugahara, H.; Gotoh, Y.; Ishikawa, J. Metal negative-ion implantation into rutile TiO_2 and enhancement of photocatalytic property under irradiation of fluorescent light. In Proceedings of the 14th International Conference on Ion Implantation Technology, Taos, NM, USA, 22–27 September 2002; Volume 2002, pp. 705–708. [CrossRef]

151. Borges, G.Á.; Elias, S.T.; da Silva, S.M.M.; Magalhães, P.O.; Macedo, S.B.; Ribeiro, A.P.D.; Guerra, E.N.S. In vitro evaluation of wound healing and antimicrobial potential of ozone therapy. *J. Cranio Maxillofac. Surg.* **2017**, *45*, 364–370. [CrossRef]

152. Huth, K.C.; Jakob, F.M.; Saugel, B.; Cappello, C.; Brand, K. Effect of ozone on oral cells compared with established antimicrobials. *Eur. J. Oral Sci.* **2006**, *114*, 435–440. [CrossRef]

153. Colombo, M.; Ceci, M.; Felisa, E.; Poggio, C.; Pietrocola, G. Cytotoxicity evaluation of a new ozonized olive oil. *Eur. J. Dent.* **2018**, *12*, 585–589. [CrossRef]

154. Kashiwazaki, J.; Nakamura, K.; Hara, Y.; Harada, R.; Wada, I.; Kanemitsu, K. Evaluation of the cytotoxicity of various hand disinfectants and ozonated water to human keratinocytes in a cultured epidermal model. *Adv. Skin Wound Care* **2020**, *33*, 313–318. [CrossRef]

155. Kuroda, K.; Azuma, K.; Mori, T.; Kawamoto, K.; Murahata, Y.; Tsuka, T.; Osaki, T.; Ito, N.; Imagawa, T.; Itoh, F.; et al. The safety and anti-tumor effects of ozonated water in vivo. *Int. J. Mol. Sci.* **2015**, *16*, 25108–25120. [CrossRef]

156. Liu, Z.; Wang, R.; Kan, F.; Jiang, F. Synthesis and characterization of TiO_2 Nanoparticles. *Asian J. Chem.* **2014**, *26*, 655–659. [CrossRef]

157. Byranvand, M.M.; Kharat, A.N.; Fatholahi, L.; Beiranvand, Z.M. A review on synthesis of nano-TiO_2 via different methods. *J. Nanostructures* **2013**, *3*, 1–9. [CrossRef]

158. Buraso, W.; Lachom, V.; Siriya, P.; Laokul, P. Synthesis of TiO_2 nanoparticles via a simple precipitation method and photocatalytic performance. *Mater. Res. Express* **2018**, *5*, 115003. [CrossRef]

159. Jeantelot, G.; Samy, O.C.; Julien, S.K.; Edy, A.H.; Anjum, D.H.; Sergei, L.; Moussab, H.; Luigi, C.; Jean-Marie, B. Morphology control of anatase TiO_2 for well-defined surface chemistry. *Phys. Chem. Chem. Phys.* **2018**, *20*, 14362–14373. [CrossRef] [PubMed]

160. Heringa, M.B.; Geraets, L.; van Eijkeren, J.C.H.; Vandebriel, R.J.; de Jong, W.H.; Oomen, A.G. Risk assessment of titanium dioxide nanoparticles via oral exposure, including toxicokinetic considerations. *Nanotoxicology* **2016**, *10*, 1515–1525. [CrossRef]

161. Zhang, D.; Qi, L.; Ma, J.; Cheng, H. Formation of crystalline nanosized titania in reverse micelles at room temperature. *J. Mater. Chem.* **2002**, *12*, 3677–3680. [CrossRef]

162. Dréno, B.; Alexis, A.; Chuberre, B.; Marinovich, M. Safety of titanium dioxide nanoparticles in cosmetics. *J. Eur. Acad. Dermatology Venereol.* **2019**, *33*, 34–46. [CrossRef]

163. Kurzmann, C.; Verheyen, J.; Coto, M.; Kumar, R.V.; Divitini, G.; Shokoohi-Tabrizi, H.A.; Verheyen, P.; de Moor, R.J.G.; Moritz, A.; Agis, H. In vitro evaluation of experimental light activated gels for tooth bleaching. *Photochem. Photobiol. Sci.* **2019**, *18*, 1009–1019. [CrossRef]

164. Shirkavad, S.; Moslehifard, E. Effect of TiO$_2$ nanoparticles on tensile strength of dental acrylic resins. *J. Dent. Res. Dent. Clin. Dent. Prospects* **2014**, *8*, 197–203. [CrossRef]

165. Meena, R.; Rani, M.; Pal, R.; Rajamani, P. Nano-TiO$_2$-induced apoptosis by oxidative stress-mediated DNA damage and activation of p53 in human embryonic kidney cells. *Appl. Biochem. Biotechnol.* **2012**, *167*, 791–808. [CrossRef]

166. Shukla, R.K.; Kumar, A.; Gurbani, D.; Pandey, A.K.; Singh, S.; Dhawan, A. TiO$_2$ nanoparticles induce oxidative DNA damage and apoptosis in human liver cells. *Nanotoxicology* **2013**, *7*, 48–60. [CrossRef] [PubMed]

167. Srivastava, R.K.; Rahman, Q.; Kashyap, M.; Singh, A.; Jain, G.; Jahan, S.; Lohani, M.; Lantow, M. A Pant. Nano-titanium dioxide induces genotoxicity and apoptosis in human lung cancer cell line, A549. *Hum. Exp. Toxicol.* **2013**, *32*, 153–166. [CrossRef] [PubMed]

168. Ekstrand-Hammarström, B.; Akfur, C.M.; Andersson, P.O.; Lejon, C.; Österlund, L.; Bucht, A. Human primary bronchial epithelial cells respond differently to titanium dioxide nanoparticles than the lung epithelial cell lines A549 and BEAS-2B. *Nanotoxicology* **2012**, *6*, 623–634. [CrossRef] [PubMed]

169. Shukla, R.K.; Sharma, V.; Pandey, A.K.; Singh, S.; Sultana, S.; Dhawan, A. ROS-mediated genotoxicity induced by titanium dioxide nanoparticles in human epidermal cells. *Toxicol. Vitr.* **2011**, *25*, 231–241. [CrossRef]

170. Xue, C.; Wu, J.; Lan, F.; Liu, W.; Yang, X.; Zeng, F.; Xu, H. Nano titanium dioxide induces the generation of ROS and potential damage in HaCaT cells under UVA irradiation. *J. Nanosci. Nanotechnol.* **2010**, *10*, 8500–8507. [CrossRef] [PubMed]

171. Gao, X.; Gao, X.; Wang, Y.; Peng, S.; Yue, B.; Fan, C.; Chen, W.; Li, X. Comparative toxicities of bismuth oxybromide and titanium dioxide exposure on human skin keratinocyte cells. *Chemosphere* **2015**, *135*, 83–93. [CrossRef]

172. Wright, C.; Iyer, A.K.V.; Wang, L.; Wu, N.; Yakisich, J.S.; Rojanasakul, Y.; Azad, N. Effects of titanium dioxide nanoparticles on human keratinocytes. *Drug Chem. Toxicol.* **2017**, *40*, 90–100. [CrossRef]

173. Kumar, S.; Meena, R.; Paulraj, R. Role of macrophage (M1 and M2) in titanium-dioxide nanoparticle-induced oxidative stress and inflammatory response in rat. *Appl. Biochem. Biotechnol.* **2016**, *180*, 1257–1275. [CrossRef]

174. El-Bassyouni, G.T.; Eshak, M.G.; Barakat, I.A.H.; Khalil, W.K.B. Immunotoxicity evaluation of novel bioactive composites in male mice as promising orthopaedic implants. *Cent. Eur. J. Immunol.* **2017**, *42*, 54–67. [CrossRef]

175. Jawaad, R.S.; Sultan, K.F.; Al-Hamadani, A.H. Synthesis of silver nanoparticles. *J. Eng. Appl. Sci.* **2014**, *9*, 586–592. [CrossRef]

176. Habouti, S.; Solterbeck, C.H.; Es-Souni, M. Synthesis of silver nano-fir-twigs and application to single molecules detection. *J. Mater. Chem.* **2010**, *20*, 5215–5219. [CrossRef]

177. Alshehri, A.H.; Jakubowska, M.; MlOzNiak, A.; Horaczek, M.; Rudka, D.; Free, C.; Carey, J.D. Enhanced electrical conductivity of silver nanoparticles for high frequency electronic applications. *ACS Appl. Mater. Interfaces* **2012**, *4*, 7007–7010. [CrossRef] [PubMed]

178. Deshmukh, S.P.; Patil, S.M.; Mullani, S.B.; Delekar, S.D. Silver nanoparticles as an effective disinfectant: A review. *Mater.Sci. Eng. C* **2019**, *97*, 954–965. [CrossRef] [PubMed]

179. Rodrigues, A.G.; Gonçalves, P.J.R.d.; Ottoni, C.A.; Ruiz, R.d.; Morgano, M.A.; de Araújo, W.L.; de Melo, I.S.; de Souza, A.O. Functional textiles impregnated with biogenic silver nanoparticles from *Bionectria ochroleuca* and its antimicrobial activity. *Biomed. Microdevices* **2019**, *21*, 1–10. [CrossRef] [PubMed]

180. Jia, C.; Zhang, Y.; Cui, J.; Gan, L. The antibacterial properties and safety of a nanoparticle-coated parquet floor. *Coatings* **2019**, *9*, 403. [CrossRef]

181. Li, Y.; Qin, T.; Ingle, T.; Yan, J.; He, W.; Yin, J.J.; Chen, T. Differential genotoxicity mechanisms of silver nanoparticles and silver ions. *Arch. Toxicol.* **2017**, *91*, 509–519. [CrossRef] [PubMed]

182. Lanone, S.; Rogerieux, F.; Geys, J.; Dupont, A.; Hoet, P. Comparative toxicity of 24 manufactured nanoparticles in human alveolar epithelial and macrophage cell lines. *Part. Fibre Toxicol.* **2009**, *6*, 14. [CrossRef]

183. AshaRani, P.V.; Mun, G.L.K.; Hande, M.P.; Valiyaveettil, S. Cytotoxicity and genotoxicity of silver nanoparticles in human cells. *ACS Nano* **2009**, *3*, 279–290. [CrossRef]

184. Albers, C.E.; Hofstetter, W.; Siebenrock, K.A.; Landmann, R.; Klenke, F.M. In vitro cytotoxicity of silver nanoparticles on osteoblasts and osteoclasts at antibacterial concentrations. *Nanotoxicology* **2013**, *7*, 30–36. [CrossRef]

185. Pauksch, L.; Hartmann, S.; Rohnke, M.; Szalay, G.; Alt, V.; Schnettler, R.; Lips, K.S. Biocompatibility of silver nanoparticles and silver ions in primary human mesenchymal stem cells and osteoblasts. *Acta Biomater.* **2014**, *10*, 439–449. [CrossRef]

186. Pérez-Díaz, M.A.; Boegli, L.; James, G.; Velasquillo, C.; Sanchez-Sanchez, R.; Martinez-Martinez, R.E.; Martínez-Castañón, G.A.; Martinez-Gutierrez, F. Silver nanoparticles with antimicrobial activities against Streptococcus mutans and their cytotoxic effect. *Mater. Sci. Eng. C* **2015**, *55*, 360–366. [CrossRef] [PubMed]

187. Salaie, R.N.; Besinis, A.; Le, H.; Tredwin, C.; Handy, R.D. The biocompatibility of silver and nanohydroxyapatite coatings on titanium dental implants with human primary osteoblast cells. *Mater. Sci. Eng. C* **2020**, *107*, 110210. [CrossRef] [PubMed]

188. Cowley, A.; Woodward, B. A healthy future: Platinum in medical applications platinum group metals enhance the quality of life of the global population. *Platin. Metals Rev.* **2011**, *55*, 98–107. [CrossRef]

189. Narayanan, R.; El-Sayed, M.A. Shape-dependent catalytic activity of platinum nanoparticles in colloidal solution. *Nano Lett.* **2004**, *4*, 1343–1348. [CrossRef]

190. Wierichs, R.J.; Meyer-Lueckel, H. Systematic review on noninvasive treatment of root caries lesions. *J. Dent. Res.* **2015**, *94*, 261–271. [CrossRef]

191. Ramirez, E.; Eradès, L.; Philippot, K.; Lecante, P.; Chaudret, B. Shape control of platinum nanoparticles. *Adv. Funct. Mater.* **2007**, *17*, 2219–2228. [CrossRef]
192. Bigall, N.C.; Härtling, T.; Klose, M.; Simon, P.; Eng, L.M.; Eychmüller, A. Monodisperse platinum nanospheres with adjustable diameters from 10 to 100 nm: Synthesis and distinct optical properties. *Nano Lett.* **2008**, *8*, 4588–4592. [CrossRef]
193. Islam, M.A.; Bhuiya, M.A.K.; Islam, M.S. A review on chemical synthesis process of platinum nanoparticles. *Asia Pacific J. Energy Environ.* **2014**, *1*, 103–116. [CrossRef]
194. Loan, T.T.; Do, L.T.; Yoo, H. Platinum nanoparticles induce apoptosis on raw 264.7 macrophage cells. *J. Nanosci. Nanotechnol.* **2018**, *18*, 861–864. [CrossRef]
195. Konieczny, P.; Goralczyk, A.G.; Szmyd, R.; Skalniak, L.; Koziel, J.; Filon, F.L.; Crosera, M.; Cierniak, A.; Zuba-Surma, E.K.; Borowczyk, J.; et al. Effects triggered by platinum nanoparticles on primary keratinocytes. *Int. J. Nanomed.* **2013**, *8*, 3963. [CrossRef]
196. Labrador-Rached, C.J.; Browning, R.T.; Braydich-Stolle, L.K.; Comfort, K.K. Toxicological implications of platinum nanoparticle exposure: Stimulation of intracellular stress, inflammatory response, and akt signaling in vitro. *J. Toxicol.* **2018**. [CrossRef] [PubMed]
197. Lin, C.-X.; Gu, J.-L.; Cao, J.-M. The acute toxic effects of platinum nanoparticles on ion channels, transmembrane potentials of cardiomyocytes in vitro and heart rhythm in vivo in mice. *Int. J. Nanomed.* **2019**, *14*, 5595–5609. [CrossRef] [PubMed]
198. Siddiqi, K.S.; Husen, A. Green synthesis, characterization and uses of palladium/platinum nanoparticles. *Nanoscale Res. Lett.* **2016**, *11*, 482. [CrossRef] [PubMed]
199. Almeer, R.S.; Ali, D.; Alarifi, S.; Alkahtani, S.; Almansour, M. Green platinum nanoparticles interaction with HEK293 cells: Cellular toxicity, apoptosis, and genetic damage. *Dose Response* **2018**, *16*. [CrossRef]
200. Şahin, B.; Aygün, A.; Gündüz, H.; Şahin, K.; Demir, E.; Akocak, S.; Şen, F. Cytotoxic effects of platinum nanoparticles obtained from pomegranate extract by the green synthesis method on the MCF-7 cell line. *Colloids Surf. B Biointerfaces* **2018**, *163*, 119–124. [CrossRef]
201. Unuofin, J.O.; Oladipo, A.O.; Msagati, T.A.M.; Lebelo, S.L.; Meddows-Taylor, S.; More, G.K. Novel silver-platinum bimetallic nanoalloy synthesized from Vernonia mespilifolia extract: Antioxidant, antimicrobial, and cytotoxic activities. *Arab. J. Chem.* **2020**, *13*, 6639–6648. [CrossRef]
202. Suh, Y.; Patel, S.; Kaitlyn, R.; Gandhi, J.; Joshi, G.; Smith, N.L.; Khan, S.A. Clinical utility of ozone therapy in dental and oral medicine. *Med. Gas. Res.* **2019**, *9*, 163–167. [CrossRef]
203. Hashimoto, M.; Sasaki, J.I.; Yamaguchi, S.; Kawai, K.; Kawakami, H.; Iwasaki, Y.; Imazato, S. Gold nanoparticles inhibit matrix metalloproteases without cytotoxicity. *J. Dent. Res.* **2015**, *94*, 1085–1091. [CrossRef]
204. Stankic, S.; Suman, S.; Haque, F.; Vidic, J. Pure and multi metal oxide nanoparticles: Synthesis, antibacterial and cytotoxic properties. *J. Nanobiotechnol.* **2016**, *14*. [CrossRef]
205. Slavin, Y.N.; Asnis, J.; Häfeli, U.O.; Bach, H. Metal nanoparticles: Understanding the mechanisms behind antibacterial activity. *J. Nanobiotechnol.* **2017**, *15*, 1–20. [CrossRef]
206. Lemire, J.A.; Harrison, J.J.; Turner, R.J. Antimicrobial activity of metals: Mechanisms, molecular targets and applications. *Nat. Rev. Microbiol.* **2013**, *11*, 371–384. [CrossRef] [PubMed]

 *nanomaterials*

Review

Nanomaterials Application in Orthodontics

Wojciech Zakrzewski [1], Maciej Dobrzynski [2,*], Wojciech Dobrzynski [3], Anna Zawadzka-Knefel [4], Mateusz Janecki [5], Karolina Kurek [6], Adam Lubojanski [1], Maria Szymonowicz [1], Zbigniew Rybak [1] and Rafal J. Wiglusz [7,8,*]

[1] Department of Experimental Surgery and Biomaterial Research, Wroclaw Medical University, Bujwida 44, 50-345 Wroclaw, Poland; wojciech.zakrzewski@student.umed.wroc.pl (W.Z.); adam.lubojanski@student.umed.wroc.pl (A.L.); maria.szymonowicz@umed.wroc.pl (M.S.); zbigniew.rybak@umed.wroc.pl (Z.R.)

[2] Department of Pediatric Dentistry and Preclinical Dentistry, Wroclaw Medical University, Krakowska 26, 50-425 Wroclaw, Poland

[3] Student Scientific Circle at the Department of Dental Materials, School of Medicine with the Division of Dentistry in Zabrze, Medical University of Silesia in Katowice, Akademicki Sq. 17, 41-902 Bytom, Poland; wojt.dobrzynski@wp.pl

[4] Department of Conservative Dentistry and Endodontics Wroclaw Medical University, Krakowska 26, 50-425 Wroclaw, Poland; anna.zawadzka-knefel@umed.wroc.pl

[5] Department of Maxillofacial Surgery, Mikulicz Radecki's University Hospital, Borowska 213, 50-556 Wroclaw, Poland; matjanecki@gmail.com

[6] Rajdent, Kozielewskiego 9, 42-200 Czestochowa, Poland; karolinakurek93@gmail.com

[7] International Institute of Translational Medicine, Jesionowa 11 St., 55–124 Malin, Poland

[8] Institute of Low Temperature and Structure Research, Polish Academy of Sciences, Okolna 2, 50-422 Wroclaw, Poland

* Correspondence: maciej.dobrzynski@umed.wroc.pl (M.D.); r.wiglusz@intibs.pl (R.J.W.); Tel.: +48-71-7840-378 (M.D.); +48-71-3954-159 (R.J.W.); Fax: +48-71-344-10-29 (R.J.W.)

Citation: Zakrzewski, W.; Dobrzynski, M.; Dobrzynski, W.; Zawadzka-Knefel, A.; Janecki, M.; Kurek, K.; Lubojanski, A.; Szymonowicz, M.; Rybak, Z.; Wiglusz, R.J Nanomaterials Application in Orthodontics. *Nanomaterials* **2021**, *11*, 337. https://doi.org/10.3390/nano11020337

Academic Editor: Maria Letizia Manca

Received: 31 December 2020
Accepted: 24 January 2021
Published: 28 January 2021

Abstract: Nanotechnology has gained importance in recent years due to its ability to enhance material properties, including antimicrobial characteristics. Nanotechnology is applicable in various aspects of orthodontics. This scientific work focuses on the concept of nanotechnology and its applications in the field of orthodontics, including, among others, enhancement of antimicrobial characteristics of orthodontic resins, leading to reduction of enamel demineralization or control of friction force during orthodontic movement. The latter one enables effective orthodontic treatment while using less force. Emphasis is put on antimicrobial and mechanical characteristics of nanomaterials during orthodontic treatment. The manuscript sums up the current knowledge about nanomaterials' influence on orthodontic appliances.

Keywords: nanomaterials; orthodontics; brackets; wires; antimicrobial effect

1. Introduction

Nanomaterials are widely used in modern clinical dentistry. They improve various properties, such as antimicrobial properties, durability of materials. These particles do not exceed 100 nm, due to they obtain a better ratio between the surface and mass. The larger the surface area of the material, the greater its reactivity. It is also easier to absorb them in the body, which can also result in high cytotoxicity [1]. Nanomaterials are used in many areas of dentistry, such as conservative dentistry, endodontics, oral, and maxillofacial surgery, periodontics, orthodontics, and prosthetics [2]. Orthodontics is a branch of dentistry dealing with the improvement of occlusal conditions and facial aesthetics in both children and adults. In cooperation with other specialists (such as dental surgeons, maxillofacial surgeons, periodontists), the orthodontist is able to significantly improve the patient's quality of life [3]. Nanotechnology is used, among others, in brackets, archives, elastomeric ligatures, orthodontic adhesives. Improving the microbicidal properties, reducing friction

67

and increasing the strength of the material are some of the advantages. However, a significant problem is the potential cytotoxicity of nanomaterials, therefore further research is needed [2].

The prolonged process of wearing orthodontic braces results in increased accumulation of dental plaque and eventually results in a greater risk of caries. Its development is generally associated with the activity of cariogenous bacteria due to prolonged dental plaque accumulation on teeth surfaces, deficiencies, avitaminosis, and diet. The demineralization process that starts the caries is called a white spot lesion (WSL), meaning, that decalcification of enamel surfaces adjacent to the orthodontic appliances is directly associated with orthodontic treatment [4]. Several studies confirm the accelerated accumulation of WLS in orthodontic treatments. Such tendency creates clinical problems leading to unacceptable esthetic alterations that, in some cases, might lead to conservative, restorative treatment. Research shows that more plaque can accumulate around composites compared to other restorative materials, which results in an increased percentage of secondary caries [5]. Moreover, resin composites do not have bacteriostatic properties.

Promising results in the prevention of pathological changes associated with orthodontic treatment are obtained through the use of nanotechnology. According to the European Commission states that: "Nanomaterial is defined as a natural, incidental, or manufactured material containing particles, in an unbound state or as an aggregate or as an agglomerate and where, for 50% or more of the particles in the number size distribution, one or more external dimensions is in the size range 1–100 nm. In specific cases and where warranted by concerns for the environment, health, safety, or competitiveness the number size distribution threshold of 50% may be replaced by a threshold between 1% and 50%" [6]. Implication of nanotechnology is beneficial to humans, it has been broadly used in the modern dentistry in in restorative dentistry as an additive nanoparticle with remineralizing properties in composite resins, dental adhesives, oral care products, in the control of bacterial biofilm as an antibacterial and antimineralizing additive in dental hygiene products such as toothpaste, mouth rinses, and composite resins. Nanotechnology is useful in the diagnosis of malignant and precancerous cavity diseases, periodontal diseases, and is also used in implantology—as a modification of the implant surface [7] and in the use of impression materials [8]. The development of technology gives better opportunities to both patient and orthodontist due to new physicochemical, mechanical and antibacterial properties of nanosized materials and can be used in coating orthodontic wires, elastomeric ligatures, and brackets, producing shape memory polymers and orthodontic bonding materials. Not only can we control biofilm formation, reduce bacterial activity and act anticariogenic, but also, through the desired tooth movement, shorten the treatment time.

There are many advantages in medicine of using nanotechnology; however, it creates many doubts regarding the safety for humans and the environment. Nanoparticles can easily penetrate tissues and can affect biological behaviors at different levels. It is necessary to conduct detailed research on the environmental and toxicological properties in order to assess the risk and lead a sustainable application of nanomaterials. The aim of this work was to describe and summarize the current use of nanoparticles and their antibacterial activity in orthodontics, including resin, brackets, and archwires.

2. Nano-Coatings in Orthodontic Archwires

Minimizing the frictional forces between the orthodontic wire and brackets has the potential to increase the desired tooth movement and thus shorten treatment time. In recent years, nanoparticles have been used as a component of dry lubricants. These solid-phase materials are capable of reducing the friction between two sliding surfaces without the need for a liquid medium. One of the many examples are Inorganic fullerene-like tungsten sulfide nanoparticles (IF-WS2) that are used as self-lubricating coatings for orthodontic stainless steel wires [9]. Friction tests simulating the performance of coated and uncoated wires were carried out on an Instron machine, scanning electron microscopy (SEM) and energy dispersive X-ray spectroscopy (EDS) analysis of the coated wires showed a clear impregnation of IF-WS2 nanoparticles in the Ni-P matrix.

Atomic force microscopy (AFM) was used as a tool to assess the surface roughness of stainless steel (SS), beta-titanium (β-Ti), and nickel-titanium(NiTi) wires [10]. The surface roughness measurement of the AFM method confirmed the fact that the roughness of the measures on the effectiveness of sliding mechanics, the corrosion behavior, and aesthetics of orthodontic arches. The influence of decontamination and clinical exposure on the modulus of elasticity, hardness and surface roughness of SS and NiTi arches, and AFM paper coupled with a nanoindenter were assessed [11]. The results of the AFM popularity assessment that the decontamination regimen and clinical exposure had no statistically significant effect on NiTi wires, but had a statistically significant effect on SS wires. In a diagnostic study, the clinical significance of statistical studies, analysis, and testing of the arch equipment on orthodontic movement is not predicted.

2.1. Nano Coatings Reducing Friction on Orthodontic Archwires

Orthodontic arches are used to generate biomechanical forces that are transmitted through the brackets to move the teeth and correct malocclusion, spacing, or crowding. They are also used for retention purposes, i.e., to keep the teeth in their current position. Currently, orthodontic arches are made of non-precious metal alloys. The most common types of wire are SS, NiTi, and β-Ti alloy wires. In the case of sliding mechanics, friction between the wire and the lock is one of the major factors influencing tooth movement. When one moving object makes contact with another, friction occurs on the contact surface, which causes resistance to the movement of the teeth. This frictional force is proportional to the force with which the contacting surfaces are pressed against each other and is governed by the interface surface characteristics (smooth/rough, chemically reactive/passive, or lubricant modified). Minimizing the frictional forces between the orthodontic wire and brackets will accelerate the desired tooth movement and thus shorten the treatment time.

NiTi substrates can be coated with cobalt and a layer of IF-WS2 nanoparticles using the electrodeposition method. The coated substrates showed friction reduction of up to 66% when compared to the uncoated ones. The results of such studies may have potential applications in reducing friction when using NiTi orthodontic wires. On the other hand, allergic reactions in patients with nickel sensitivity may be the disadvantage of introducing nickel into this type of coating. Therefore, the effect of such NiP coatings on stainless steel and NiTi wires should be assessed for biocompatibility in animal models and further human trials.

2.2. Delivering Nanoparticles from an Elastomeric Ligature

Elastomeric ligatures can serve as a support scaffold to deliver nanoparticles that can be anti-cariogenic or anti-inflammatory. They may also carry embedded antibiotic drug molecules. The release of anti-cariogenic fluoride from elastomeric ligatures has already been described in the literature [12,13]. Research has shown that fluoride release is characterized by an initial burst of fluoride in the first few days followed by a logarithmic fall. The whole process is effective against common enamel demineralization around the orthodontic bracket during treatment [14].

2.3. Shape Memory Polymers (SMP) in Orthodontics

In the last decade, there has been a growing interest in the production of aesthetic orthodontic wires to complement brackets in the color of the teeth. Shape memory polymers (SMPs) are materials that can remember equilibrium shapes and then manipulate and fix them into a temporary or dormant shape under certain temperature and stress conditions. They can later relax to their original, stress-free state under thermal, electrical, or environmental conditions. This relaxation is related to the elastic deformation stored in the previous manipulation. Recovery of SMP into equilibrium shape can be accompanied by an appropriate and prescribed force, useful for orthodontic tooth movement, or a macroscopic change in shape that is useful in ligation mechanisms. Due to the ability of SMP to have two shapes, these devices meet requirements unattainable by modern orthodontic materials, allowing the orthodontist to insert them into the patient's mouth more easily and comfortably [15].

When placed in the oral cavity, these polymers can be activated by body temperature or light-activated photoactive nanoparticles thereby causing tooth movement. SMP orthodontic wires can provide an improvement over traditional orthodontic materials as they provide lighter, more consistent forces which, in turn, can cause less pain to patients. Also, SMP materials are transparent, stainable, and stain-resistant, providing the patient with a more aesthetic apparatus during treatment. High percent elongation of the SMP apparatus (up to about 300%) allows for the application of continuous forces over a large range of tooth movement, and thus, fewer patient visits [16,17]. Future directions of research on shape—nanocomposite polymers with memory for the production of aesthetic orthodontic wires may have interesting potential in the research of orthodontic biomaterials.

2.4. Control of Oral Biofilms during Orthodontic Treatment

Nanoparticles have a larger surface area to volume ratio (per unit mass) compared to non-nano scale particles, interacting more closely with microbial membranes and providing a much larger surface area for antimicrobial activity. In particular, metal nanoparticles with a size of 1–10 nm showed the highest biocidal effect on bacteria [18]. Silver has a long history of use in medicine as an antibacterial agent [19]. The antimicrobial properties of nanoparticles have been exploited through the mechanism of joining dental materials with nanoparticles or coating the surface with nanoparticles to prevent adhesion of microbes to reduce biofilm formation [20,21]. It was found that resin composites containing fillers implanted with silver ions that release silver ions have an antibacterial effect on oral *streptococci* [22].

Ahn et al. [16] compared an experimental composite adhesive (ECA) containing silica nanofillers and silver nanoparticles with two conventional composite adhesives and a resin-modified glass ionomer (RMGI) to investigate the surface characteristics, physical properties, and antimicrobial activity against cariogenic *streptococci*. The results suggest that the ECAs had rougher surfaces than conventional adhesives due to the addition of silver nanoparticles. Bacterial adhesion to ECA was lower than to traditional adhesives, which was not affected by saliva. Bacterial suspensions containing ECA show slower growth of bacteria than those containing conventional adhesives. There is no significant difference in the shear bond strength and fracture strength of the bond between ECA and conventional adhesives.

3. Bracket Materials

The development of technology for the production of orthodontic materials and products provides better opportunities for patients with functional, health, and aesthetic results. It also improves the daily technical performance of the orthodontist. To perform their function properly, the brackets should have good biocompatibility, correct hardness, and strength, smooth archwire slot to reduce frictional resistance, smooth surface to reduce plaque deposition, should be precisely manufactured for each tooth, have high corrosion resistance, and ionic release [23].

Orthodontic braces are manufactured by three main methods which may be used in combination: Casting, injection molding, and milling from different types of material including metal, plastics, ceramics, and combinations.

Among the compositions of metallic, we can distinguish stainless steel, non-nickel steel, low-nickel stainless, cobalt–chromium alloys, titanium, and its alloys, gold alloys, and platinum alloys [24].

Metallic materials and their alloys are characterized by high mechanical parameters, usually better than ceramics or polymers. The surface in contact with the wire should have a relatively high modulus of elasticity to minimize the disbursement of energy transmitted by the wire from inexpedient plastic deformation and difficult enough to minimize expenditure wear caused by the movement of the activated wire. On the other hand, the base of the bracket must be sufficiently deformable to facilitate removal during treatment completion [24].

Stainless steel is a metallic alloy commonly used in the production of orthodontics brackets, due to its low cost, higher modulus of elasticity, and good biomechanical properties [25]. It can be classified as austenitic, martensitic, ferritic, duplex (austenitic-ferritic), and precipitation-hardening. The most commonly used alloys in orthodontic brackets are 303, 304, 316, 317, 17-4 PH [26]. Conventional type 316 L austenitic stainless steel is composed of %wt: Iron: Balance, manganese: 2.0, chromium: 16–18, nickel: 10–14, molybdenum: 2–3, and traces of phosphorus, sulfur, and carbon [27]. Although this alloy works well in clinical use, signs of corrosion have been observed.

17-4 PH alloy demonstrates improvement in corrosion resistance, frictional behavior, and cytotoxicity, more than austenitic stainless steels 303, 204, and 316/316 L [26]. Nickel stabilizes the austenitic phase; the anti-corrosive properties and the ductility are improved while the addition of chromium facilitates the formation of a passive anti-corrosion coating [28].

Although allergenic, cytotoxic, and mutagenic their content in the brackets is so small that their use is safe. Since the occurrence of adverse reactions, it was considered that exposure to these elements should be kept to a minimum. This resulted in the introduction of various non-nickel or very low nickel content stainless steel which is more resistant to corrosion and does not release nickel into the oral cavity. Compared to 316 L, Alloy 2205 is harder and less corrosive [29].

Titanium was used in the construction of brackets as a material with a proven lack of allergenicity and increased corrosion resistance. The many current dental and medical applications have made titanium the obvious choice of all the available components.

Commercially pure titanium grade 4 and Ti-6Al-4V alloy are the most widely used types for manufacturing orthodontic brackets, The different methods of obtaining the brackets result in significant differences in physical, mechanical, and bulk material properties. Corrosion resistance is achieved due to the presence of a thin passive protective layer made of titanium oxide. This layer is more stable than its counterpart chrome oxide on stainless steel [30]. Gold-coated brackets were introduced as an alternative to steel and titanium brackets. They are plated with 300 micro inches of 24 karat gold, therefore, have significantly brighter appearance. Moreover, they have better mechanical properties compared with conventional brackets made of stainless steel alloys. Gold alloy brackets are introduced as highly anticorrosive and the first choice for patients allergic to nickel (Ni) [30]. Significant side effects have not been observed clinically.

Additionally, nano-sized gold particles can be used on orthodontic appliances e.g., aligners, to increase its antibacterial activity, by preventing biofilm formation as can be seen in Figure 1. Both the gingiva and teeth are covered by aligners for almost the entire day, which is a risk factor for plaque accumulation. Gold particles also show positive biocompatibility both in vitro and in vivo.

A suitable substitute for stainless steel brackets are those coated with a platinum layer. Platinum has been found as a material totally compatible in the oral environment. Its alloys are five times more resistant to abrasion than gold and compared to stainless steel, they have excellent corrosion resistance, a harder surface which reduces friction and improves the mechanics of sliding. As a combination of the platinum layer and the unique implantation process, a barrier has been created that protects against the diffusion of nickel, cobalt, and chromium.

Similar electrochemical properties, including excellent corrosion resistance, to that of platinum brackets, are demonstrated by those made of cobalt chrome steel [31]. Regarding friction resistance, cobalt–chromium brackets are comparable, but have slightly less friction than stainless steel brackets when used with stainless steel wires; however, cobalt–chromium brackets offer more friction than titanium brackets with both stainless steel and beta-titanium wires [32].

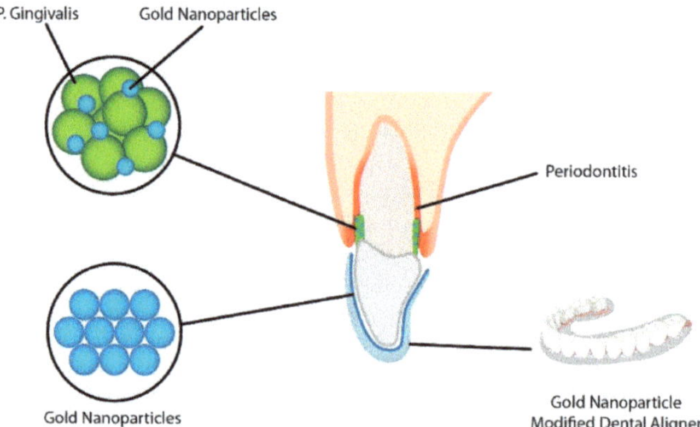

Figure 1. Aligners coated with modified gold nanoparticles causing enhanced antibacterial activity against *Porphyromonas gingivalis*. Due to presence of the coated aligner, the number of bacterial cells was decreased, causing increased biofilm formation prevention.

Although metal brackets exhibit excellent mechanical properties and provide many clinical advantages the issue of aesthetics remains a challenge. Elements made from ceramics and plastics have been widely used in clinical orthodontics.

The first plastic brackets appeared in the early 1970s and were made of acrylic, then polycarbonate, but unfortunately problems related to them were quickly noticed. They had a tendency to water sorption, change color upon contact with the ultraviolet light and some food or drinks [33].

There has been observed an increased adhesion of pathogens like *Streptococcus mutans* and *Candida albicans*. In order to eliminate problems and improve their properties the following solutions are possible: Reinforcement with other materials such as ceramic or fiberglass fillers and/or metal slots, chemical modification of the polymer and alternative polymers for instance urethane dimethacrylate, high-density polyethylene, and EBP [34]. Research shows that compared to stainless steel brackets, plastic brackets are only suitable for clinical use if they have a metal slot [35].

An important issue is the biocompatibility of plastic materials, especially in terms of cytotoxic effects of particle- and fiber-reinforced polycarbonate orthodontic brackets in fibroblast and breast cancer cells through the activation of mitochondrial cell death mechanisms [36].

Although polycarbonate brackets with metal reinforced slots demonstrate a significantly lesser degree of deformation, followed by pure polyurethane, pure polycarbonate, and fiberglass reinforced polycarbonate brackets torque problems still exist. Ceramic reinforced polycarbonate brackets showed the highest deformation under torque stresses [37].

Polyoxymethylene brackets were found to be harder and less rough. Unfortunately, this material is also unattractive due to the opacity and milky color. Moreover, it appears to release some formaldehyde over time.

There is still a search for an ideal polymer that would combine the optical properties of translucency and the mechanical properties of stiffness, resistance to water absorption, and degradation. The introduction of new materials should ensure this does not release toxic compounds, in particular leaching of monomer Bis-GMA (bisphenol A-glycidyl methacrylate), TEGDMA (triethylene glycol dimethacrylate) [38]. The advantage of polymer brackets, as in the case of those from stainless steel, is the ease and safety of removing them from the tooth.

Among the brackets ensuring excellent aesthetic and optimum stable properties, we also include those made of ceramics. Their advantages are high rigidity and abrasion resistance as well as biocompatibility, and they are free from discoloration. Ceramic brackets are usually composed of aluminum oxides. There are two varieties currently available polycrystalline and monocrystalline (Saphire) forms, depending on their method of production. Another category is the polycrystalline Zirconia which has been offered as an alternative to alumina ceramic [39]. Polycrystalline zirconia brackets have the greatest toughness amongst all ceramics however are very opaque and can exhibit intrinsic colors. The monocrystal alumina brackets, which are noticeably clearer and consequently more aesthetic, along with having higher strength, than the polycrystalline alumina brackets, show low fracture toughness, due to the lack of internal grain boundaries, the presence of pores, and machining damage from milling [26]. Ceramic materials have some disadvantages associated with iatrogenic enamel damage due to their hardness, bonding and debonding, Frictional resistance. Orthodontists may experience problems with bracket breakage and fracture resistance, particularly when trying the ligature or fracture from archwires forces.

4. Nanomaterials in Orthodontics

Nanomaterials versatility allows them to be used in many situations during orthodontic clinical treatment, as can be seen in Table 1.

Table 1. Nanomaterials application in dentistry.

Nanomaterial	Method of Use	Application	References
Silver NPs (AgNPs)	Applied as a coating agent on titanium	Implants	[40,41]
Zinc oxide NPs (ZnONPs)	Incorporated into dental resins	Resin composite adhesives	[42,43]
Chitosan NPs	Conjugated with silver nanoparticles	Resin composites adhesives	[44,45]
Copper (I) oxide NPs (Cu$_2$ONPs)	Antimicrobial effect in resin adhesives	Resin composites adhesives	[46]
Titanium (IV) oxide NPs (TiO$_2$NPs)	Nanotubes on titanium surfaces and incorporated with ZnONPs	Implants	[47,48]
Gold NPs (AuNPs)	Modified gold nanoparticles (AuDAPT) coated onto orthodontic aligners	Antimicrobial coated aligner	[49]
Carbonate hydroxyapatite nanocrystal	Antibacterial and antidemineralizing properties	Toothpastes, mouthwashes and composite resins	[50]
Amorphous Calcium Phosphate (ACP)	Antibacterial and antidemineralizing properties	Antibacterial and antidemineralizing properties	[51]
Novel Poly(l-lactic acid) (PLLA)/Multi-walled carbon nanotubes (MWNTs)/hydroxyapatite (HA) nanofibrous scaffolds	Polymer solution FOR entire-tooth regeneration	Dental Surface applications	[52]
Bioactive peptide—Amphiphile nanofibers	Branched peptide Amphiphile molecules containing the peptide motif Arg-Gly-Asp, or "RGD"	Dental surface applications	[53]

Friction is one of the major factors present during retraction or alignment of teeth during orthodontic treatment. One of the methods to overcome high friction is the application of higher forces during treatment. Such action can have one significant disadvantage-undesirable anchorage loss [54]. On the other hand, there are other methods of overcoming unwanted friction, including alteration of the bracket design or wire shape and size. At last, there is a possibility of nanoparticle coating addition. To benefit from the antibacterial properties of nanoparticles, there are two main strategies in orthodontics to reduce biofilm formation. One strategy focuses on coating the surface of orthodontic brackets or wires with nanoparticles [55]. The other is about combining nanoparticles with orthodontic adhesives or acrylic materials. The advantages of nanocomposite materials include excellent optical properties, easy handling, and excellent polishability [24]. Moreover, nanofillers can reduce the surface roughness of orthodontic adhesives, which is one of the most important factors in bacterial adhesion [25], as can be seen in Figure 2.

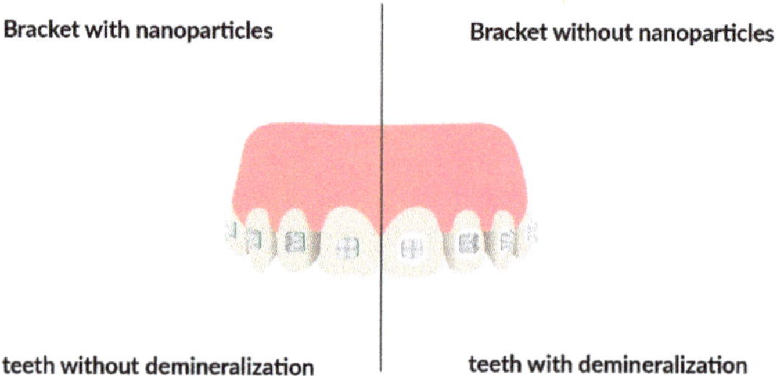

Figure 2. Comparison of teeth demineralization development with and without nanoparticles' covered brackets.

4.1. Silver Nanoparticles (AgNPs) Coating

Some studies have proposed silver nanoparticles as the most effective type of metal nanoparticles for preventing the growth of *Streptococcus mutans* [56]. Recently, silver nanoparticles (AgNPs) have been shown to be materials with excellent anti-microbial properties in a wide variety of microorganisms. In the orthodontic field, studies have incorporated AgNPs (17 nm) into orthodontic elastomeric modules, orthodontic brackets, and wires, and others, against a wide variety of bacterial species concluding that these orthodontic appliances with AgNPs could potentially combat the dental biofilm decreasing the incidence of dental enamel demineralization during and after the orthodontic treatments [57,58]. AgNPs can significantly inhibit the bacterial adherence of the *S. mutans* strain on the surfaces of the orthodontic bracket and wire appliances finding that the smaller AgNP samples demonstrated statistically to have the most important *S. mutans* antiadherence activities for orthodontic brackets and wires when compared to NiTi (nickel–titanium) and SS (stainless steel wires) [59]. It is also confirmed by several studies, that coverage of AgNPs in human dentin prevents biofilm formation on the surface of the dentin, together with bacterial growth inhibition [58,60,61]. In order for AGNPs to be a stable suspension able to limit the agglomeration, they should have zeta potential values ranging between +30 and −30 mV [62,63]. Bürgers et al. [64] confirms, that smaller AgNPs have the ability to release more silver ions, which promotes their antimicrobial effect, while the histological effect of AgNPs generally focuses on inhibition of microbial metabolism, leading to impaired production of extracellular polysaccharides and specific bacterial processes leading to its general dysfunction [65]. These studies confirm, that AgNP-coated brackets can help to decrease the spot lesions appearance during orthodontic treatment, and may be even useful in compromised patients with immune deficiency, diabetes, or elevated risk of endocarditis [66]. In addition to silver, many other nanoparticles like chitosan, copper, zinc, hydroxyapatite, and silicon dioxide can be added to composites in order to reduce bacterial activity and growth.

4.2. Chitosan

Chitosan is a naturally acquired polysaccharide that is formed by the deacetylation of chitin. It is a non-toxic, biodegradable, biocompatible, and has antibacterial properties [67], on *Agregatibacter actinomycetemcomitans*, *Porphyromonas Gingivalis*, and *Streptococcus mutans* [68,69]. Chitosan additionally has inhibiting action against fungi. This material's application as an antibacterial chemical agent in mouthwashes is limited due to its reduced solubility in water. Nonetheless, its characteristics are highly desirable in dental materials. Chitosan could be maintained inside the materials in the oral cavity due to its insolubility in water. Histologically, inhibition is caused by inactivation of the enzyme, the substitution of lipopolysaccharides, metal ions, and formation of acidic polymer like teichoic acid. Chitosan, due to its low solubility and melting temperature, can be maintained in the oral cavity for a long period of time, unlike CHX which is released and disappears in the early phase.

4.3. Copper Oxide

It was proved by Yassaei et al. [70], that no significant difference was found between silver and copper oxide (CuO) nanoparticles, but it was noted that a curing time increased with the use of copper material when compared to the silver one. The former is cheaper and additionally both physically and chemically more stable than the latter. CuO nanoparticles affect *Streptococcus mutans* bacteria in a similar way as silver particles do [56]. It was confirmed in other studies [4], that copper and copper-zinc nanoparticles had a significant inhibitory effect on the studied microbes. According to other studies, CuO is able to decrease biofilm formation from 70 up to 80% [71]. Moreover, the similar results were achieved when CuO particles were incorporated into adhesive materials [72]. Additionally, nanoparticles like CuO can act as nano-fillers and enhance the shear bond strength of adhesive.

4.4. Nitrogen-Doped Titanium Dioxide (N-Doped TiO$_2$) Brackets

The activation of N-doped TiO$_2$ leads to the formation of OH. Free radicals, superoxide ions (O$_2$), hydrogen peroxide (H$_2$O$_2$), and peroxyl radicals (HO$_2$). These chemicals exert antimicrobial activity, also reacting with lipids, enzymes, and proteins. According to Poosti et al. [73], TiO$_2$ nanoparticles of size 21 ± 5 nm can be blended to light cure orthodontic composite paste in 1, 2, and 3% and all these concentrations have similar antibacterial effects. Salehi et al. [74] proved, that nitrogen-doped TiO$_2$ brackets have shown better antimicrobial activity when compared to the uncoated stainless steel brackets. Adding TiO$_2$ to adhesives enhances its antibacterial activity without compromising its mechanical properties [75]. Nitrogen-doped TiO$_2$ brackets were also reported to present antibacterial activity against normal oral pathogenic bacteria [76].

4.5. Zinc Oxide (ZnO)

It has been observed, that as the concentration of ZnO increases, the antimicrobial activity also increases, followed by shear bond strength reduction. It is important to underline, that ZnO and CuO coated brackets have been observed with better antimicrobial characteristics on Streptococcus mutans than when the brackets were coated with CuO nanoparticles alone [77]. Kachoei et al. [78], Behroozian et al. [79] and Goto et al. [80] proved, that following ZnO nanoparticle coating, the frictional forces be-tween archwires and brackets significantly decreased. Because of that effect, these na-noparticles offer new opportunities in overcoming the unwanted friction forces, better anchorage control, and reduced risk of resorption.

5. Relationship between the Orthodontic Arch and Bracket Materials

We use various brackets and arches in orthodontic treatment. The most popular materials from which the locks are stainless steel, titanium, ceramics, and plastic. The materials that arches are usually made of are: Stainless steel, nickel-titanium alloy, chrome-cobalt steel, and titanium–molybdenum alloy. Between the arch and the orthodontic bracket, we can observe the phenomenon of friction, which makes it difficult to move the bracket along the arch. Friction is one of the crucial forces in orthodontics. It acts against the traction force (TF), which can be seen in Figure 3.

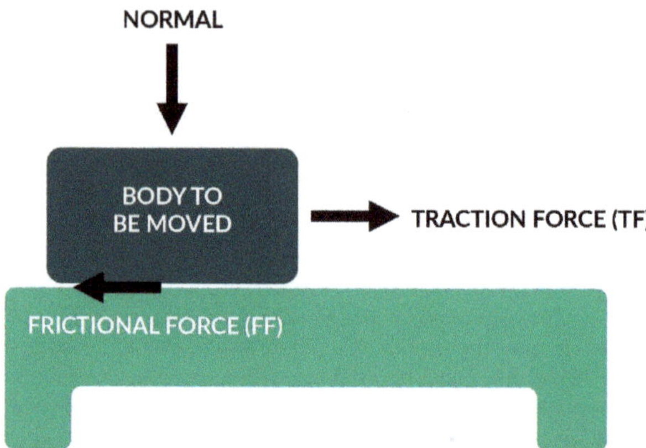

Figure 3. Different forces acting over a body under traction on top of a surface. Body to be moved, traction force (TF), friction force (FF), contact surface (CS).

The friction observed with orthodontic sliding mechanics is a clinical challenge for orthodontists—the high levels of friction can reduce the effectiveness of the mechanics, reduce the efficiency of tooth movement and further complicate anchorage control [80]. One of the main goals of orthodontic manufacturing companies is to look for new products that would generate less friction during sliding mechanics. One of them is the use of nanoparticles. There are two variables that influence the friction generated during orthodontic treatment: Mechanical and biological [81].

Mechanical factors mainly include the material of the bow and bracket. The gold standard of materials for performing sliding is the combination of stainless steel brackets and arches. Based on the research by Kusy and Whitley, the friction force is influenced by the shape and size of the arc. They claim that the friction is greater in larger diameter arches [82]. Several studies show that rectangular wires cause more friction than round wires [83]. The friction also depends on the material of the arc. It has been shown that a SS wire pulled through an SS lock produces the least resistance. NiTi wires produce a little greater friction, while titanium–molybdenum (TMA) alloys the largest (Frank and Nikolai showed that NiTi wire has less friction than SS wire) [84]. Another aspect considered in terms of the friction force is the material of the bracket and the type of the bracket. Kusy et al. [85] compared the friction level of stainless steel and titanium brackets. Titanium showed a greater coefficient of friction. Based on research [86], ceramic brackets produce almost twice as much friction as SS brackets. The new, self-ligating type of brackets appears to cause less friction, but this idea still requires scientific confirmation.

It appears that the main biological factor influencing friction is the presence of saliva which, depending on the type of bracket and arch, can act as a lubricant or as a "glue". Its action will therefore increase or decrease friction. Baker investigated the effect of saliva on friction and concluded that human saliva reduces the friction force by 15–19% [87]. The correct composition and amount of saliva are therefore important in maintaining the correct treatment. Debris that can resist on the surface of orthodontic arches also appears to be a significant variable that can increase friction during orthodontic treatment. After 8 weeks of use on orthodontic arches, significant deposits of biofilm were registered. The described nanomaterials affecting the number of bacteria can reduce their number, indirectly affecting the condition of saliva and reducing the amount of plaque on orthodontic elements. Using them could prevent increased frictional forces.

According to the studies, exposure to the oral cavity for one month can cause a significant slowdown in orthodontic movement (in this case the NiTi arches were tested) due to the accumulation of biofilm [88]. Additionally, the study suggests that the acidic pH produced by the bacteria present in the plaque increases the roughness of the arc and thus the friction between the wire and the bracket [89,90]. One of the ways to create unfavorable conditions for plaque accumulation is to try to include in orthodontic treatment the use of nanoparticles having a proven bacteriostatic effect. Properly-applied particles can also improve the mechanical factors by reducing the friction coefficient at the arc-lock interface.

6. Microbial Colonization Associated with Different Kinds of FOAs.

Fixed orthodontic appliances inhibit oral hygiene and create new retentive areas for plaque and debris as can be seen in Figure 4. It could increase the carriage of microbes and subsequent infection and it is one of the common problems that should be avoided in orthodontic treatment.

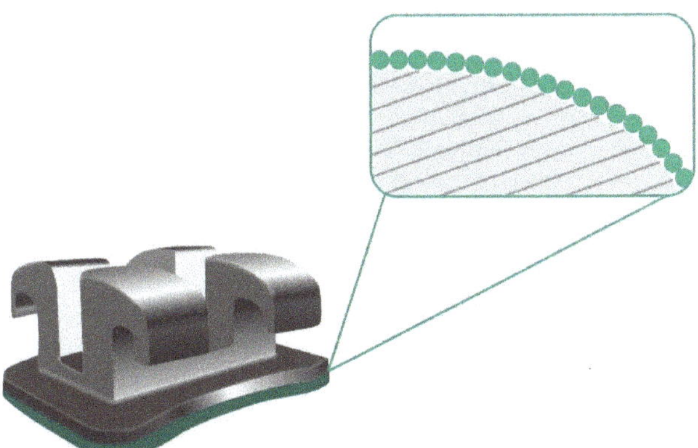

Figure 4. Orthodontic bracket covered by a nano-sized film.

The most common site for bacterial adhesion and biofilm formation is at the bracket adhesive-enamel junction, an area that is difficult to clean with daily brushing. The plaque that accumulates around orthodontic brackets often results in enamel decalcification, white spot formation, and dental caries adjacent to brackets. It is also difficult to remove microbial growth around orthodontic appliances. Its adherence to the fixed appliance is largely contributed by the bracket material and also the design of orthodontic brackets and ligating method [90,91]. The quantity and the quality of the plaque are influenced by many factors, including surface roughness, and surface-free energy [92]. Electrostatic attractions and van der Waal forces influence the adhesion of microorganisms to surfaces too [93]. Many types of braces are used in orthodontics. Bonded brackets have many advantages over bands such as better aesthetics, ease of placement, and removal and accessibility for oral hygiene [94].

7. Introduction of Nanofillers or NP (Silver, TiO$_2$) to Orthodontic Adhesives

Orthodontic adhesives showed a higher capacity to retain cariogenic *streptococci* than bracket materials. Previous short-term (24-h) in vitro studies demonstrated comparable or lower and still acceptable shear strength when nano-filled adhesives were used to fix orthodontic brackets.

Compared to traditional orthodontic adhesives, the use of nanofillers reduced the surface roughness of the adhesive; however, this was not true when silver NP was added to this mixture. Nevertheless, evaluation of the long-term effect of nanofiber adhesives on preventing enamel demineralization during orthodontic treatment, particularly around brackets and under orthodontic bands, has not yet been investigated.

Silver has been found to have antimicrobial activity against gram-positive/negative bacteria, fungi, protozoa, some viruses, and strains resistant to antibiotics [95,96], as well as cariogenic *Streptococcus mutans* [97]. Resin composites containing fillers implanted with silver ions had antibacterial properties against oral *streptococci* [22]. The addition of NP silver significantly reduces the adhesion of cariogenic *streptococci* to orthodontic adhesive compared to traditional adhesives, without compromising physical properties (shear bond strength). Adding TiO_2, SiO_2, or NP silver to acrylic orthodontic materials' cold-curing acrylic resins is common during the manufacture of removable orthodontic appliances such as expanders, fixers, and functional appliances which are mainly made of polymethyl methacrylate (PMMA). Compared to natural teeth, bacterial plaque adheres to acrylic resin braces with a larger surface area [98], which may lead to the development of caries-forming flora in the oral cavity. *Candida* Stomatitis is also an inflammation of the oral mucosa characterized by erythema (reddened areas), especially on the palate mucosa [99,100], which sometimes occurs under dentures (denture stomatitis) devices, or fixers.

CA is an opportunistic pathogen, and *Candida* is carried in the oral cavity in 25–75% of the studied populations [101]. A relationship has been suggested between the presence of a removable acrylic apparatus and the *Candida* carrier state, as well as low saliva pH [102]. In one study, the incidence of CA carriers before treatment with removable appliances was 39%; this number increased to 79% after 9 months and after treatment, and decreased to 14% after treatment. Similarly, orthodontic appliances placed on tooth tissues favored a greater proliferation of CA compared to dental appliances. The increase in *Candida* proliferation in people wearing removable appliances is probably due to protection against the natural and mechanical removal of saliva and the defense system [101].

Controlling CA proliferation under removable acrylic appliances can potentially prevent the development of orthodontic stomatitis. It is essential to find alternative therapies to eliminate CA that are tolerant to conventional antifungal drugs [100]. Investigation of the antimicrobial properties of NP acrylic materials and their use in mobile appliances is at an early stage and is limited to in vitro models. Sodagar et al. [54] investigated the changes in the bending strength of PMMA acrylic resin after adding TiO_2 (0.5%) and SiO_2 (1%) nanoparticles. The inclusion of NP in acrylic resin adversely affected the flexural strength of the final product and this effect was correlated with the concentration of NP [103]. However, a variable was observed after the addition of silver nanoparticles to the acrylic liquid of the two PMMA resins.

The mature dental plaque is composed of glucans and various microorganisms, the most common of them is *S. mutans* (the most cariogenic) and *Candida albicans*. Researchers like Shrinivaasan Nambi Rammohan, Ahn, Papaioannou, Fournier, or Brusca explored the relationship between CFUs (*S. mutans* alone, *C. albicans* alone, *S. mutans*, and *C. albicans* in combination) on surfaces of different kinds of orthodontic materials. When *S. mutans* was evaluated alone Shrinivaasan et al. [104], Papaioannou et al. [105], Fournier et al. [106], and Brusca et al. [107] found no obvious difference in the adhesion of *S. mutans* to stainless steel, plastic, and ceramic brackets. Quite different results were obtained Ahn et al. [108] There was a greater number of CFUs on stainless steel brackets than on plastic and ceramic brackets. Titanium and gold brackets showed lesser CFUs than stainless steel brackets. In the case of CA was evaluated alone, titanium brackets had the greatest of CFUs number because of the characteristics rough surface of these brackets [108] and gold brackets had the least number of CFUs because of the inert properties of gold. Plastic and ceramic brackets revealed a greater adherence than stainless steel brackets [107]. When *S. mutans* and *C. albicans* were evaluated in combination the clinical situation was different than an individual examination of these microorganisms and showed an antagonistic relationship at least in the initial growth but in the established plaque, they rather seem to exert a synergistic effect. For plastic and ceramic brackets, there was a greater number of CFUs and for metal brackets was the least [107].

Summarizing, titanium had some antibacterial properties but was not effective against the fungi. They grow by hyphae formation and the rough surface helped the increased levels of *C. albicans* [109]. Gold brackets revealed a decreased number of CFUs *S. mutans* and *C. albicans* and it could be inert properties of gold. Plastic and ceramic brackets showed greater levels of CFUs when *C. albicans* were studied alone and in combination with *S. mutans*. On composite yeasts exhibited numerous cell elongations which help in the adhesion mechanism and formation of pseudohyphae. Metallic brackets increase the level of bacterial adhesion compared with ceramic brackets because of the highest critical surface tension (greater surface energy). Stainless steel had an increased potential for microorganism attachment [110]. Properly, the material with high surface free energy will attract more bacteria than material with low surface free energy [110].

8. Nanomaterials in Orthodontics and Their Use in the Nearest Future

Nanoparticles are increasingly involved in dentistry [111]. They are used more often in conservative dentistry, endodontics [112,113], and prosthetics [111], where they become an integral part of treatment. They are used in irrigating solutions, filling materials and alloy in prosthetics. Their dynamic development should also include other fields of somatology, such as orthodontics. Currently we try find a ways to improve the value of mechanical orthodontic appliances. The use of nanomaterials partially solves this problem. The improvement of the biomechanical value of the orthodontic locks and arches, as well as the interference with the bacterial flora by nanomaterials seem worth developing. In the near future, adding nanoparticles to the materials of appliances will be the gold standard, improving the quality of orthodontic treatment. In addition to determining the basic components of the components of an orthodontic appliance, it will also be necessary to use appropriate proportions of nanoparticles in alloys. In the future, nanoparticles will also partly solve the problem of increased demineralization during treatment, which could reduce the number of complications. The use of the described particles also gives better control of the anchorage. Better control results in better, more predictable treatment [114], which reduces the stress of the orthodontist and increases patient satisfaction [115]. It will be possible to more accurately pursue the goals set at the beginning of treatment—during orthodontic diagnostics. A more thorough treatment will result in a better quality of life for the patient after treatment. The level of compatibility remains a challenge for nanoparticles in the future [116]. Further research is required to determine the safety of their use. Overcoming this problem makes it possible to easily increase the quality of orthodontic treatment. The use of nanoparticles will also reduce the described number of complications during orthodontic treatment, which will result in limiting the performance of additional procedures to eliminate complications. It also reduces treatment time, which reduces the cost of treatment. The shorter treatment time also allows more patients to be healed.

9. Materials in Orthodontics and Their Use in the Nearest Future

The future of nanotechnology in orthodontics has potential to develop in a number of additional applications as well including shape-memory polymers, self-healing materials, self-cleaning materials, biometric adhesives, tooth movement using orthodontic nanobots, and nano-changes on the surfaces of temporary anchorage devices (TADs) to increase their retention but still allow them to be removed when no longer needed [117,118].

Shape memory polymers, such as dual shape materials, belong to the group of "actively moving", which can change shape from one to the other. Orthodontics can use low stiffness transparent polymer arcs that can transform into arcs with a specific modulus of elasticity when exposed to a heat or light for example. With this procedure, it is possible to increase the effectiveness of treatment and aesthetics [119]. Self-healing materials that can repair themselves similar to biological systems. Hybrid materials have been developed, made of micro-ducts containing liquids or dissolved therapeutic agents. These materials can be used in the production of locks and orthodontic arches. A breach of the buckle or wire causes the nanobubble to burst and expose the monomer to the environment, thereby filling the resulting rupture gap with the described therapeutic agents [120].

Biometric adhesives—It is an enamel-friendly bonding mechanism for orthodontic appliances. The process takes place due to the formation of localized van der Waals forces [121]. This action ensures a strong bond between the materials without the use of a chemical substance. This material is often named "geckel". It acts as like a sticky note and exhibits strong, reversible adhesion in air and in water [122]. In orthodontics, such a procedure would ensure adequate bond strength without prior conditioning of the enamel. Self-cleaning materials have been developed, by using appropriate materials, increase the safety of using orthodontic appliances. The idea was taken from aircraft, where planes are covered with a titanium oxide nanocoating. A super-hybrid layer of hydrofluoric acid forms on the surface to prevent contamination. Photocatalytic activity resulting from the reaction of titanium oxide with light has attracted attention in orthodontic materials [123]. They try to find how inducing a reaction on the alloy of Ni-Ti archwires. By appropriate procedure—thickening the titanium oxide layer, electrolytic treatment and applying heat, it is possible to obtain a crystalline structure of rutile (titanium dioxide) on the surface of the materials [124].

10. Conclusions

Nowadays, nanotechnology plays an important role in the dental field since it has the potential to bring significant innovations and benefits. The recent positive results are a stimulus for future research, especially regarding orthodontics. The range of research including orthodontic bonding materials, covering of brackets and wires, as well as their antimicrobial characteristics has a huge potential. The review focused on scientific works concerning the use of nanoparticles in orthodontics that has been published in the literature over the last few years. The physicochemical properties gained by nano-sized materials have augmented the efficiency of orthodontic treatment. In this review, due to indicating the main types of literature reviews and referring to key studies showed that the physicochemical properties gained by nano-sized materials have augmented the efficiency of orthodontic treatment. Information can be implemented by scientists and doctors involved in the orthodontic therapy is included.

This review has also showed that the nanomaterials application regarding mechanical and antibacterial properties in orthodontics [55]. Nanoparticles can be successfully added to acrylic resins, cements, or orthodontic adhesives to prevent enamel demineralization during orthodontic treatment. Their versatility in clinical orthodontics can be seen in Table 1.

This review marked, that control and coordinated management of orthodontic treatment is crucial. Dental materials often present limitations during orthodontic treatment, but recently, nanotechnology and science have helped to partially solve some of the limitations. Nanomaterials can successfully reduce friction between the wire and the bracket, which may influence the orthodontic treatment. They are also useful in increasing the antimicrobial characteristics of materials used during treatment. Adding nanoparticles to the adhesives can increase their mean shear bond strength. This review provides several perspectives for the development of use nanomaterials in orthodontic.

Firstly, it is necessary to improve and search for new opportunities in overcoming the unwanted friction forces, better anchorage control, reducing the risk of resorption all should be based on evidence-based medicine and research generating stronger evidence.

Secondly, it is necessary to monitor the treatment of patients who use orthodontic nanomaterials due to the specificity of the oral cavity environment, which is dynamically changing. Biocompatibility and cytotoxicity are important considerations when using new bioactive materials. In the available literature, the knowledge about adverse effects resulting from the use of nanomaterials in orthodontics is limited. Despite the undoubted advantages of nanomaterials, knowledge about them is still incomplete and should be verified and carefully assessed, and the potential benefits should corresponded with the risk.

The application of nanomaterials in dentistry, especially in orthodontics is anticipated to grow further, and an interdisciplinary approach focusing on expertise in dentistry and nanomaterial science is required. The future in orthodontics will benefit greatly through nanotechnology.

Author Contributions: Conceptualization: M.D., W.Z., and R.J.W.; Methodology: W.Z., M.D., and R.J.W.; Software: W.Z., M.D., W.D. and R.J.W.; Validation: M.D. and R.J.W.; Formal analysis: W.Z.,

M.D., M.S., Z.R., and R.J.W.; Investigation: W.Z., M.D. W.D., Z.R., M.S., and R.J.W.; Data curation: W.Z., M.D., W.D., A.Z.-K., M.J., K.K., A.L., Z.R., M.S., and R.J.W.; Writing—original draft preparation: W.Z., M.D., W.D., A.Z.-K., M.J. K.K., and. R.J.W.; Writing—review and editing: W.Z., M.D., and R.J.W.; Supervision: M.D. and R.J.W.; Project administration: M.D. and R.J.W.; Funding acquisition: R.J.W. All authors have read and agreed to the published version of the manuscript.

Funding: The authors would like to acknowledge financial support from the National Science Centre (NCN) within the project "Preparation and characterization of biocomposites based on nanoapatites for theranostics" (No. UMO-2015/19/B/ST5/01330).

Informed Consent Statement: Not applicable.

Data Availability Statement: Not applicable.

Conflicts of Interest: The authors declare no conflict of interest.

References

1. Song, W.; Ge, S. Application of Antimicrobial Nanoparticles in Dentistry. *Molecules* **2019**, *24*, 1033. [CrossRef] [PubMed]
2. Sharan, J.; Singh, S.; Lale, S.V.; Mishra, M.; Koul, V.; Kharbanda, O.P. Applications of nanomaterials in dental science: A review. *J. Nanosci. Nanotechnol.* **2017**, *17*, 2235–2255. [CrossRef] [PubMed]
3. Gkantidis, N.; Christou, P.; Topouzelis, N. The orthodontic-periodontic interrelationship in integrated treatment challenges: A systematic review. *J. Oral Rehabil.* **2010**, *37*, 377–390. [CrossRef] [PubMed]
4. Behnaz, M.; Dalaie, K.; Mirmohammadsadeghi, H.; Salehi, H.; Rakhshan, V.; Aslani, F. Shear bond strength and adhesive remnant index of orthodontic brackets bonded to enamel using adhesive systems mixed with tio 2 nanoparticles. *Dental Press J. Orthod.* **2018**, *23*, 43.e1–43.e7. [CrossRef] [PubMed]
5. Sevinç, B.A.; Hanley, L. Antibacterial activity of dental composites containing zinc oxide nanoparticles. *J. Biomed. Mater. Res. Part B Appl. Biomater.* **2010**, *94*, 22–31. [CrossRef]
6. Boverhof, D.R.; Bramante, C.M.; Butala, J.H.; Clancy, S.F.; Lafranconi, W.M.; West, J.; Gordon, S.C. Comparative assessment of nanomaterial definitions and safety evaluation considerations. *Regul. Toxicol. Pharmacol.* **2015**, *73*, 137–150. [CrossRef] [PubMed]
7. Chieruzzi, M.; Pagano, S.; Moretti, S.; Pinna, R.; Milia, E.; Torre, L.; Eramo, S. Nanomaterials for tissue engineering in dentistry. *Nanomaterials* **2016**, *6*, 134. [CrossRef]
8. Feng, X.; Chen, A.; Zhang, Y.; Wang, J.; Shao, L.; Wei, L. Application of dental nanomaterials: Potential toxicity to the central nervous system. *Int. J. Nanomed.* **2015**, *10*, 3547–3565. [CrossRef]
9. Redlich, M.; Katz, A.; Rapoport, L.; Wagner, H.D.; Feldman, Y.; Tenne, R. Improved orthodontic stainless steel wires coated with inorganic fullerene-like nanoparticles of WS2 impregnated in electroless nickel-phosphorous film. *Dent. Mater.* **2008**, *24*, 1640–1646. [CrossRef]
10. Bourauel, C.; Fries, T.; Drescher, D.; Plietsch, R. Surface roughness of orthodontic wires via atomic force microscopy, laser specular reflectance, and profilometry. *Eur. J. Orthod.* **1998**, *20*, 79–92. [CrossRef]
11. Alcock, J.P.; Barbour, M.E.; Sandy, J.R.; Ireland, A.J. Nanoindentation of orthodontic archwires: The effect of decontamination and clinical use on hardness, elastic modulus and surface roughness. *Dent. Mater.* **2009**, *25*, 1039–1043. [CrossRef] [PubMed]
12. Doherty, U.B.; Benson, P.E.; Higham, S.M. Fluoride-releasing elastomeric ligatures assessed with the in situ caries model. *Eur. J. Orthod.* **2002**, *24*, 371–378. [CrossRef] [PubMed]
13. Miura, K.K.; Ito, I.Y.; Enoki, C.; Elias, A.M.; Matsumoto, M.A.N. Anticariogenic effect of fluoride-releasing elastomers in orthodontic patients. *Braz. Oral Res.* **2007**, *21*, 228–233. [CrossRef] [PubMed]
14. Nalbantgil, D.; Oztoprak, M.O.; Cakan, D.G.; Bozkurt, K.; Arun, T. Prevention of demineralization around orthodontic brackets using two different fluoride varnishes. *Eur. J. Dent.* **2013**, *7*, 41–47. [CrossRef] [PubMed]
15. Jung, Y.C.; Cho, J.W. Application of shape memory polyurethane in orthodontic. *J. Mater. Sci. Mater. Med.* **2010**, *21*, 2881–2886. [CrossRef]
16. Meng, Q.; Hu, J. A review of shape memory polymer composites and blends. *Compos. Part A Appl. Sci. Manuf.* **2009**, *40*, 1661–1672. [CrossRef]
17. Leng, J.; Lan, X.; Liu, Y.; Du, S. Shape-memory polymers and their composites: Stimulus methods and applications. *Prog. Mater. Sci.* **2011**, *56*, 1077–1135. [CrossRef]
18. Allaker, R.P. Critical review in oral biology & medicine: The use of nanoparticles to control oral biofilm formation. *J. Dent. Res.* **2010**, *89*, 1175–1186. [CrossRef]
19. Hill, W. *Argyria; the Pharmacology of Silver*; The Williams & Wilkins Company: Baltimore, MD, USA, 1939.
20. Moolya, N.; Sharma, R.; Shetty, A.; Gupta, N.; Gupta, A.; Jalan, V. Orthodontic bracket designs and their impact on microbial profile and periodontal disease: A clinical trial. *J. Orthod. Sci.* **2014**, *3*, 125. [CrossRef]
21. Monteiro, D.R.; Gorup, L.F.; Takamiya, A.S.; Ruvollo-Filho, A.C.; de Camargo, E.R.; Barbosa, D.B. The growing importance of materials that prevent microbial adhesion: Antimicrobial effect of medical devices containing silver. *Int. J. Antimicrob. Agents* **2009**, *34*, 103–110. [CrossRef]

22. Yamamoto, K.; Ohashi, S.; Aono, M.; Kokubo, T.; Yamada, I.; Yamauchi, J. Antibacterial activity of silver ions implanted in SiO2 filler on oral streptococci. *Dent. Mater.* **1996**, *12*, 227–229. [CrossRef]

23. Oh, K.T.; Choo, S.U.; Kim, K.M.; Kim, K.N. A stainless steel bracket for orthodontic application. *Eur. J. Orthod.* **2005**, *27*, 237–244. [CrossRef] [PubMed]

24. Eliades, T.; Zinelis, S.; Bourauel, C.; Eliades, G. Manufacturing of Orthodontic Brackets: A Review of Metallurgical Perspectives and Applications. *Recent Patents Mater. Sci.* **2010**, *1*, 135–139. [CrossRef]

25. Ogiński, T.; Kawala, B.; Mikulewicz, M.; Antoszewska-Smith, J. A Clinical Comparison of Failure Rates of Metallic and Ceramic Brackets: A Twelve-Month Study. *BioMed Res. Int.* **2020**, *2020*. [CrossRef] [PubMed]

26. Iijima, M.; Zinelis, S.; Papageorgiou, S.N.; Brantley, W.; Eliades, T. Orthodontic brackets. In *Orthodontic Applications of Biomaterials*; Elsevier: New York, NY, USA, 2017; pp. 75–96.

27. Zinelis, S.; Sifakakis, I.; Katsaros, C.; Eliades, T. Microstructural and mechanical characterization of contemporary lingual orthodontic brackets. *Eur. J. Orthod.* **2014**, *36*, 389–393. [CrossRef] [PubMed]

28. Jithesh, C.; Venkataramana, V.; Penumatsa, N.; Reddy, S.N.; Poornima, K.Y.; Rajasigamani, K. Comparative evaluation of nickel discharge from brackets in artificial saliva at different time intervals. *J. Pharm. Bioallied Sci.* **2015**, *7*, S587–S593. [CrossRef]

29. Platt, J.A.; Guzman, A.; Zuccari, A.; Thornburg, D.W.; Rhodes, B.F.; Oshida, Y.; Moore, B.K. Corrosion behavior of 2205 duplex stainless steel. *Am. J. Orthod. Dentofac. Orthop.* **1997**, *112*, 69–79. [CrossRef]

30. Orthodontic Brackets Selection, Placement and Debonding. Available online: https://www.researchgate.net/publication/275582644_ORTHODONTIC_BRACKETS_SelectionPlacement_and_Debonding (accessed on 21 September 2020).

31. Schiff, N.; Dalard, F.; Lissac, M.; Morgon, L.; Grosgogeat, B. Corrosion resistance of three orthodontic brackets: A comparative study of three fluoride mouthwashes. *Eur. J. Orthod.* **2005**, *27*, 541–549. [CrossRef]

32. Nair, S.; Janardhanam, P.; Padmanabhan, R. Evaluation of the effect of bracket and archwire composition on frictional forces in the buccal segments. *Indian J. Dent. Res.* **2012**, *23*, 203. [CrossRef]

33. Ali, O.; Makou, M.; Papadopoulos, T.; Eliades, G. Laboratory evaluation of modern plastic brackets. *Eur. J. Orthod.* **2012**, *34*, 595–602. [CrossRef]

34. Matsui, S.; Umezaki, E.; Komazawa, D.; Otsuka, Y.; Suda, N. Evaluation of mechanical properties of esthetic brackets. *J. Dent. Biomech.* **2015**, *6*. [CrossRef] [PubMed]

35. Sadat-Khonsari, R.; Moshtaghy, A.; Schlegel, V.; Kahl-Nieke, B.; Möller, M.; Bauss, O. Die Verformung von Kunststoffbrackets unter Torque-belastung: Eine Vergleichsstudie. *J. Orofac. Orthop.* **2004**, *65*, 26–33. [CrossRef] [PubMed]

36. Retamoso, L.B.; Luz, T.B.; Marinowic, D.R.; Machado, D.C.; De Menezes, L.M.; Freitas, M.P.M.; Oshima, H.M.S. Cytotoxicity of esthetic, metallic, and nickel-free orthodontic brackets: Cellular behavior and viability. *Am. J. Orthod. Dentofac. Orthop.* **2012**, *142*, 70–74. [CrossRef]

37. Russell, J.S. Current products and practice: Aesthetic orthodontic brackets. *J. Orthod.* **2005**, *32*, 146–163. [CrossRef] [PubMed]

38. Kloukos, D.; Pandis, N.; Eliades, T. Bisphenol-A and residual monomer leaching from orthodontic adhesive resins and polycarbonate brackets: A systematic review. *Am. J. Orthod. Dentofac. Orthop.* **2013**, *143*. [CrossRef]

39. Elekdag-Türk, S.; Yilmaz (née Huda Ebulkbash), H. Ceramic Brackets Revisited. In *Current Approaches in Orthodontics*; IntechOpen: London, UK, 2019.

40. Venugopal, A.; Muthuchamy, N.; Tejani, H.; Anantha-Iyenga-Gopalan; Lee, K.P.; Lee, H.J.; Kyung, H.M. Incorporation of silver nanoparticles on the surface of orthodontic microimplants to achieve antimicrobial properties. *Korean J. Orthod.* **2017**, *47*, 3–10. [CrossRef]

41. Padovani, G.C.; Feitosa, V.P.; Sauro, S.; Tay, F.R.; Durán, G.; Paula, A.J.; Durán, N. Advances in Dental Materials through Nanotechnology: Facts, Perspectives and Toxicological Aspects. *Trends Biotechnol.* **2015**, *33*, 621–636. [CrossRef]

42. Kasraei, S.; Sami, L.; Hendi, S.; AliKhani, M.-Y.; Rezaei-Soufi, L.; Khamverdi, Z. Antibacterial properties of composite resins incorporating silver and zinc oxide nanoparticles on Streptococcus mutans and Lactobacillus. *Restor. Dent. Endod.* **2014**, *39*, 109. [CrossRef]

43. Arun, D.; Adikari Mudiyanselage, D.; Gulam Mohamed, R.; Liddell, M.; Monsur Hassan, N.M.; Sharma, D. Does the Addition of Zinc Oxide Nanoparticles Improve the Antibacterial Properties of Direct Dental Composite Resins? A Systematic Review. *Materials (Basel)* **2020**, *14*, 40. [CrossRef]

44. Targino, A.G.R.; Flores, M.A.P.; Dos Santos, V.E.; De Godoy Bené Bezerra, F.; De Luna Freire, H.; Galembeck, A.; Rosenblatt, A. An innovative approach to treating dental decay in children. A new anti-caries agent. *J. Mater. Sci. Mater. Med.* **2014**, *25*, 2041–2047. [CrossRef]

45. Kalaivani, R.; Maruthupandy, M.; Muneeswaran, T.; Hameedha Beevi, A.; Anand, M.; Ramakritinan, C.M.; Kumaraguru, A.K. Synthesis of chitosan mediated silver nanoparticles (Ag NPs) for potential antimicrobial applications. *Front. Lab. Med.* **2018**, *2*, 30–35. [CrossRef]

46. Vargas-Reus, M.A.; Memarzadeh, K.; Huang, J.; Ren, G.G.; Allaker, R.P. Antimicrobial activity of nanoparticulate metal oxides against peri-implantitis pathogens. *Int. J. Antimicrob. Agents* **2012**, *40*, 135–139. [CrossRef] [PubMed]

47. Liu, W.; Su, P.; Chen, S.; Wang, N.; Ma, Y.; Liu, Y.; Wang, J.; Zhang, Z.; Li, H.; Webster, T.J. Synthesis of TiO2 nanotubes with ZnO nanoparticles to achieve antibacterial properties and stem cell compatibility. *Nanoscale* **2014**, *6*, 9050–9062. [CrossRef] [PubMed]

48. Sodagar, A.; Akhoundi, M.S.A.; Bahador, A.; Jalali, Y.F.; Behzadi, Z.; Elhaminejad, F.; Mirhashemi, A.H. Effect of TiO2 nanoparticles incorporation on antibacterial properties and shear bond strength of dental composite used in orthodontics. *Dental Press J. Orthod.* **2017**, *22*, 67–74. [CrossRef] [PubMed]

49. Zhang, M.; Liu, X.; Xie, Y.; Zhang, Q.; Zhang, W.; Jiang, X.; Lin, J. Biological Safe Gold Nanoparticle-Modified Dental Aligner Prevents the Porphyromonas gingivalis Biofilm Formation. *ACS Omega* **2020**, *5*, 18685–18692. [CrossRef]

50. Roveri, N.; Battistella, E.; Foltran, I.; Foresti, E.; Iafisco, M.; Lelli, M.; Palazzo, B.; Rimondini, L. Synthetic biomimetic carbonate-hydroxyapatite nanocrystals for enamel remineralization. In Proceedings of the Advanced Materials Research; Trans Tech Publications: Cham, Switzerland, 2008; Volume 47–50, Part 2. pp. 821–824.

51. Chen, C.; Weir, M.D.; Cheng, L.; Lin, N.J.; Lin-Gibson, S.; Chow, L.C.; Zhou, X.; Xu, H.H.K. Antibacterial activity and ion release of bonding agent containing amorphous calcium phosphate nanoparticles. *Dent. Mater.* **2014**, *30*, 891–901. [CrossRef]

52. Ladd, M.R.; Lee, S.J.; Stitzel, J.D.; Atala, A.; Yoo, J.J. Co-electrospun dual scaffolding system with potential for muscle-tendon junction tissue engineering. *Biomaterials* **2011**, *32*, 1549–1559. [CrossRef]

53. Huang, Z.; Sargeant, T.D.; Hulvat, J.F.; Mata, A.; Bringas, P.; Koh, C.Y.; Stupp, S.I.; Snead, M.L. Bioactive nanofibers instruct cells to proliferate and differentiate during enamel regeneration. *J. Bone Miner. Res.* **2008**, *23*, 1995–2006. [CrossRef]

54. Batra, P. Nanoparticles and their Applications in Orthodontics. *Adv. Dent. Oral Health* **2016**, *2*. [CrossRef]

55. Borzabadi-Farahani, A.; Borzabadi, E.; Lynch, E. Nanoparticles in orthodontics, a review of antimicrobial and anti-caries applications. *Acta Odontol. Scand.* **2014**, *72*, 413–417. [CrossRef]

56. Bapat, R.A.; Chaubal, T.V.; Joshi, C.P.; Bapat, P.R.; Choudhury, H.; Pandey, M.; Gorain, B.; Kesharwani, P. An overview of application of silver nanoparticles for biomaterials in dentistry. *Mater. Sci. Eng. C* **2018**, *91*, 881–898. [CrossRef] [PubMed]

57. Hernández-Gómora, A.E.; Lara-Carrillo, E.; Robles-Navarro, J.B.; Scougall-Vilchis, R.J.; Hernández-López, S.; Medina-Solís, C.E.; Morales-Luckie, R.A. Biosynthesis of silver nanoparticles on orthodontic elastomeric modules: Evaluation of mechanical and antibacterial properties. *Molecules* **2017**, *22*, 1407. [CrossRef] [PubMed]

58. Mhaske, A.R.; Shetty, P.C.; Bhat, N.S.; Ramachandra, C.S.; Laxmikanth, S.M.; Nagarahalli, K.; Tekale, P.D. Antiadherent and antibacterial properties of stainless steel and NiTi orthodontic wires coated with silver against Lactobacillus acidophilus—an in vitro study. *Prog. Orthod.* **2015**, *16*. [CrossRef]

59. Espinosa-Cristóbal, L.F.; López-Ruiz, N.; Cabada-Tarín, D.; Reyes-López, S.Y.; Zaragoza-Contreras, A.; Constandse-Cortéz, D.; Donohué-Cornejo, A.; Tovar-Carrillo, K.; Cuevas-González, J.C.; Kobayashi, T. Antiadherence and antimicrobial properties of silver nanoparticles against streptococcus mutans on brackets and wires used for orthodontic treatments. *J. Nanomater.* **2018**, *2018*. [CrossRef]

60. Besinis, A.; Hadi, S.D.; Le, H.R.; Tredwin, C.; Handy, R.D. Antibacterial activity and biofilm inhibition by surface modified titanium alloy medical implants following application of silver, titanium dioxide and hydroxyapatite nanocoatings. *Nanotoxicology* **2017**, *11*, 327–338. [CrossRef] [PubMed]

61. Besinis, A.; De Peralta, T.; Handy, R.D. Inhibition of biofilm formation and antibacterial properties of a silver nano-coating on human dentine. *Nanotoxicology* **2014**, *8*, 745–754. [CrossRef] [PubMed]

62. Tuan, T.Q.; Van Son, N.; Dung, H.T.K.; Luong, N.H.; Thuy, B.T.; Van Anh, N.T.; Hoa, N.D.; Hai, N.H. Preparation and properties of silver nanoparticles loaded in activated carbon for biological and environmental applications. *J. Hazard. Mater.* **2011**, *192*, 1321–1329. [CrossRef] [PubMed]

63. Vanitha, G.; Rajavel, K.; Boopathy, G.; Veeravazhuthi, V.; Neelamegam, P. Physiochemical charge stabilization of silver nanoparticles and its antibacterial applications. *Chem. Phys. Lett.* **2017**, *669*, 71–79. [CrossRef]

64. Bürgers, R.; Eidt, A.; Frankenberger, R.; Rosentritt, M.; Schweikl, H.; Handel, G.; Hahnel, S. The anti-adherence activity and bactericidal effect of microparticulate silver additives in composite resin materials. *Arch. Oral Biol.* **2009**, *54*, 595–601. [CrossRef]

65. Espinosa-Cristóbal, L.F.; Martinez-Castanon, G.A.; Téllez-Déctor, E.J.; Niño-Martínez, N.; Zavala-Alonso, N.V.; Loyola-Rodríguez, J.P. Adherence inhibition of Streptococcus mutans on dental enamel surface using silver nanoparticles. *Mater. Sci. Eng. C* **2013**, *33*, 2197–2202. [CrossRef]

66. PRIME PubMed | Nanosilver Coated Orthodontic Brackets: In Vivo Antibacterial Properties and Ion Release. Available online: https://wwww.unboundmedicine.com/medline/citation/26787659/Nanosilver_coated_orthodontic_brackets:_in_vivo_antibacterial_properties_and_Ion_release_ (accessed on 13 July 2020).

67. Kim, J.-S.; Shin, D.-H. Inhibitory effect on Streptococcus mutans and mechanical properties of the chitosan containing composite resin. *Restor. Dent. Endod.* **2013**, *38*, 36. [CrossRef] [PubMed]

68. Inhibitory Effect of Water-Soluble Chitosan on Growth of Streptococcus Mutans—PubMed. Available online: https://pubmed.ncbi.nlm.nih.gov/14964411/ (accessed on 14 July 2020).

69. Ikinci, G.; Şenel, S.; Akincibay, H.; Kaş, S.; Erciş, S.; Wilson, C.G.; Hincal, A.A. Effect of chitosan on a periodontal pathogen Porphyromonas gingivalis. *Int. J. Pharm.* **2002**, *235*, 121–127. [CrossRef]

70. Yassaei, S.; Nasr, A.; Zandi, H.; Motallaei, M.N. Comparison of antibacterial effects of orthodontic composites containing different nanoparticles on Streptococcus mutans at different times. *Dental Press J. Orthod.* **2020**, *25*, 52–60. [CrossRef] [PubMed]

71. Eshed, M.; Lellouche, J.; Matalon, S.; Gedanken, A.; Banin, E. Sonochemical coatings of ZnO and CuO nanoparticles inhibit streptococcus mutans biofilm formation on teeth model. *Langmuir* **2012**, *28*, 12288–12295. [CrossRef]

72. The Effect of CuO Nanoparticles on Antimicrobial Effects and Shear Bond Strength of Orthodontic Adhesives—PubMed. Available online: https://pubmed.ncbi.nlm.nih.gov/29492409/ (accessed on 16 July 2020).

73. Poosti, M.; Ramazanzadeh, B.; Zebarjad, M.; Javadzadeh, P.; Naderinasab, M.; Shakeri, M.T. Shear bond strength and antibacterial effects of orthodontic composite containing TiO2 nanoparticles. *Eur. J. Orthod.* **2013**, *35*, 676–679. [CrossRef]

74. Salehi, P.; Babanouri, N.; Roein-Peikar, M.; Zare, F. Long-term antimicrobial assessment of orthodontic brackets coated with nitrogen-doped titanium dioxide against Streptococcus mutans. *Prog. Orthod.* **2018**, *19*, 35. [CrossRef]

75. Ahn, S.J.; Lee, S.J.; Kook, J.K.; Lim, B.S. Experimental antimicrobial orthodontic adhesives using nanofillers and silver nanoparticles. *Dent. Mater.* **2009**, *25*, 206–213. [CrossRef]

76. Cao, B.; Wang, Y.; Li, N.; Liu, B.; Zhang, Y. Preparation of an orthodontic bracket coated with an nitrogen-doped TiO2-xNy thin film and examination of its antimicrobial performance. *Dent. Mater. J.* **2013**, *32*, 311–316. [CrossRef]

77. Kachoei, M.; Eskandarinejad, F.; Divband, B.; Khatamian, M. The effect of zinc oxide nanoparticles deposition for friction reduction on orthodontic wires. *Dent. Res. J. (Isfahan)* **2013**, *10*, 499–505. [CrossRef]

78. Behroozian, A.; Kachoei, M.; Khatamian, M.; Divband, B. The effect of ZnO nanoparticle coating on the frictionalresistance between orthodontic wires and ceramic brackets. *J. Dent. Res. Dent. Clin. Dent. Prospects* **2016**, *10*, 106–111. [CrossRef]

79. Goto, M.; Kasahara, A.; Tosa, M. Low-friction coatings of zinc oxide synthesized by optimization of crystal preferred orientation. *Tribol. Lett.* **2011**, *43*, 155–162. [CrossRef]

80. Rossouw, P.E. Friction: An overview. *Semin. Orthod.* **2003**, *9*, 218–222. [CrossRef]

81. Influence of Fluid Media on the Frictional Coefficients in Orthodontic Sliding | Request PDF. Available online: https://www.researchgate.net/publication/248866046_Influence_of_fluid_media_on_the_frictional_coefficients_in_orthodontic_sliding (accessed on 13 September 2020).

82. Kusy, R.P.; Whitley, J.Q. Influence of archwire and bracket dimensions on sliding mechanics: Derivations and determinations of the critical contact angles for binding. *Eur. J. Orthod.* **1999**, *21*, 199–208. [CrossRef] [PubMed]

83. Drescher, D.; Bourauel, C.; Schumacher, H.A. Frictional forces between bracket and arch wire. *Am. J. Orthod. Dentofac. Orthop.* **1989**, *96*, 397–404. [CrossRef]

84. Frank, C.A.; Nikolai, R.J. A comparative study of frictional resistances between orthodontic bracket and arch wire. *Am. J. Orthod.* **1980**, *78*, 593–609. [CrossRef]

85. Kusy, R.P.; Whitley, J.Q.; Ambrose, W.W.; Newman, J.G. Evaluation of titanium brackets for orthodontic treatment: Part I. The passive configuration. *Am. J. Orthod. Dentofac. Orthop.* **1998**, *114*, 558–572. [CrossRef]

86. Pratten, D.H.; Popli, K.; Germane, N.; Gunsolley, J.C. Frictional resistance of ceramic and stainless steel orthodontic brackets. *Am. J. Orthod. Dentofac. Orthop.* **1990**, *98*, 398–403. [CrossRef]

87. Baker, K.L.; Nieberg, L.G.; Weimer, A.D.; Hanna, M. Frictional changes in force values caused by saliva substitution. *Am. J. Orthod. Dentofac. Orthop.* **1987**, *91*, 316–320. [CrossRef]

88. Sapata, D.M.; Ramos, A.L.; Sábio, S.; Normando, D.; Pascotto, R.C. Evaluation of biofilm accumulation on and deactivation force of orthodontic Ni-Ti archwires before and after exposure to an oral medium: A prospective clinical study. *J. Dent. Res. Dent. Clin. Dent. Prospects* **2020**, *14*, 41–47. [CrossRef]

89. Lin, J.; Han, S.; Zhu, J.; Wang, X.; Chen, Y.; Vollrath, O.; Wang, H.; Mehl, C. Influence of fluoride-containing acidic artificial saliva on the mechanical properties of Nickel-Titanium orthodontics wires. *Indian J. Dent. Res.* **2012**, *23*, 591–595. [CrossRef]

90. Wichelhaus, A.; Geserick, M.; Hibst, R.; Sander, F.G. The effect of surface treatment and clinical use on friction in NiTi orthodontic wires. *Dent. Mater.* **2005**, *21*, 938–945. [CrossRef] [PubMed]

91. Sukontapatipark, W.; El-Agroudi, M.A.; Selliseth, N.J.; Thunold, K.; Selvig, K.A. Bacterial colonization associated with fixed orthodontic appliances. A scanning electron microscopy study. *Eur. J. Orthod.* **2001**, *23*, 475–484. [CrossRef] [PubMed]

92. Hägg, U.; Kaveewatcharanont, P.; Samaranayake, Y.H.; Samaranayake, L.P. The effect of fixed orthodontic appliances on the oral carriage of Candida species and Enterobacteriaceae. *Eur. J. Orthod.* **2004**, *26*, 623–629. [CrossRef] [PubMed]

93. Anhoury, P.; Nathanson, D.; Hughes, C.V.; Socransky, S.; Feres, M.; Chou, L.L. Microbial profile on metallic and ceramic bracket materials. *Angle Orthod.* **2002**, *72*. [CrossRef]

94. Boyd, R.L.; Baumrind, S. Periodontal considerations in the use of bonds or bands on molars in adolescents and adults. *Angle Orthod.* **1992**, *62*. [CrossRef]

95. Stobie, N.; Duffy, B.; McCormack, D.E.; Colreavy, J.; Hidalgo, M.; McHale, P.; Hinder, S.J. Prevention of Staphylococcus epidermidis biofilm formation using a low-temperature processed silver-doped phenyltriethoxysilane sol-gel coating. *Biomaterials* **2008**, *29*, 963–969. [CrossRef]

96. Abstract—Europe PMC. Available online: https://europepmc.org/article/med/19339161 (accessed on 9 December 2020).

97. Hernández-Sierra, J.F.; Ruiz, F.; Cruz Pena, D.C.; Martínez-Gutiérrez, F.; Martínez, A.E.; de Jesús Pozos Guillén, A.; Tapia-Pérez, H.; Martínez Castañón, G. The antimicrobial sensitivy of Streptococcus mutans to nanoparticles of silver, zinc oxide, and gold. *Nanomed. Nanotechnol. Biol. Med.* **2008**, *4*, 237–240. [CrossRef]

98. Radford, D.R.; Challacombe, S.J.; Walter, J.D. Denture plaque and adherence of Candida albicans to denture-base materials in vivo and in vitro. *Crit. Rev. Oral Biol. Med.* **1999**, *10*, 99–116. [CrossRef]

99. Monteiro, D.R.; Gorup, L.F.; Takamiya, A.S.; de Camargo, E.R.; Filho, A.C.R.; Barbosa, D.B. Silver Distribution and Release from an Antimicrobial Denture Base Resin Containing Silver Colloidal Nanoparticles. *J. Prosthodont.* **2012**, *21*, 7–15. [CrossRef]

100. Spampinato, C.; Leonardi, D. Candida infections, causes, targets, and resistance mechanisms: Traditional and alternative antifungal agents. *BioMed Res. Int.* **2013**, *2013*. [CrossRef]

101. Hibino, K.; Wong, R.W.K.; HÄgg, U.; Samaranayake, L.P. The effects of orthodontic appliances on Candida in the human mouth. *Int. J. Paediatr. Dent.* **2009**, *19*, 301–308. [CrossRef] [PubMed]
102. Arendorf, T.; Addy, M. Candidal carriage and plaque distribution before, during and after removable orthodontic appliance therapy. *J. Clin. Periodontol.* **1985**, *12*, 360–368. [CrossRef] [PubMed]
103. Sodagar, A.; Bahador, A.; Khalil, S.; Saffar Shahroudi, A.; Zaman Kassaee, M. The effect of TiO2 and SiO2 nanoparticles on flexural strength of poly (methyl methacrylate) acrylic resins. *J. Prosthodont. Res.* **2013**, *57*, 15–19. [CrossRef] [PubMed]
104. Juvvadi, S.; Rammohan, S.; Gandikota, C.; Challa, P.; Manne, R.; Mathur, A. Adherence of Streptococcus mutans and Candida albicans to different bracket materials. *J. Pharm. Bioallied Sci.* **2012**, *4*, 212. [CrossRef] [PubMed]
105. O'Sullivan, J.M.; Jenkinson, H.F.; Cannon, R.D. Adhesion of Candida albicans to oral streptococci is promoted by selective adsorption of salivary proteins to the streptococcal cell surface. *Microbiology* **2000**, *146*, 41–48. [CrossRef] [PubMed]
106. Forsberg, C.M.; Brattström, V.; Malmberg, E.; Nord, C.E. Ligature wires and elastomeric rings: Two methods of ligation, and their association with microbial colonization of streptococcus mutans and iactobacilli. *Eur. J. Orthod.* **1991**, *13*, 416–420. [CrossRef]
107. Brusca, M.I.; Chara, O.; Sterin-Borda, L.; Rosa, A.C. Influence of different orthodontic brackets on adherence of microorganisms in vitro. *Angle Orthod.* **2007**, *77*, 331–336. [CrossRef]
108. Lim, B.S.; Lee, S.J.; Lee, J.W.; Ahn, S.J. Quantitative analysis of adhesion of cariogenic streptococci to orthodontic raw materials. *Am. J. Orthod. Dentofac. Orthop.* **2008**, *133*, 882–888. [CrossRef]
109. Colony forming Unit Levels of Salivary Lactobacilli and Streptococcus Mutans in Orthodontic Patients—PubMed. Available online: https://pubmed.ncbi.nlm.nih.gov/16302600/ (accessed on 20 September 2020).
110. Desoet, J.J.; Van Loveren, C.; Lammens, A.J.; Pavićič, M.; Hamburg, C.; Ten Cate, J.M.; De Graaff, J. Differences in cariogenicity between fresh isolates of streptococcus sobrinus and streptococcus mutans. *Caries Res.* **1991**, *25*, 116–122. [CrossRef]
111. Neel, A.; Bozec, L.; Perez, R.A.; Kim, H.-W.; Knowles, J.C. Nanotechnology in dentistry: Prevention, diagnosis, and therapy. *Int. J. Nanomed.* **2015**, *10*, 6371. [CrossRef]
112. Shrestha, A.; Kishen, A. Antibacterial Nanoparticles in Endodontics: A Review. *J. Endod.* **2016**, *42*, 1417–1426. [CrossRef] [PubMed]
113. Raura, N.; Garg, A.; Arora, A.; Roma, M. Nanoparticle technology and its implications in endodontics: A review. *Biomater. Res.* **2020**, *24*, 21. [CrossRef] [PubMed]
114. Roberts-Harry, D.; Sandy, J. Orthodontics. Part 9: Anchorage control and distal movement. *Br. Dent. J.* **2004**, *196*, 255–263. [CrossRef] [PubMed]
115. Stress in Dentistry—PubMed. Available online: https://pubmed.ncbi.nlm.nih.gov/9828615/ (accessed on 14 January 2021).
116. Biocompatibility and Toxicity of Nanoparticles and Nanotubes. Available online: https://www.hindawi.com/journals/jnm/2012/548389/ (accessed on 14 January 2021).
117. Subramani, K.; Huja, S.; Kluemper, G.T.; Morford, L.; Hartsfield, J.K. Nanotechnology in Orthodontics-1: The Past, Present, and a Perspective of the Future. In *Nanobiomaterials in Clinical Dentistry*; Elsevier Inc.: New York, NY, USA, 2012; pp. 231–247. ISBN 9781455731275.
118. Eliades, T. Orthodontic material applications over the past century: Evolution of research methods to address clinical queries. *Am. J. Orthod. Dentofac. Orthop.* **2015**, *147*, S224–S231. [CrossRef]
119. Lendlein, A.; Jiang, H.; Jünger, O.; Langer, R. Light-induced shape-memory polymers. *Nature* **2005**, *434*, 879–882. [CrossRef]
120. Cordier, P.; Tournilhac, F.; Soulié-Ziakovic, C.; Leibler, L. Self-healing and thermoreversible rubber from supramolecular assembly. *Nature* **2008**, *451*, 977–980. [CrossRef]
121. Berengueres, J.; Saito, S.; Tadakuma, K. Structural properties of a scaled gecko foot-hair. *Bioinspir. Biomim.* **2007**, *2*, 1–8. [CrossRef]
122. Lee, H.; Lee, B.P.; Messersmith, P.B. A reversible wet/dry adhesive inspired by mussels and geckos. *Nature* **2007**, *448*, 338–341. [CrossRef]
123. Özyildiz, F.; Uzel, A.; Hazar, A.S.; Güden, M.; Ölmez, S.; Aras, I.; Karaboz, İ. Photocatalytic antimicrobial effect of TiO2 anatase thin-film–coated orthodontic arch wires on 3 oral pathogens. *Turkish J. Biol.* **2014**, *38*, 289–295. [CrossRef]
124. Brantley, W.A.; Eliades, T. The Role of Biomedical Engineers in the Design and Manufacture of Customized Orthodontic Appliances. In *Integrated Clinical Orthodontics*; John Wiley and Sons: New York, NY, USA, 2013; pp. 366–379. ISBN 9781444335972.

 *nanomaterials*

Article

The Comprehensive Approach to Preparation and Investigation of the Eu^{3+} Doped Hydroxyapatite/poly(L-lactide) Nanocomposites: Promising Materials for Theranostics Application

Katarzyna Szyszka [1], Sara Targonska [1], Malgorzata Gazinska [2], Konrad Szustakiewicz [2] and Rafal J. Wiglusz [1,3,*]

[1] Institute of Low Temperature and Structure Research, Polish Academy of Science, ul. Okolna 2, 50-422 Wroclaw, Poland
[2] Wroclaw University of Science and Technology, Polymer Engineering and Technology Division, Wyb. Wyspianskiego 27, 50-370 Wroclaw, Poland
[3] Centre for Advanced Materials and Smart Structures, Polish Academy of Sciences, Okolna 2, 50-950 Wroclaw, Poland
* Correspondence: r.wiglusz@intibs.pl; Tel.: +48-071-3954-159

Received: 3 July 2019; Accepted: 6 August 2019; Published: 10 August 2019

Abstract: In response to the need for new materials for theranostics application, the structural and spectroscopic properties of composites designed for medical applications, received in the melt mixing process, were evaluated. A composite based on medical grade poly(L-lactide) (PLLA) and calcium hydroxyapatite (HAp) doped with Eu^{3+} ions was obtained by using a twin screw extruder. Pure calcium Hap, as well as the one doped with Eu^{3+} ions, was prepared using the precipitation method and then used as a filler. XRPD (X-ray Powder Diffraction) and IR (Infrared) spectroscopy were applied to investigate the structural properties of the obtained materials. DSC (Differential Scanning Calorimetry) was used to assess the Eu^{3+} ion content on phase transitions in PLLA. The tensile properties were also investigated. The excitation, emission spectra as well as decay time were measured to determine the spectroscopic properties. The simplified Judd–Ofelt (J-O) theory was applied and a detailed analysis in connection with the observed structural and spectroscopic measurements was made and described.

Keywords: calcium hydroxyapatite nanopowders; rare earth ions; poly(L-lactide); nanocomposites; twin screw extrusion

---

1. Introduction

Regenerative medicine comes in many forms, promising to develop new biomedical treatments for people who suffer due to the burden of trauma, congenital defects and degenerative diseases. Recently, the greatest effort has been put in seeking materials that could help and accelerate the regenerative process by stimulating the body's own repair mechanisms to functionally heal previously irreparable tissues or organs [1]. Nowadays, the materials with great biological properties, broadly used in medicine, are calcium phosphates and poly(L-lactide) (PLLA) [2].

The biocompatible and biodegradable poly(L-lactide) is one of the most popular polymers produced from renewable raw materials [3]. It can enhance the adhesion and elasticity of composite materials, therefore, it is broadly used in medical and biomedical applications, i.e., for vascular stent production [4], for drug delivery systems [5], for tissue engineering [6], and as surgical sutures, implants and screws. Moreover, it is used in such fields as water purification by oil adsorption [7,8], food packaging [9] or photocatalysis degradation [10]. Poly(L-lactide) can be physically modified by doping it with fillers like bioceramic: β-TCP [11] or chitosan to facilitate cell adhesion, reduce the amount of degradation products, improve cell proliferation, increase hydrophilicity and improve bending between bones and polymeric implants [12]. Among the fillers, phosphates are particularly important, because they are the main inorganic component of vertebrate bones (bone and teeth) [13]. One of the most important materials in regenerative medicine is hydroxyapatite (hereafter: HAp) [14–16], which is biocompatible, bioactive and able to form a chemical bond with living tissues. Hydroxyapatite has been broadly investigated as a carrier meant for drug and gene delivery, cellular imaging and biosensing [17,18]. Moreover, it is relatively simple to obtain and modify. A formation of composites based on a PLLA and HAp matrix leads to combining their own advantages and overcoming their disadvantages aimed at bone tissue engineering scaffolds. During degradation of the PLLA/HAp material in the human body, the phosphates are gradually released and transformed into natural bone tissues that can raise the osseointegration of the composite calcium phosphate bioceramics [19].

Composites and nanocomposites based on bioresorbable thermoplastic aliphatic polyesters, i.e., poly(L-lactide) (PLLA), polycaprolactone (PCL) or copolymers thereof (like PGLA) and calcium hydroxyapatites have been widely studied due to their potential applications in regenerative medicine, especially for bone tissue engineering [12,20–27]. Usually, these systems can be used as implants for small bone defects or as scaffolds [28,29].

Many ternary systems based on PLLA/HAp and a third, bioactive component have also been investigated. Many components have been used as antibacterial dopants for PLLA, i.e., chitosan [30,31] or nano silver [32]. Lately, many articles have reported the modification of hydroxyapatite through the introduction of inorganic functional agents like silver [33–36], zinc [34,36], gold [34,35], copper [35,36] into HAp to achieve antibacterial properties or enhance osteoblast adhesion and proliferation [37]. Furthermore, nanomaterials based on lanthanide compounds are great candidates for biofluorescence probes due to the narrow emission bands, long lifetimes, low photobleaching and relatively low toxicity in comparison with conventional organic dyes. Their long emission lifetimes enable using time-resolved spectroscopy for eliminating the fluorescence of tissue [38–41]. The last-mentioned properties are especially important and can be used in theranostics for "personalized medicine" with a dual diagnostic and therapeutic function for in vivo imaging applications [42,43].

In this study, we present the synthesis and physico-chemical properties of composites based on bioresorbable PLLA and Eu^{3+} doped hydroxyapatite obtained through twin-screw, co-rotating micro-extrusion as compostes aimed at theranostic application (bone regeneration promotion and bio-imaging possibility).

2. Materials and Methods

2.1. Materials

Poly(L-lactide) Resomer L210s (PLLA) supplied by Evonik (Darmstadt, Germany) and synthetic hydroxyapatite were used in our research. Calcium hydroxyapatite and Eu^{3+} doped hydroxyapatite were prepared in our lab according to the procedures described below. As starting substrates, $Ca(NO_3)_2 \cdot 4H_2O$ ($\geq$99% Acros Organics, Schwerte, Germany), $(NH_4)_2HPO_4$ ($\geq$99.0% Fluka, Bucharest, Romania), Eu_2O_3 (99.99% Alfa Aesar, Karlsruhe, Germany), $NH_3 \cdot H_2O$ (99% Avantor Performance Materials Poland S.A., Gliwice, Poland), and HNO_3 (ultrapure Avantor Performance Materials Poland S.A., Gliwice, Poland) were used.

2.2. Synthesis of HAp and Eu^{3+}-Doped HAp Powders

Nanocrystalline powders of pure $Ca_{10}(PO_4)_6(OH)_2$ and that activated by Eu^{3+} ions were synthesized by using the precipitation method. The concentration of dopant ions was set to 1 mol%, 3 mol%, 5 mol% Eu^{3+} in a ratio to the overall molar content of calcium cations as the following process. The stoichiometric amount of Eu_2O_3 was digested in an excess of HNO_3 to receive water-soluble europium nitrate, and then the europium nitrate hydrates were re-crystallized three times in order to eliminate of the HNO_3 excess. Afterwards, the stoichiometric amount of calcium nitrate and europium nitrate was dissolved in deionized water and mixed together. Then the stoichiometric amount of $(NH_4)_2HPO_4$ was added to the previous mixture, leading to the fast precipitation of the intermediate product. The pH value of the suspension was adjusted to 10 by ammonia. The reaction mixture was heated and stirred for 3 h. Subsequently, the obtained products were washed several times with de-ionized water and dried at 70 °C for 24 h. As the final products, three types of europium-doped hydroxyapatites (1 mol% Eu^{3+}:HAp, 3 mol% Eu^{3+}:HAp, 5 mol% Eu^{3+}:HAp) were obtained. Calcium hydroxyapatite (HAp) was also prepared as a reference sample.

2.3. Preparation of the PLLA/Eu^{3+}-Doped Hydroxyapatite Composites

PLLA/Eu^{3+}:HAp composites having 10 wt.% of the fillers were prepared using the Thermo Scientific Process 11 (Waltham, MA, USA) co-rotating twin-screw micro-extruder (D = 20 mm, L/D = 40) with screw rotation speed of 200 min^{-1} and barrel temperature profile of 200–180 °C (from hopper to die) in a nitrogen atmosphere. The extruder screw geometry was presented in our previous work [44]. The composites were extruded in a one-step process. PLLA- and Eu^{3+}-doped HAp were dried in 80 °C under vacuum for 4 h before compounding. After extrusion, the composites were cooled down in the air and pelletized.

The composites were formed into foils through the casting extrusion technique using the Ultra Micro Cast Film extruder (Labtech Engineering, Sweden/Thailand) having flat die with a width of 75 mm, a conical screw with a diameter ranging from 18 to 8 mm (from hopper to die), L/d = 24, temperature of extrusion of 200 °C and screw speed of 100 rpm. Using these parameters, foils with a thickness of ~100 µm were obtained. The obtained materials are summarized in Table 1.

Table 1. Summary of the investigated materials.

Symbol	PLLA (wt.%)	Hap (wt.%)
HAp	0	100
1 mol% Eu^{3+}:HAp	0	100
3 mol% Eu^{3+}:HAp	0	100
5 mol% Eu^{3+}:HAp	0	100
PLLA	100	0
PLLA/HAp	90	10
PLLA/1 mol% Eu^{3+}:HAp	90	10
PLLA/3 mol% Eu^{3+}:HAp	90	10
PLLA/5 mol% Eu^{3+}:HAp	90	10

2.4. Characterization

Powder X-ray diffraction patterns were measured in the 2θ range of 2–60° by using a Rigaku Ultima IV (Tokyo, Japan) X-ray diffractometer equipped with Ni-filtered Cu Kα_1 radiation (Kα_1 = 1.54060 Å, U = 40 kV, I = 30 mA). The experimental XRD patterns were compared with the standards obtained from the Inorganic Crystal Structure Database (ICSD) and analyzed. The degree of crystallinity from XRD patterns was calculated using deconvolution made in Origin 8.0.

High-resolution transmission electron microscopy (HRTEM) images were done by a Philips CM-20 SuperTwin microscope (Eindhoven, The Netherlands), operating at 200 kV. Sample was prepared by dispersing a small amount of specimen in methanol and putting a droplet of the suspension on a copper microscope gird covered with carbon.

The surface morphology and element mapping of the nanocomposites were observed with a scanning electron microscope equipped with energy dispersive spectroscopy FEI Nova NanoSEM 230

(Hillsboro, OR, USA) with an EDS spectrometer (EDAX Genesis XM4) at an acceleration voltage of 18 kV and spot 3.0. Before observation, a layer of graphite was sprayed uniformly over the samples.

IR spectra were acquired using a Thermo Scientific Nicolet iS10 FT-IR Spectrometer (Waltham, MA, USA) equipped with an Automated Beamsplitter exchange system (iS50 ABX containing a DLaTGS KBr detector), a built-in all-reflective diamond ATR module (iS50 ATR), Thermo Scientific Polaris™ and a HeNe laser as an IR radiation source. IR spectra were recorded at 295 K temperature in the 4000–500 cm^{-1} range in KBr pellets with a spectral resolution of 2 cm^{-1}.

The emission, excitation spectra, and luminescence kinetics were recorded using an FLS980 fluorescence spectrometer (Edinburgh Instruments, Kirkton Campus, UK). A 450 W xenon lamp was used as an excitation source. The radiation from the lamp was filtered by a 300 mm monochromator equipped with holographic grating (1800 grooves per mm, blaze 250 nm) for the emission and excitation spectra. A microsecond flashlamp (μF2) was used for the measurements of luminescence kinetics, whereas a Hamamatsu R928P photomultiplier was used as a detector. All the emission and excitation spectra were corrected according to the apparatus characteristics and the excitation source intensity. The luminescence kinetics was recorded at 616 nm according to electric dipole transition ($^5D_0 \rightarrow {^7F_2}$). The powders were placed in quarc tube and the composites were placed directly in the holder in the spectrometer.

The temperatures of phase transitions were evaluated using the DSC1 STARe Differential Scanning Calorimeter System from Mettler-Toledo (Giessen, Germany). The research was conducted in a nitrogen environment for a temperature range of 25 ÷ 200 °C. The heating/cooling rate was 5 K/min and the gas flow rate remained at 20 mL/min (two thermal cycles).

Samples for mechanical tests were cut out from the obtained foils, from the middle region having ~100 μm, using a 5A type cutter (PN-EN ISO 527-2). Tensile properties tests were conducted on a universal Instron 5966 (Norwood, MA, USA) machine with the speed of 1 mm/min (for Young's modulus) and 10 mm/min (for tensile strength and strain at break). Before measurements, samples were kept in 80 °C for 2 h to reduce internal stresses.

3. Results

3.1. Morphology

The morphology of the received hydroxyapatite nanopowders doped with 3 mol% of Eu^{3+} was examined using the TEM and Selected Area Electron Diffraction (SAED) techniques. The received nanoapatite is nanocrystalline, loosely aggregated, and its shape is irregular (see Figure 1). The crystal phase purity of the obtained hydroxyapatite was additionally proved by a SAED analysis. The particles size distribution based on SEM images of pure HAp and 3 mol% Eu^{3+}:HAp particles, as well as representative SEM images of powders were presented on Figure S1.

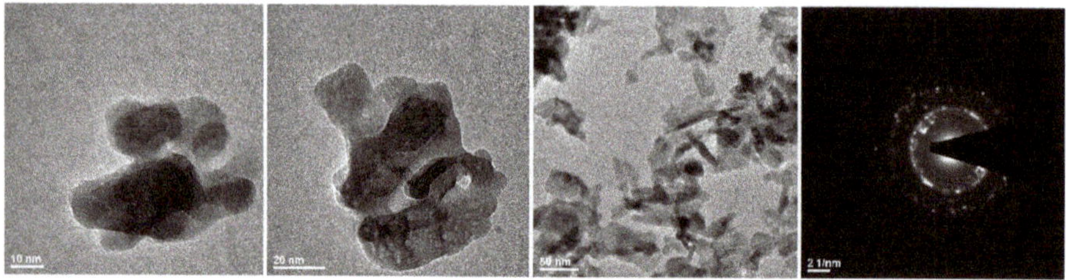

Figure 1. Representative TEM images and a SAED image of the 3 mol% Eu^{3+}:$Ca_{10}(PO_4)_6(OH)_2$.

The SEM pictures of the PLLA/HAp composites are shown in Figure 2, Figures S2 and S3. All the materials consist of 90 wt.% of PLLA and 10 wt.% of hydroxyapatite. The main difference between the materials is the Eu^{3+} content in the hydroxyapatite. The pictures indicate a random distribution of fillers in all the systems. The hydroxyapatite particle size in all the composites ranges from a few to over one hundred nanometers. The Eu^{3+} addition does not affect the distribution of HAp in the composites. The EDAX picture (Figure S4) also shows the distribution of basic elements (Ca, P) in the PLLA/5 mol% Eu^{3+}:HAp. Additionally, the figure presents the distribution of Eu^{3+} in the system.

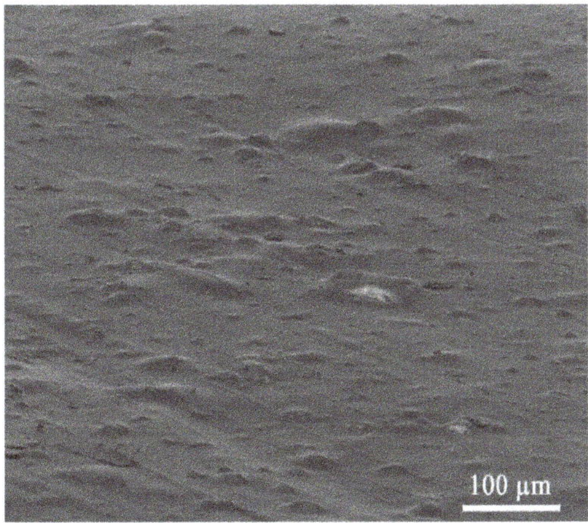

Figure 2. SEM image of composite surface for PLLA/HAp.

3.2. Structural Analysis of Eu^{3+}-Doped Composites

The crystal phase purity of $Ca_{10}(PO_4)_6(OH)_2$ nanocrystals doped with x Eu^{3+} ions (where x = 1, 3, 5 mol%) was checked with the powder XRD technique and was compared with the reference standard of the hexagonal $Ca_{10}(PO_4)_6(OH)_2$ lattice ascribed to the $P6_3/m$ space group [45] (ICSD-180315) (see Figure S5). The diffraction patterns of the Eu^{3+}:$Ca_{10}(PO_4)_6(OH)_2$ nanoparticles embedded into the poly(L-lactide) composite having the function of optically active ions concentration obtained with the extrusion method are presented in Figure 3. The neat poly(L-lactide) crystallized in the orthorhombic α'-polylactide with $P2_12_12_1$ space group. In this case, the degree of crystallinity was approximately 11% (Figure S6a). The deconvolution of XRD curves of the hydroxyapatite-doped PLLA, which revealed amorphous PLLA in all the investigated cases, was performed (see Figure S6b–e). Interestingly, mesophase with an amount in the range of 1.9% to 5.9% was also found in PLLA doped with HAp. The highest mesophase content was found in the systems having the highest Eu^{3+} content (3 and 5 mol%).

Infrared spectra were recorded for the entire materials to get a deeper insight into the structure. The IR spectra of HAp consist of typical active vibrational bands of phosphate and hydroxyl groups (see Figure S7).

The IR spectra of the PLLA/HAp composites comprise vibrational bands related to the hydroxyapatite and poly(L-lactide) (see Figure 4). The typical poly(L-lactide) active stretching and banding vibrations of the $-CH_3$ and $-CH_2$ groups are observed at 2848.3 and 2915.4 cm^{-1}, respectively. The most intense peak at 1749.1 cm^{-1} is connected with $-C = O$ stretching vibration. The symmetric and asymmetric stretching of the C-C(=O)-O group was observed at 1452.1 and 1150 cm^{-1}. Peaks at this range can be connected with ester groups existing in polylactide and lactide molecules. The 1380.7 cm^{-1} and 1361.0 cm^{-1} peaks were assigned to the scissor vibration $\delta_s(-CH_3)$ group and the bending vibration of the $\delta_1(-CH_3)$ group. The vibration of the ester group ($-C-O-$) derived from poly(L-lactide) molecules can also be clearly observed at 1180.7 cm^{-1}. Peaks lying at higher energy corresponded to the vibration of groups belonging to HAp. Regarding the infrared spectra of pure hydroxyapatite, it was possible to determine all of the functional groups of hydroxyapatite. In the case of polylactide composites, the position of phosphate bands is shifted slightly. The triply degenerate ν_3 antisymmetric stretching of the phosphate groups is observed at 1039.4 cm^{-1} (ν_3) and 1082.8 cm^{-1} (ν_3). The ratio intensity of these peaks in reference to nanopowders (Figure 4) is different. In the composite, the peak around 1080 cm^{-1} is more intense, because the ester group vibration (-C-O-) of poly(L-lactide) and the phosphate group vibration ($v_3(PO_4^{3-})$) of hydroxyapatite are overlapped. The symmetric stretching (v_1) of the PO_4^{3-} groups was assigned to the peak at 955.5 cm^{-1}. The spectra of all obtained composites show great similarity and are in agreement with the literature data [46–48].

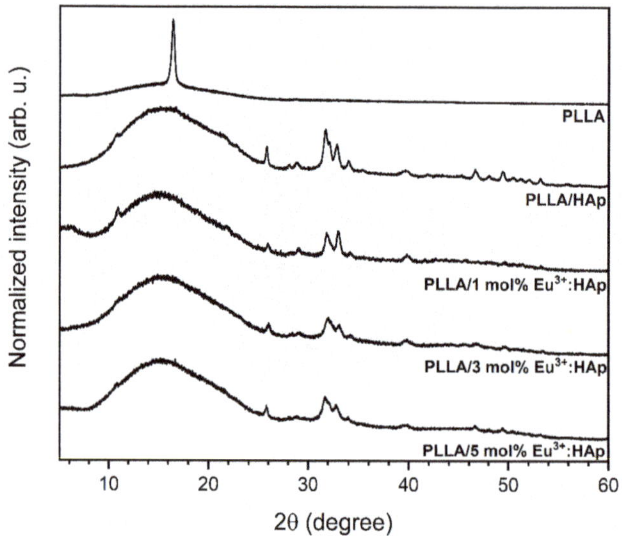

Figure 3. X-ray diffraction patterns of PLLA/x mol% Eu^{3+}:HAp composites (where x = 0–5) obtained with the extrusion method.

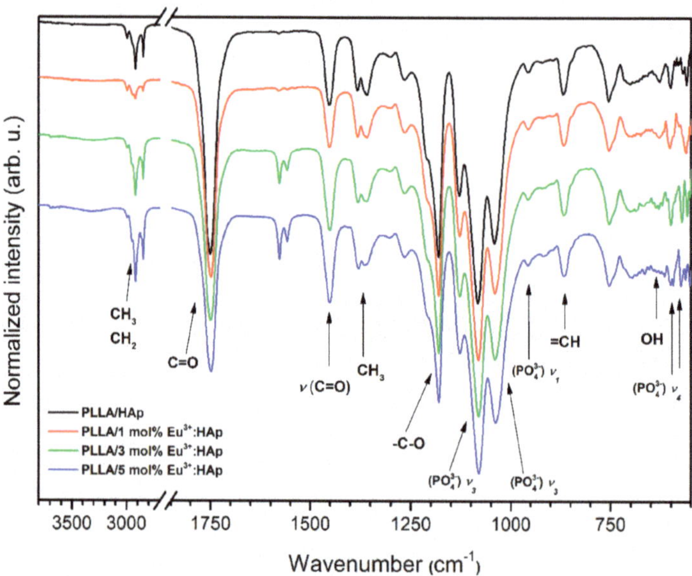

Figure 4. IR spectra of PLLA/x mol% Eu^{3+}:HAp composites (where x = 0–5) obtained via extrusion in situ.

3.3. Thermal Properties

More information about the supramolecular structure of PLLA in the composites was obtained from DSC measurements. Upon the first heating, the DSC curves of PLLA and the composites are presented in Figure 5a, characteristic phase transitions are visible, such as glass transition with T_g at around 61°C, cold crystallization at the mean temperature range of 80 °C ÷ 120 °C, and melting with preceding small exothermic effect corresponding to alpha'-alpha reorganization [49,50]. The thermal parameters estimated form the first heating DSC curves are collected in Table S1.

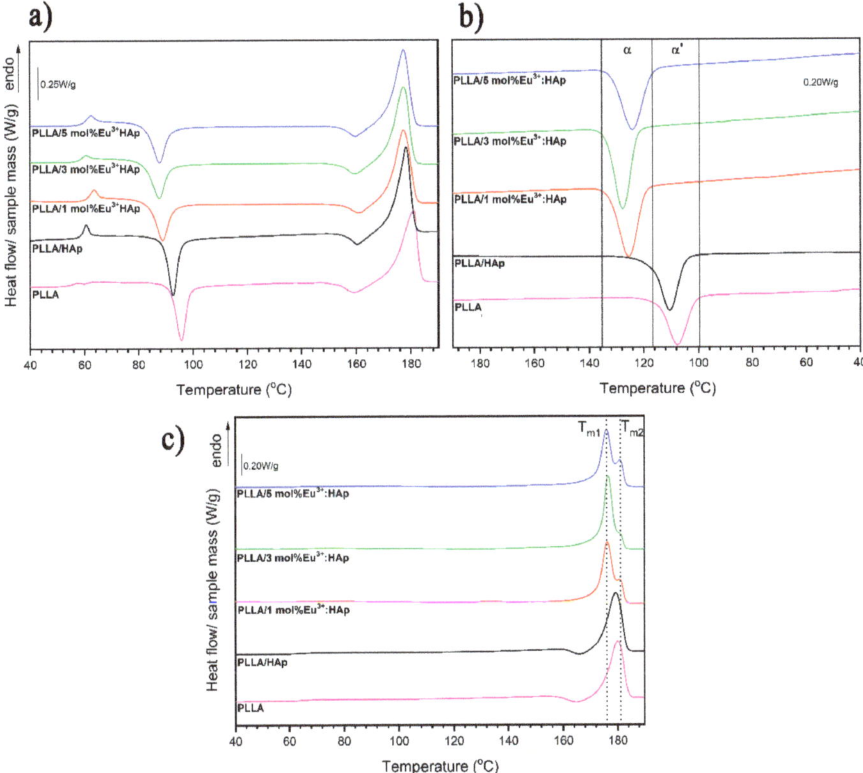

Figure 5. The first heating (**a**), cooling (**b**) and second heating (**c**) DSC curves of PLLA and PLLA/x mol% Eu^{3+}:HAp composites (where x = 0–5).

The influence of HAp particles and Eu^{3+} ions doping HAp on melt crystallization of PLLA can be analyzed based on the cooling DSC curves presented in Figure 5b and Table S2. For comparison of the cooling DSC curves, it is clearly visible that melt crystallization exotherms of PLLA in the analyzed samples have different temperature ranges. Neat PLLA and PLLA with unmodified HAp crystallize upon cooling at lower temperature ranges than PLLA with HAp modified with Eu^{3+}. The Eu^{3+} ions have an influence on crystallization kinetics of PLLA. In the presence of Eu^{3+}, PLLA crystallizes faster. Moreover, it is known that at lower temperatures, it crystallizes into alpha and alpha' crystals whereas at higher temperatures, PLLA crystallizes in the alpha form [49]. It can be expected that the presence of Eu^{3+} has an influence on the crystalline form of PLLA that forms on melt crystallization. The verification of this assumption can be found in the second heating DSC curves (Figure 5c, Table S3).

3.4. Spectroscopic Properties

The excitation emission spectra of all the obtained materials were measured at room temperature, with a recording emission wavelength at 616 nm, which responds to the maximum of the most intense electric dipole transition ($^5D_0 \rightarrow {}^7F_2$). The spectra were corrected to the intensity of the excitation source and normalized to the most intense transition. All spectra contain characteristic sharp lines associated with the intraconfigurational 4f-4f transitions of the Eu^{3+} ions, as well as an intense and broad band related to the ligand-to-metal charge transfer (CT) $O^{2-} \rightarrow Eu^{3+}$ transition located in the UV region. The excitation emission spectra presented in Figure S8 belong to Eu^{3+}:HAp nanopowders and in Figure 6 belong to PLLA/Eu^{3+}:HAp composites. The 4f orbitals of lanthanide ions are well isolated by the external 5s, 5p and 5d shells and well protected against the influence of the crystal field. Due to this feature, the barycenters of the f-f lanthanide electron transitions are weakly affected by the ligand field, and the position of these peaks remains almost independent of the host lattice structure [51,52]. The narrow peaks observed at 299.2 nm (33 422 cm^{-1}) were ascribed to the $^7F_0 \rightarrow {}^5F_{(4,3,2,1)}$, $^3P_0$ transitions, at 319.7 nm (31 279 cm^{-1}) to $^7F_0 \rightarrow {}^5H_{(6,5,4,7,3)}$, at 363.4 nm (27 518 cm^{-1}) to $^7F_0 \rightarrow {}^5D_4$, $^5L_8$ at 377.0 nm (26 525 cm^{-1}) to $^5L_8$, $^7F_0 \rightarrow G_2$, $^5L_7$, $^5G_3$, at 394.4 nm (25 355 cm^{-1}) to $^7F_0 \rightarrow {}^5L_6$, at 416.2 nm (24 027 cm^{-1}) to $^7F_0 \rightarrow {}^5D_3$ at 465.9 nm (21 464 cm^{-1}) to $^7F_0 \rightarrow {}^5D_2$, as well as at 527.0 nm (18 975 cm^{-1}) to $^7F_0 \rightarrow {}^5D_1$. The allowed CT transition is strongly affected by electron-lattice coupling, and the peak position depends on the surrounding symmetry of the ion. In the case of PLLA/HAp composites, the CT maximum is located at 253.5 nm (39 448 cm^{-1}). It is well known that the substitution of divalent Ca^{2+} by trivalent Eu^{3+} cations requires the charge compensation mechanism. The two mechanisms are well known for the apatite host lattice and have been described previously in [51,53,54].

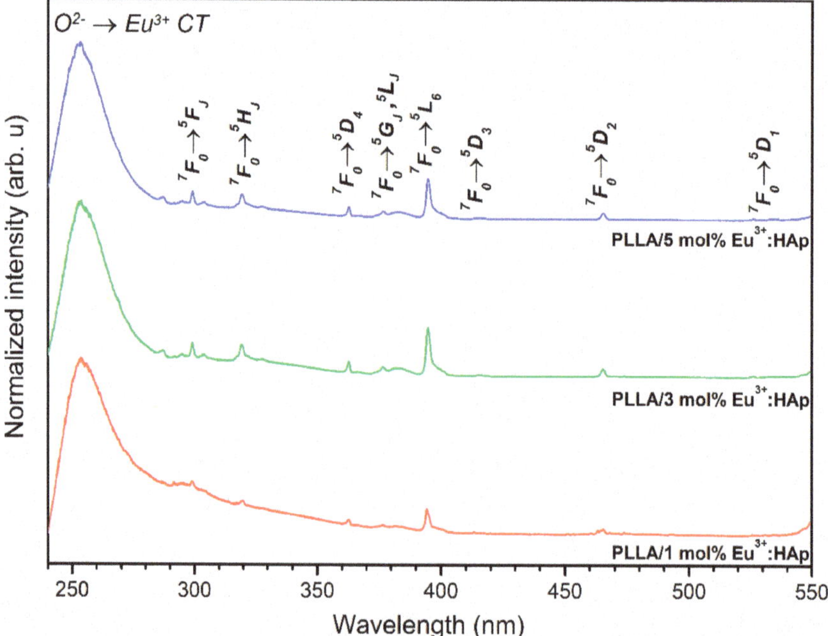

Figure 6. Excitation spectra of x mol% Eu^{3+}:HAp nanoparticles (where x = 1–5) incorporated into PLLA composites.

The room temperature emission spectra of the entire nanopowders and nanocomposites were measured in the spectral range of 500–750 nm under 394.5 nm excitation as a function of optically active ions concentration. The emission spectra of Eu^{3+}:HAp nanopowders are collected in Figure S9 and the emission spectra of PLLA/Eu^{3+}:HAp composites are collected in Figure 7. The spectra were normalized to the $^5D_0 \rightarrow {}^7F_1$ transition. As can be seen, the emission spectra of Eu^{3+} ions are composed of five bands related to the $^5D_0 \rightarrow {}^7F_{0,1,2,3,4}$ transitions occurring respectively at ca. 573.5 nm (17 437 cm^{-1}), 590.0 nm (16 949 cm^{-1}), 616.2 nm (16 228 cm^{-1}), 653.2 nm (15 309 cm^{-1}) and 700.9 nm (14 267 cm^{-1}). The most important transitions in the study of the structural and spectroscopic properties of Eu^{3+} ions are the $^5D_0 \rightarrow {}^7F_{0,1,2}$ transitions.

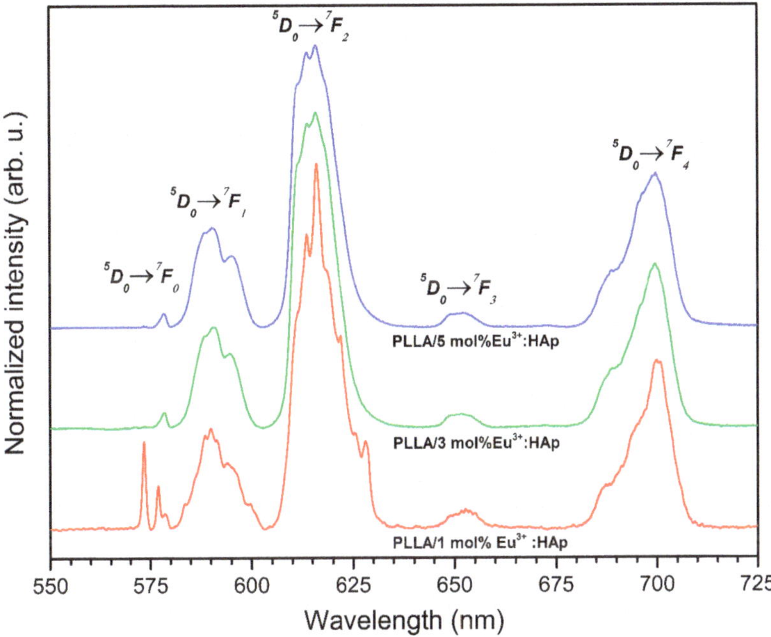

Figure 7. Emission spectra of PLLA/x mol% Eu^{3+}:HAp composites (where x = 0–5) incorporated into poly(L-lactide) composites obtained by using the extrusion method.

The ratio of the integral intensities of the $^5D_0 \rightarrow {}^7F_2$ electric dipole transition to the $^5D_0 \rightarrow {}^7F_1$ magnetic dipole transition is needed to evaluate the asymmetry of the coordination polyhedron of europium(III) ions and the variations in the local point symmetry. This is possible due to the fact that the intensity of the $^5D_0 \rightarrow {}^7F_2$ transition is very sensitive to even small changes in the local environment of Eu^{3+} ions in the crystal field, whereas the intensity of the $^5D_0 \rightarrow {}^7F_1$ transition is nearly independent of those influences. If an Eu^{3+} ion is located in a centrosymmetric site, the only permitted transition is the magnetic one. In the opposite case, the electric dipole transition is dominant. The ratio of the relative emission intensities (R) is defined by the equation:

$$R = \frac{\int {}^5D_0 \rightarrow {}^7F_2}{\int {}^5D_0 \rightarrow {}^7F_1} \tag{1}$$

The higher the ratio between these transitions is, the less centrosymmetric the local environment around Eu^{3+} ions becomes. The impact of the Eu^{3+} ions concentration in powders and composites on the R values are presented in Table 2. It is difficult to observe a straight tendency in the case of both materials. With the influence of the Eu^{3+} ions concentration, the R factor increases, so the Eu^{3+} ions local environment became more distorted, but in the case of 5 mol% Eu^{3+}:HAp, it decreased in both powders and in composites. Comparing both materials, the R value was higher in composites, which indicates that in these materials, the surroundings of the Eu^{3+} ions is more distorted.

Table 2. Decay rates of radiative (A_{rad}), non-radiative (A_{nrad}) and total (A_{tot}) processes of $^5D_0 \rightarrow {}^7F_J$ transitions, luminescence lifetimes (τ), intensity parameters (Ω_2, Ω_4), quantum efficiency (η) and asymmetry ratio (R) of the powders and composites.

Sample	A_{rad} (s^{-1})	A_{nrad} (s^{-1})	A_{tot} (s^{-1})	T (ms)	Ω_2 (10^{-20} cm^2)	Ω_4 (10^{-20} cm^2)	η (%)	R
x mol% Eu^{3+}:HAp Powders								
1 mol% Eu^{3+}	160.05	305.07	465.12	2.15	4.0441	1.0793	34.41	2.874
3 mol% Eu^{3+}	179.94	391.49	571.43	1.75	4.6633	1.3196	31.49	3.314
5 mol% Eu^{3+}	155.55	387.93	543.48	1.84	3.7878	1.2680	28.62	2.692
PLLA/x mol% Eu^{3+}:HAp Composites								
1 mol% Eu^{3+}	230.30	414.86	645.16	1.55	4.9835	4.5359	35.70	3.542
3 mol% Eu^{3+}	218.28	517.02	735.29	1.36	4.5349	4.5455	29.69	3.223
5 mol% Eu^{3+}	201.85	561.50	763.36	1.31	4.0362	4.3208	26.44	2.868

The simplified Judd–Ofelt approach was implemented to provide the intensity parameters Ω_2 and Ω_4, as well as to have a deeper insight into the structure. The results of this approach are shown in Table 2. The value of the Ω_2 parameter indicates some changes in the distortion of europium coordination polyhedra caused by such factors as ions concentration, annealing temperature, etc., and could be related to an increase of the $Eu^{3+} - O^{2-}$ bond covalency. It is worth noting that the value of the Ω_2 parameter has the same tendency as the R factor in powders and composites. The value of the Ω_4 factor supplies some information about changes in the electron density around Eu^{3+} cations. This value cannot be straightly correlated with changes in the Eu^{3+} ions symmetry, but it can add some information about the electron density variations of the O^{2-} anions surrounding that influences the CT band position.

To determine the detailed spectroscopic properties of nanopowders and nanocomposites, the luminescence kinetics was analyzed. The luminescence life times were recorded at room temperature. The materials were excited by a 394.5 nm line and monitored at 618 nm corresponding to the $^5D_0 \rightarrow {}^7F_2$ transition. The luminescence kinetic curves of the Eu^{3+}:HAp nanopowders are shown in Figure S10, and for the Eu^{3+}:Ca$_{10}$(PO$_4$)$_6$(OH)$_2$ nanoparticles embedded into poly(L-lactide) composites, they are shown in Figure 8.

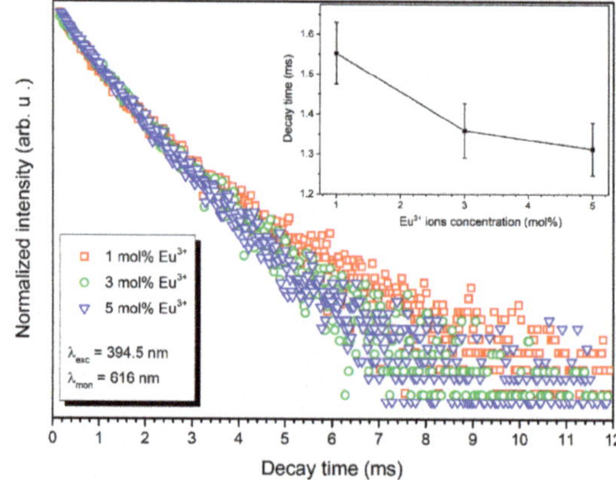

Figure 8. Emission kinetics of the Eu^{3+}:Ca$_{10}$(PO$_4$)$_6$(OH)$_2$ nanoparticles embedded into poly(L-lactide) composites obtained by using the extrusion method. Insert: The decay time as a function of Eu^{3+} ions concentration.

3.5. Tensile Properties

In Figure 9, we presented the tensile results for PLLA-based composites with 10 wt.% of HAp with different Eu^{3+} ions concentration. The parameters are very important factors in designing 3D scaffolds for bone tissue engineering. From among three basic methods (polymerization in situ, solution-casting and melt mixing), we chose melt mixing-extrusion in a twin-screw extruder. This method ensures good filler distribution in the polymer. The preparation method is the most important in the case of nanocomposites preparation.

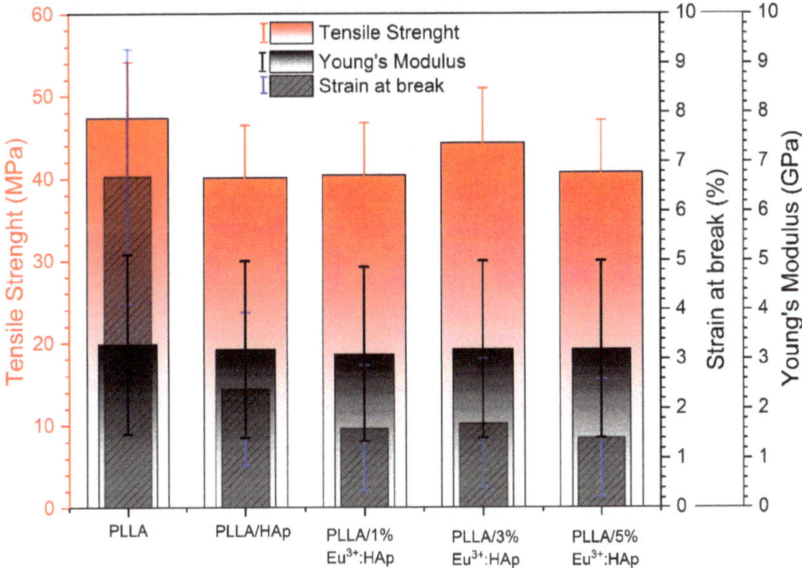

Figure 9. Tensile properties for PLLA/HAp composites: tensile strength (red left axis), strain at break (right axis) and Young's modulus (right axis).

4. Discussion

4.1. Thermal Properties

The strongest differences in thermal properties of PLLA and the composites concern cold crystallization. In the presence of HAp particles, the onset of cold crystallization (T_{cc}^{onset}), as well as that of peak temperature (T_{cc}), is about 3 °C lower compared to neat PLLA. This result indicates the well-known nucleating activity of HAp particles on PLLA cold crystallization [50]. It is worth underlining that the stronger effectiveness of nucleation activity towards cold crystallization is exhibited by HAp particles doped with Eu^{3+} ions. The lowest T_{cc}^{onset} was registered for the PLLA/3 mol% Eu^{3+}:HAp composite, the cold crystallization onset is about 9.4 °C lower than for neat PLLA.

Moreover, the enthalpy of cold crystallization (ΔH_{cc}) is lower in the case of the PLLA/Eu^{3+}:HAp composite than for neat PLLA and PLLA with the unmodified HAp particles composite. The lower values of cold crystallization enthalpy prove the higher crystallinity of PLLA in the presence of Eu^{3+}:HAp. The presence of Eu^{3+} ions and its molar content have a strong influence on the crystallinity degree of PLLA in composites.

The lack of exothermic effects of the alpha'-alpha transition in the second heating DSC curves in the case of composites with Eu^{3+} confirms that, upon subsequent cooling, PLLA crystallized in the alpha form. By contrast, neat PLLA and PLLA with unmodified HAp crystallized upon cooling as a mixture of alpha and alpha′ crystals, which is designated by the presence of an exothermic peak at 165 °C corresponding to the alpha′-alpha reorganization.

The PLLA crystallizes faster and in alpha and alpha′ form in the presence of Eu^{3+} ions embedded in the apatite. The HAp doped with Eu^{3+} ions shown high nucleation activity towards cold crystallization as well as high crystallinity degree of PLLA in composites.

4.2. Spectroscopic Properties

The red emission was observed from the Eu^{3+}:HAp powders and the PLLA/Eu^{3+}:HAp composites. The analysis of the $^5D_0 \rightarrow {}^7F_0$ transition can provide data about the number of crystallographic sites substituted by Eu^{3+} ions into the host structure. Additionally, the existence of this transition approves that the Eu^{3+} ions are located at a low-symmetry environment, and it is observed only if the Eu^{3+} ions occupy sites with a local symmetry of C_n, C_{nv} or C_s. The $^5D_0 \rightarrow {}^7F_0$ transition is split into three components located at 573.5 nm (17 437 cm^{-1}) and 577.2 nm (17 325 cm^{-1}) and 578.6 nm (17 283 cm^{-1}) in the case of Eu^{3+}:HAp nanopowders. The Eu^{3+} ions occupy three different types of crystallographic sites, which is well known in the literature on apatite host matrixes [51–54]. The $^5D_0 \rightarrow {}^7F_1$ and $^5D_0 \rightarrow {}^7F_2$ transitions consist of many overlapped Stark components. The most intense emission was observed for the hypersensitive $^5D_0 \rightarrow {}^7F_2$ electronic transition. Its intensity is highly influenced by the local symmetry of the Eu^{3+} ions as well as the type of ligands rather than the intensities of the other electronic transitions. This transition is notably used to calculate the asymmetry of the Eu^{3+} site. In the case of europium(III) emission in the PLLA/1 mol% Eu^{3+}:HAp composite, it is visible that the $^5D_0 \rightarrow {}^7F_0$ transition is split into three components, but in the case of higher concentration, it is possible to distinguish only a single line. Surprisingly, the major difference is visible in the intensity of the $^5D_0 \rightarrow {}^7F_4$ transition which is much more intense in the case of the composite in the entire range of optically active ions concentration.

The decay profiles are non-exponential. Since the physical meaning of multi-exponential fitting is complicated to explain, the lifetime values were calculated from the effective emission decay times using the following equation:

$$\tau_m = \frac{\int_0^\infty tI(t)dt}{\int_0^\infty I(t)dt} \cong \frac{\int_0^{t^{max}} tI(t)dt}{\int_0^{t^{max}} I(t)dt} \tag{2}$$

where $I(t)$ is the luminescence intensity at time t corrected for the background, and the integrals are calculated over the range of $0 < t < t^{max}$, where $t^{max} >> \tau_m$. The value of decay time is shorter in the case of the Eu^{3+} ions-doped hydroxyapatite embedded into poly(L-lactide). This observation could be connected with different response of the Eu^{3+} ions present on the surface in nanopowders than in composites. The decay times shortened with an increase of Eu^{3+} ions concentration in HAp embedded into PLLA.

4.3. Tensile Properties

The tensile strength of the PLLA/HAp composite is ~40.1 MPa, which is comparable to all the three PLLA/Eu^{3+}:HAp composites (40.4–44.3 MPa with a standard deviation 2.4–2.9 MPa). In this case, no influence of Eu^{3+} ions on the tensile strength of the composites was observed. However, tensile strength of all the composites is lower compared to neat PLLA. The effect of mechanical properties reduction was described earlier [55] for PLLA doped with cellulose nanocrystals or carbon nanotubes but also for hydroxyapatite [16]. The reason for this effect is polymer–filler adhesion/interaction. In the research, neat calcium hydroxyapatite without any surface modification has been used. In this case, adhesion is rather poor and is not improved by the Eu^{3+} ions incorporation. Another reason is the filler content. It has been shown before that, in the extruded composites, the higher the content of hydroxyapatite, the lower the tensile strength [16]. Young's modulus all the values for the composites are in the range of 3.1–3.2 GPa with a standard deviation of 0.1–0.3 GPa. When comparing the strain at break parameter, the reduction for PLLA/HAp (2.4%) as compared to neat PLLA (6.7%) can be observed along with further reduction of PLLA/Eu^{3+}:HAp (1.4–1.7%) composites as compared to the PLLA/HAp system. The last effect is very slight (0.7–1.0%); however, the measurement error ranges from 0.1% to 0.2%. The tensile strength, Young's modulus and strain at break parameters value are the highest in the case of neat PLLA and are comparable to each other for all PLLA/HAp composites.

5. Conclusions

In the present research, pure as well as Eu^{3+}-doped hydroxyapatite were synthesized by using the precipitation method and were confirmed by the XRD patterns analysis. The poly(L-lactide)/europium(III)-doped hydroxyapatite composite foils were successfully fabricated using the twin-screw co-rotating micro-extrusion technique in the weight ratio 9:1. The obtained composites were found to have similar properties to neat PLLA. Differential Scanning Calorimetry revealed some differences in thermal properties of PLLA and the composites. In the presence of HAp particles, the cold crystallization is lower in comparison with neat PLLA. This result indicated a well-known nucleating activity of HAp particles towards the cold crystallization. Moreover, stronger effectiveness of nucleation activity towards cold crystallization was shown in the case of HAp particles doped with Eu^{3+} ions. The HAp nanopowders and composites doped with Eu^{3+} ions are characterized by red emission under UV radiation with the $^5D_0 \rightarrow {}^7F_2$ transition as the most intense. The luminescence kinetics shows a non-exponential curve indicating that more than one emitting center is present in the investigated materials. The value of luminescence decays is lower in the case of the Eu^{3+} ions-doped hydroxyapatite embedded into poly(L-lactide).

The PLLA/Eu^{3+}:HAp composites were obtained as prospective candidates to theranostic applications (therapy and diagnostics) due to support of bone healing by hydroxyapatite and bio-imaging possibility of Eu^{3+} ions.

Supplementary Materials: The Supplementary Materials are available online at http://www.mdpi.com/2079-4991/9/8/1146/s1.

Author Contributions: R.J.W. conceived and designed the experiments as well as analyzed all data; K.S. (Katarzyna Szyszka) and S.T. performed the experiments as well as analyzed all data; M.G. and K.S. (Konrad Szustakiewicz) contributed reagents/materials/analysis tools; all authors contributed to the writing of the paper.

Funding: This work was financially supported by the National Science Centre (NCN) under the projects 'Preparation and characterization of nanoapatites doped with rare earth ions and their biocomposites' (No. UMO-2012/05/E/ST5/03904) and 'Preparation and characterization of biocomposites based on nanoapatites for theranostics' (No. UMO-2015/19/B/ST5/01330).

Acknowledgments: The authors would like to thank Aneta Ciupa-Litwa for IR measurements and Damian Szymanski for the assistance with SEM-EDX measurements and Alicja Borowiecka for synthesis of the studied powders as well as Bartlomiej Kryszak for performing XRD analysis and help by deconvolution of WAXS patterns.

Conflicts of Interest: The authors declare no conflict of interest.

References

1. Chen, F.-M.; Liu, X. Advancing biomaterials of human origin for tissue engineering. *Prog. Polym. Sci.* **2016**, *53*, 86–168. [CrossRef] [PubMed]

2. Terukina, T.; Saito, H.; Tomita, Y.; Hattori, Y.; Otsuka, M. Development and effect of a sustainable and controllable simvastatin-releasing device based on PLGA microspheres/carbonate apatite cement composite: In vitro evaluation for use as a drug delivery system from bone-like biomaterial. *J. Drug Deliv. Sci. Technol.* **2017**, *37*, 74–80. [CrossRef]

3. Tsuji, H. Poly(lactic acid) stereocomplexes: A decade of progress. *Adv. Drug Deliv. Rev.* **2016**, *107*, 97–135. [CrossRef] [PubMed]

4. Puppi, D.; Pirosa, A.; Lupi, G.; Erba, P.A.; Giachi, G.; Chiellini, F. Design and fabrication of novel polymeric biodegradable stents for small caliber blood vessels by computer-aided wet-spinning. *Biomed. Mater.* **2017**, *12*, 035011. [CrossRef] [PubMed]

5.   Zhu, Y.; Pyda, M.; Cebe, P. Electrospun fibers of poly(L-lactic acid) containing lovastatin with potential applications in drug delivery. *J. Appl. Polym. Sci.* **2017**, *134*, 45287. [CrossRef]

6.   Ji, X.; Yang, W.; Wang, T.; Mao, C.; Guo, L.; Xiao, J.; He, N. Coaxially electrospun core/shell structured poly(L-lactide) acid/chitosan nanofibers for potential drug carrier in tissue engineering. *J. Biomed. Nanotechnol.* **2013**, *9*, 1672–1678. [CrossRef] [PubMed]

7.   Liang, J.W.; Prasad, G.; Wang, S.C.; Wu, J.L.; Lu, S.G. Enhancement of the Oil Absorption Capacity of Poly(Lactic Acid) Nano Porous Fibrous Membranes Derived via a Facile Electrospinning Method. *Appl. Sci.* **2019**, *9*, 1014. [CrossRef]

8.   le Phuong, H.A.; Ayob, N.A.I.; Blanford, C.F.; Rawi, N.F.M.; Szekely, G. Nonwoven Membrane Supports from Renewable Resources: Bamboo Fiber Reinforced Poly(Lactic Acid) Composites. *ACS Sustain. Chem. Eng.* **2019**, *7*, 11885–11893. [CrossRef]

9.   Muller, J.; González-Martínez, C.; Chiralt, A. Combination of Poly(lactic) Acid and Starch for Biodegradable Food Packaging. *Materials* **2017**, *10*, 952. [CrossRef]

10. Shi, J.; Zhang, L.; Xiao, P.; Huang, Y.; Chen, P.; Wang, X.; Gu, J.; Zhang, J.; Chen, T. Biodegradable PLA Nonwoven Fabric with Controllable Wettability for Efficient Water Purification and Photocatalysis Degradation. *ACS Sustain. Chem. Eng.* **2018**, *6*, 2445–2452. [CrossRef]

11. Prabhakaran, M.P.; Venugopal, J.; Ramakrishna, S. Electrospun nanostructured scaffolds for bone tissue engineering. *Acta Biomater.* **2009**, *5*, 2884–2893. [CrossRef]

12. Okamoto, M.; John, B. Synthetic biopolymer nanocomposites for tissue engineering scaffolds. *Prog. Polym. Sci.* **2013**, *38*, 1487–1503. [CrossRef]

13. Dorozhkin, S.V. Calcium orthophosphates in dentistry. *J. Mater. Sci. Mater. Med.* **2013**, *24*, 1335–1363. [CrossRef]

14. Pietrasik, J.; Szustakiewicz, K.; Zaborski, M.; Haberko, K. Hydroxyapatite: An environmentally friendly filler for elastomers. *Mol. Cryst. Liq. Cryst.* **2008**, *483*, 172–178. [CrossRef]

15. Szustakiewicz, K.; Stępak, B.; Antończak, A.J.; Maj, M.; Gazińska, M.; Kryszak, B.; Pigłowski, J. Femtosecond laser-induced modification of PLLA/hydroxypatite composite. *Polym. Degrad. Stab.* **2018**, *149*, 152–161. [CrossRef]

16. Smieszek, A.; Marycz, K.; Szustakiewicz, K.; Kryszak, B.; Targonska, S.; Zawisza, K.; Watras, A.; Wiglusz, R.J. New approach to modification of poly (l-lactic acid) with nano-hydroxyapatite improving functionality of human adipose-derived stromal cells (hASCs) through increased viability and enhanced mitochondrial activity. *Mater. Sci. Eng. C* **2019**, *98*, 213–226. [CrossRef]

17. Wei, Y.; He, Y.; Li, X.; Chen, H.; Deng, X. Cellular uptake and delivery-dependent effects of Tb^{3+}-doped hydroxyapatite nanorods. *Molecules* **2017**, *22*, 1043. [CrossRef]

18. Curtin, C.M.; Cunniffe, G.M.; Lyons, F.G.; Bessho, K.; Dickson, G.R.; Duffy, G.P.; O'Brien, F.J. Innovative collagen nano-hydroxyapatite scaffolds offer a highly efficient non-viral gene delivery platform for stem cell-mediated bone formation. *Adv. Mater.* **2012**, *24*, 749–754. [CrossRef]

19. Yuan, Q.; Wu, J.; Qin, C.; Xu, A.; Zhang, Z.; Lin, S.; Ren, X.; Zhang, P. Spin-coating synthesis and characterization of Zn-doped hydroxyapatite/polylactic acid composite coatings. *Surf. Coat. Technol.* **2016**, *307*, 461–469. [CrossRef]

20. Wang, X.; Song, G.; Lou, T. Fabrication and characterization of nano-composite scaffold of PLLA/silane modified hydroxyapatite. *Med. Eng. Phys.* **2010**, *32*, 391–397. [CrossRef]

21. Ajdukovic, Z. Repair of Bone Tissue Affected by Osteoporosis with Hydroxyapatite-Poly-L-lactide (HAp-PLLA) With and Without Blood Plasma. *J. Biomater. Appl.* **2005**, *20*, 179–190. [CrossRef]

22. Nishida, Y.; Domura, R.; Sakai, R.; Okamoto, M.; Arakawa, S.; Ishiki, R.; Salick, M.R.; Turng, L.S. Fabrication of PLLA/HA composite scaffolds modified by DNA. *Polymer* **2015**, *56*, 73–81. [CrossRef]

23. Park, K.; Kim, J.J. Effect of Surface-activated PLLA Scaffold on Apatite Formation in Simulated Body Fluid. *J. Bioact. Compat. Polym.* **2010**, *25*, 27–39. [CrossRef]

24. Sabir, M.I.; Xu, X.; Li, L. A review on biodegradable polymeric materials for bone tissue engineering applications. *J. Mater. Sci.* **2009**, *44*, 5713–5724. [CrossRef]

25. Junchao, W.; Yanfeng, D.; Yiwang, C.; Xuesi, C. Mechanical and thermal properties of polypeptide modified hydroxyapatite/poly(L-lactide) nanocomposites. *Sci. China Chem.* **2011**, *54*, 431–437. [CrossRef]

26. Wilberforce, S.I.J.; Finlayson, C.E.; Best, S.M.; Cameron, R.E. The influence of hydroxyapatite (HA) microparticles (m) and nanoparticles (n) on the thermal and dynamic mechanical properties of poly-L-lactide. *Polymer* **2011**, *52*, 2883–2890. [CrossRef]

27. Cho, Y.S.; Hong, M.W.; Jeong, H.-J.; Lee, S.J.; Kim, Y.Y.; Cho, Y.S. The fabrication of well-interconnected polycaprolactone/hydroxyapatite composite scaffolds, enhancing the exposure of hydroxyapatite using the wire-network molding technique. *J. Biomed. Mater. Res. Part B Appl. Biomater.* **2017**, *105*, 2315–2325. [CrossRef]

28. Duan, B.; Wang, M.; Zhou, W.Y.; Cheung, W.L.; Li, Z.Y.; Lu, W.W. Three-dimensional nanocomposite scaffolds fabricated via selective laser sintering for bone tissue engineering. *Acta Biomater.* **2010**, *6*, 4495–4505. [CrossRef]

29. Szustakiewicz, K.; Gazińska, M.; Kryszak, B.; Grzymajło, M.; Pigłowski, J.; Wiglusz, R.J.; Okamoto, M. The influence of hydroxyapatite content on properties of poly(L-lactide)/hydroxyapatite porous scaffolds obtained using thermal induced phase separation technique. *Eur. Polym. J.* **2019**, *113*, 313–320. [CrossRef]

30. Zhang, C.Y.; Zhang, C.L.; Wang, J.F.; Lu, C.H.; Zhuang, Z.; Wang, X.P.; Fang, Q.F. Fabrication and in vitro investigation of nanohydroxyapatite, chitosan, poly(L-lactic acid) ternary biocomposite. *J. Appl. Polym. Sci.* **2013**, *127*, 2152–2159. [CrossRef]

31. Niu, X.; Feng, Q.; Wang, M.; Guo, X.; Zheng, Q. Porous nano-HA/collagen/PLLA scaffold containing chitosan microspheres for controlled delivery of synthetic peptide derived from BMP-2. *J. Control. Release* **2009**, *134*, 111–117. [CrossRef]

32. Zhu, S.; Sun, H.; Geng, H.; Liu, D.; Zhang, X.; Cai, Q.; Yang, X. Dual functional polylactide–hydroxyapatite nanocomposites for bone regeneration with nano- silver being loaded via reductive polydopamine. *RSC Adv.* **2016**, *6*, 91349–91360. [CrossRef]

33. Popa, C.L.; Ciobanu, C.S.; Voicu, G.; Vasile, E.; Chifiriuc, M.C.; Iconaru, S.L.; Predoi, D. Influence of Thermal Treatment on the Antimicrobial Activity of Silver-Doped Biological Apatite. *Nanoscale Res. Lett.* **2015**, *10*, 502. [CrossRef]

34. Mocanu, A.; Furtos, G.; Rapuntean, S.; Horovitz, O.; Flore, C.; Garbo, C.; Danisteanu, A.; Rapuntean, G.; Prejmerean, C. Tomoaia-Cotisel, Synthesis; Characterization and antimicrobial effects of composites based on multi-substituted hydroxyapatite and silver nanoparticles. *Appl. Surf. Sci.* **2014**, *298*, 225–235. [CrossRef]

35. Banerjee, S.; Bagchi, B.; Bhandary, S.; Kool, A.; Hoque, N.A.; Thakur, P.; Das, S. A facile vacuum assisted synthesis of nanoparticle impregnated hydroxyapatite composites having excellent antimicrobial properties and biocompatibility. *Ceram. Int.* **2017**, *44*, 1066–1077. [CrossRef]

36. Marques, C.F.; Olhero, S.; Abrantes, J.C.C.; Marote, A.; Ferreira, S.; Vieira, S.I.; Ferreira, J.M.F. Biocompatibility and antimicrobial activity of biphasic calcium phosphate powders doped with metal ions for regenerative medicine. *Ceram. Int.* **2017**, *43*, 15719–15728. [CrossRef]

37. Hickey, D.J.; Ercan, B.; Sun, L.; Webster, T.J. Adding MgO nanoparticles to hydroxyapatite-PLLA nanocomposites for improved bone tissue engineering applications. *Acta Biomater.* **2015**, *14*, 175–184. [CrossRef]

38. Nikolaev, A.; Kolesnikov, I.; Frank-Kamenetskaya, O.; Kuz'mina, M. Europium concentration effect on characteristics and luminescent properties of hydroxyapatite nanocrystalline powders. *J. Mol. Struct.* **2017**, *1149*, 323–331. [CrossRef]

39. Wiglusz, R.J.; Kedziora, A.; Lukowiak, A.; Doroszkiewicz, W.; Strek, W. Hydroxyapatites and Europium(III) doped hydroxyapatites as a carrier of silver nanoparticles and their antimicrobial activity. *J. Biomed. Nanotechnol.* **2018**, *8*, 605–612. [CrossRef]

40. Tesch, A.; Wenisch, C.; Herrmann, K.-H.; Reichenbach, J.R.; Warncke, P.; Fischer, D.; Müller, F.A. Luminomagnetic Eu^{3+}- and Dy^{3+}-doped hydroxyapatite for multimodal imaging. *Mater. Sci. Eng. C* **2017**, *81*, 422–431. [CrossRef]

41. Syamchand, S.S.; Sony, G. Europium enabled luminescent nanoparticles for biomedical applications. *J. Lumin.* **2015**, *165*, 190–215. [CrossRef]

42. Chen, M.H.; Yoshioka, T.; Ikoma, T.; Hanagata, N.; Lin, F.H.; Tanaka, J. Photoluminescence and doping mechanism of theranostic Eu^{3+}/Fe^{3+} dual-doped hydroxyapatite nanoparticles. *Sci. Technol. Adv. Mater.* **2014**, *15*, 055005. [CrossRef]

43. Hasna, K.; Kumar, S.S.; Komath, M.; Varma, M.R.; Jayaraj, M.K.; Kumar, K.R. Synthesis of chemically pure, luminescent Eu^{3+} doped HAp nanoparticles: A promising fluorescent probe for in vivo imaging applications. *Phys. Chem. Chem. Phys.* **2013**, *15*, 8106. [CrossRef]

44. Szustakiewicz, K.; Gazińska, M.; Kiersnowski, A.; Pigłowski, J. Polyamide 6/organomontmorillonite nanocomposites based on waste materials. *Polimery* **2011**, *56*, 397–400. [CrossRef]

45. Balan, E.; Delattre, S.; Roche, D.; Segalen, L.; Morin, G.; Guillaumet, M.; Blanchard, M.; Lazzeri, M.; Brouder, C.; Salje, E.K.H. Line-broadening effects in the powder infrared spectrum of apatite. *Phys. Chem. Miner.* **2011**, *38*, 111–122. [CrossRef]

46. Mofokeng, J.P.; Luyt, A.S.; Tábi, T.; Kovács, J. Comparison of injection moulded, natural fibre-reinforced composites with PP and PLA as matrices. *J. Thermoplast. Compos. Mater.* **2012**, *25*, 927–948. [CrossRef]

47. Ignjatović, N.; Savić, V.; Najman, S.; Plavšić, M.; Uskoković, D. A study of HAp/PLLA composite as a substitute for bone powder, using FT-IR spectroscopy. *Biomaterials* **2011**, *22*, 571–575. [CrossRef]

48. Rasselet, D.; Ruellan, A.; Guinault, A.; Miquelard-Garnier, G.; Sollogoub, C.; Fayolle, B. Oxidative degradation of polylactide (PLA) and its effects on physical and mechanical properties. *Eur. Polym. J.* **2014**, *50*, 109–116. [CrossRef]

49. Androsch, R.; Schick, C.; di Lorenzo, M.L. Melting of conformationally disordered crystals (α' phase) of poly(L-lactic acid). *Macromol. Chem. Phys.* **2014**, *215*, 1134–1139. [CrossRef]

50. Wilberforce, S.I.J.; Finlayson, C.E.; Best, S.M.; Cameron, R.E. A comparative study of the thermal and dynamic mechanical behaviour of quenched and annealed bioresorbable poly-L-lactide/α-tricalcium phosphate nanocomposites. *Acta Biomater.* **2011**, *7*, 2176–2184. [CrossRef]

51. Sobierajska, P.; Pazik, R.; Zawisza, K.; Renaudin, G.; Nedelec, J.-M.; Wiglusz, R.J. Effect of lithium substitution on the charge compensation, structural and luminescence properties of nanocrystalline $Ca_{10}(PO_4)_6F_2$ activated with Eu^{3+} ions. *CrystEngComm* **2016**, *18*, 3447–3455. [CrossRef]

52. Zawisza, K.; Strzep, A.; Wiglusz, R.J. Influence of annealing temperature on the spectroscopic properties of hydroxyapatite analogues doped with Eu^{3+}. *New J. Chem.* **2017**, *41*, 9990–9999. [CrossRef]

53. Zawisza, K.; Wiglusz, R.J. Preferential site occupancy of Eu^{3+} ions in strontium hydroxyapatite nanocrystalline - $Sr_{10}(PO_4)_6(OH)_2$ - Structural and spectroscopic characterisation. *Dalton Trans.* **2017**, *46*, 3265–3275. [CrossRef]

54. Zawisza, K.; Sobierajska, P.; Renaudin, G.; Nedelec, J.-M.; Wiglusz, R.J. Effects of crystalline growth on structural and luminescence properties of $Ca_{(10-3x)}Eu_{2x}(PO_4)_6F_2$ nanoparticles fabricated by using a microwave driven hydrothermal process. *CrystEngComm* **2017**, *19*, 6936–6949. [CrossRef]

55. Armentano, I.; Puglia, D.; Luzi, F.; Arciola, C.R.; Morena, F.; Martino, S.; Torre, L. Nanocomposites based on biodegradable polymers. *Materials* **2018**, *11*, 795. [CrossRef]

Article

SmS/EuS/SmS Tri-Layer Thin Films: The Role of Diffusion in the Pressure Triggered Semiconductor-Metal Transition

Andreas Sousanis, Dirk Poelman and Philippe F. Smet *

Lumilab, Department of Solid State Sciences, Ghent University, Krijgslaan 281/S1, 9000 Gent, Belgium; andreas.sousanis@ugent.be (A.S.); dirk.poelman@ugent.be (D.P.)
* Correspondence: philippe.smet@ugent.be

Received: 16 September 2019; Accepted: 20 October 2019; Published: 24 October 2019

Abstract: While SmS thin films show an irreversible semiconductor-metal transition upon application of pressure, the switching characteristics can be modified by alloying with other elements, such as europium. This manuscript reports on the resistance response of tri-layer SmS/EuS/SmS thin films upon applying pressure and on the correlation between the resistance response and the interdiffusion between the layers. SmS thin films were deposited by e-beam sublimation of Sm in an H_2S atmosphere, while EuS was directly sublimated by e-beam from EuS. Structural properties of the separate thin films were first studied before the deposition of the final nanocomposite tri-layer system. Piezoresistance measurements demonstrated two sharp resistance drops. The first drop, at lower pressure, corresponds to the switching characteristic of SmS. The second drop, at higher pressure, is attributed to EuS, partially mixed with SmS. This behavior provides either a well-defined three or two states system, depending on the degree of mixing. Depth profiling using x-ray photoelectron spectroscopy (XPS) revealed partial diffusion between the compounds upon deposition at a substrate temperature of 400 °C. Thinner tri-layer systems were also deposited to provide more interdiffusion. A higher EuS concentration led to a continuous transition as a function of pressure. This study shows that EuS-modified SmS thin films are possible systems for piezo-electronic devices, such as memory devices, RF (radio frequency) switches and piezoresistive sensors.

Keywords: SmS; EuS; semiconductor-metal transition; structural properties; piezoresistivity; interdiffusion; rare earths; thin films

---

1. Introduction

Materials science plays a significant role in the fabrication of new devices based on chemical compounds with specific properties, providing unprecedented device capabilities. Such a device is the piezoelectronic transistor (PET) [1–4], which can exploit the pressure-induced semiconductor to metal transition (SMT) of certain compounds. Examples are Sm chalcogenides [5,6], Mott insulators, as well as other material oxides (e.g., Sr_2IrO_4) [7]. Specifically, SmS features a hysteretic pressure-induced SMT [6,8,9] at around 0.65 GPa, where the resistance drops significantly. In single crystals, the system returns to the semiconducting state upon release of the pressure, while for thin films thermal annealing or a tensile force is needed. Although this behavior was mainly studied in bulk crystals [6,10], we recently managed to observe hysteretic resistance loops in SmS thin films [11]. Another approach to provide a tunable, hysteretic piezoresistive behavior could be the use of alloyed systems [12,13] that shift the energy bands of the primary material (SmS), without inducing the metallic state, leading to reversible switching characteristics upon releasing force. In order to achieve this, SmS can be alloyed with similar materials that have a somewhat wider band gap. This slightly opens the band gap and shifts the SMT to higher pressures, in comparison to pure SmS. In Figure 1, we see a qualitative representation of the hysteretic discontinuous resistance change in SmS as a function of pressure, as observed in single crystals. This drop can be explained as follows. At atmospheric pressure, the material possesses its semiconducting high resistance state (HRS), which changes to a metallic state (low resistance state, or LRS) when pressure is applied. The required pressure strongly depends on the gap between the *4f* states of Sm ions and the *5d(t$_{2g}$)* degenerate conduction band (CB) [14]. SmS shows a smaller gap (~0.15 eV), than SmSe (transition pressure = 2 GPa) and SmTe (transition pressure = 4.5 GPa), resulting in an isostructural transition at lower pressures (at around 0.65 GPa) [15]. The substitution of SmS with, for instance, wide band gap rare earth-based materials (e.g., EuS or YbS) [16,17], promotes a shift of the *5d* band of the alloyed system to higher energies, with the increase of the substituent further increasing the Sm *4f-5d* band gap. Nevertheless, there are other elements substituting for Sm, which directly promote a chemically triggered transition to the metallic state at ambient conditions. Such elements, for example, are Gd and Y [12,13,17,18], both decreasing the phase transition to even lower pressure. This tunability of the piezoresistive response promotes SmS and alloyed SmS as possible candidates for memory and RF (radio frequency) switching devices [1–4,19].

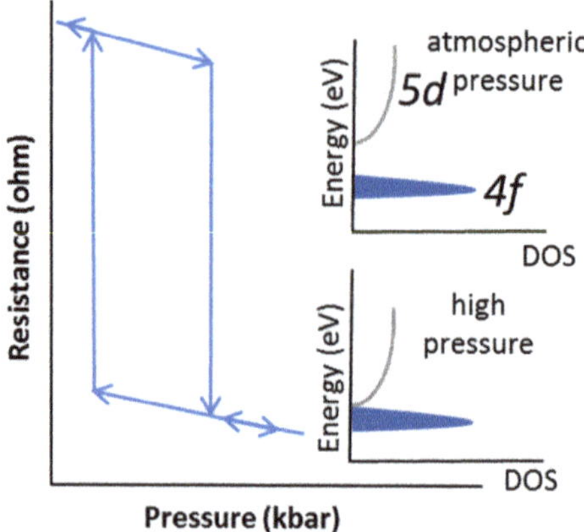

Figure 1. Schematic representation of the piezoresistive response of single crystal SmS. At the critical pressure, a change from the high-resistive to the low-resistive state occurs, due to the pressure induced shift of the *5d* conduction band towards the *4f* states of Sm^{2+} (inset). The top inset shows the gap between the *4f* states and *5d* conduction band at atmospheric pressure. The bottom inset demonstrates the closing of the gap, upon application of pressure. Upon release of the pressure, the SmS returns to the high-resistive state.

In this work, we report on the piezoresistance response of a SmS/EuS/SmS tri-layer thin film system. This tri-layer is an experimentally easy and controlled way to study the alloyed system Sm$_{1-x}$Eu$_x$S. Since the individual films are very thin, interdiffusion between the layers is expected, especially at elevated temperatures during deposition or after post-deposition thermal annealing. We first present some of the basic properties of the separate compounds before showing the structural properties of the tri-layer system. We used two different substrate temperatures to study the diffusion between the layers upon deposition. Thermal post-annealing was performed to further investigate diffusion. The resistance drop, and its subsequent rise after pressure release, confirms that this material system is a promising candidate for several strain-based sensing devices. The resistance response strongly depends on the mixing between the layers. The as-deposited tri-layers at 250 °C showed a three states system behavior, while deposition at 400 °C can lead to a conventional system with two states.

2. Materials and Methods

An e-beam evaporator (model: Leybold Univex 450, Leybold GmbH, Germany) was used to deposit SmS on Corning (1737F) glass and 6 inch Si (100) wafers. Sm metal (Smart Elements GmbH, Vienna, Austria, 99.99%) was used as the target material and deposited at a rate of typically 0.8 nm/s under a reactive H_2S (Praxair Inc., Danbury, CT, USA, 99.8%) flow, leading to a pressure of 1×10^{-5} mbar during the deposition. The base pressure of the system was approximately 2×10^{-6} mbar. Details on the deposition conditions required for obtaining stoichiometric and well-crystallized SmS are described elsewhere [20]. EuS was first synthesized starting from Eu_2O_3 (Alfa Aesar, Thermo Fisher GmbH, Germany, 99.99%) powder. Eu_2O_3 was placed in a tube furnace for 2 h under H_2S flow, at 1000 °C. The produced sulfurized powder consisted of EuS, as confirmed by XRD (not shown). Then, by using a hydraulic press, EuS pellets were prepared as target material for the deposition of EuS thin films by e-beam evaporation under identical H_2S flow as for the deposition of SmS. Resistivity values were calculated from the sheet resistance measured using a four probe setup. In order to electrically insulate the thin films from the Si substrate, dedicated samples were deposited on top of a 600 nm Al_2O_3 (Alfa Aesar, Thermo Fisher GmbH, Germany, 99.99%) thin film, prepared by e-beam evaporation. For electrical measurements where the resistivity was measured across the thin film, iridium bottom and top electrodes (thickness 50 nm) were deposited by e-beam evaporation, starting from an Ir slug (Aldrich Chemistry BVBA, Belgium, 99.9%), with the substrate at room temperature. The bottom electrodes were blanket layers deposited over the entire substrate, whereas the top electrodes were deposited through a mask, yielding circular electrodes with a diameter of 1.5 mm. The SMT was introduced in SmS by rubbing the thin film surface with a round-shaped metal tip, without visibly damaging the thin film surface. The reverse MST (metal to semiconductor transition) was induced by thermal annealing at 400 °C in vacuum. The semiconducting and metallic SmS state will be referred to as S-SmS/HRS (high resistance state) and M-SmS/LRS (low resistance state), respectively.

Structural characterization of the fabricated thin films was carried out via X-ray diffraction (XRD) using a standard powder diffractometer (D8 with Ni-filtered $CuK\alpha_1$ radiation, $\lambda = 0.154059$ nm, Bruker AXS GmbH, Karlsruhe, Germany). In situ high temperature X-ray diffraction (HTXRD) patterns were also measured using a Bruker D8 Discover system (equally using $CuK\alpha_1$ radiation) with an integrated annealing chamber, able to support several atmospheric conditions. In the latter case, a linear detector was used, which allowed for collecting diffraction patterns in seconds, without moving the sample or detector. SEM analysis was carried out in an FEI electron microscope (Quanta FEG 200, Hillsboro, Oregon, USA), with a point resolution of 1.7 nm at 20 kV. Ultraviolet-visible (UV-Vis) spectra were recorded at room temperature in the specular reflectance geometry (V-W method) with a Varian Cary 500 UV-Vis spectrophotometer (Agilent, Santa Clara, CA, USA) in the wavelength range of 200–800 nm. Piezoresistance measurements were performed using a homemade device, similar to reported devices [21]. Finally, we used X-ray photoelectron spectroscopy (XPS) to detect and confirm the presence and diffusion of the components in the multi-layered structures. The used set-up was an ESCA S-probe VG (Thermo Fisher GmbH, Germany) with an $Al(K_a)$ source (1486.6 eV). The base pressure of the system was 5×10^{-10} mbar, while the pressure during Ar-sputtering increased up to 2×10^{-7} mbar. The sputter time was 25 s per step and the measurement time after each step was 5 min, unless mentioned otherwise. In all studies of the tri-layers, the Sm $3d_{5/2}$, S $2p$, O $1s$, and Eu $3d_{5/2}$ peaks were used for the calculation of elemental concentrations.

3. Results

3.1. Individual SmS and EuS Thin Films

EuS thin films with a thickness of 25 nm are largely transparent in the visible region, although the UV-Vis spectra show a broad absorption band between 1.8 and 2.8 eV, caused by the transition from the $4f^7$ ground state to the $4f^6 5d$ configuration in Eu^{2+} (Figure 2a). This is in line with EuS being a natural ferromagnetic semiconductor [22], showing an optical band gap of 1.65 eV for thin film

EuS [23]. EuS thin films with a thickness of 25 nm deposited at 250 °C on Si (100) wafer show a good crystallinity with preferential growth orientation of the (200) planes parallel to the substrate. Notice that SmS and EuS have the same rock salt lattice structure, with almost identical lattice constants for S-SmS and EuS (5.970 Å and 5.968 Å respectively). Consequently, standard XRD analysis cannot be used to discriminate between both materials.

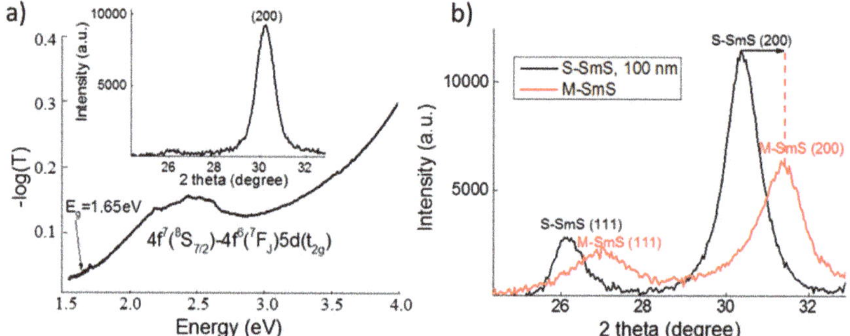

Figure 2. Basic properties of the deposited SmS and EuS thin films. (**a**) Optical (on glass) and structural (inset, on Si wafer) fingerprint of a 25 nm as-deposited EuS. (**b**) XRD patterns showing the structural behavior in the as-deposited (black line) S-SmS, as well as in the metallic state (red line), after rubbing. The lattice planes are indicated. In both thin film depositions, the substrate temperature was 250 °C.

Applying pressure to these EuS thin films does not introduce any transition. This is different for the SmS thin films, where moderate pressure can provide the SMT at room temperature. Usually, soft polishing is used to induce M-SmS [24,25]. Here, gently rubbing the sample surface leads to the SMT, with the accompanying reduction in lattice constant. This is demonstrated by the corresponding XRD peaks shifting to higher 2θ values (Figure 2b), in accordance with previous investigations [26]. The 100 nm S-SmS thin films show resistivity values in the range from 1.5 to 5 × 10^{-1} Ωcm (as derived from the sheet resistance), while the rubbed 100 nm M-SmS layers demonstrated 3–4 × 10^{-3} Ωcm. These values are comparable to previous investigations [27]. The as-deposited EuS thin films showed a sheet resistance of about 1.7 MΩ (25 nm EuS thin film), which is almost one order of magnitude lower, relative to previous investigations on highly optimized insulating EuS thin films (the sheet resistance is higher than 20 MΩ for thicknesses between 20 and 200 nm) [28].

In the SmS/EuS/SmS tri-layers, we studied the diffusion process in this nanocomposite system in order to tune the piezoresistive behavior. The substrate temperature, post-deposition annealing, as well as the thickness of the diffusive layer (EuS) relative to the full tri-layer thickness were changed to influence the layer interdiffusion.

3.2. Deposition and Annealing of Tri-Layers

Figure 3 shows an XRD plot of a 35/30/35 nm SmS/EuS/SmS tri-layer deposited at 250 °C, before (black curve) and after rubbing (red curve). After rubbing the sample surface, we observe two peaks for the (200) lattice plane, one related to EuS and one to M-SmS. Hence, at a deposition temperature of 250 °C, the individual layers do not thoroughly mix. XPS analysis, discussed below, confirms that the thin films are not mixed when deposited at 250 °C.

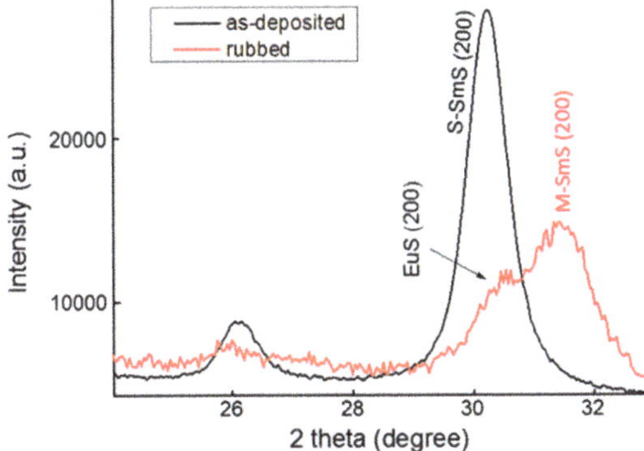

Figure 3. XRD patterns of an as-deposited SmS/EuS/SmS (35/30/35 nm) tri-layer deposited at 250 °C (black curve), and after rubbing (red curve).

In order to probe the influence of temperature on the structural evolution of the tri-layer thin films, in situ HTXRD was performed up to 800 °C (with a heating rate of 10 °C/s). Figure 4a shows the in situ HTXRD patterns of an as-deposited (at 250 °C) 35/30/35 nm SmS/EuS/SmS tri-layer under 20% O_2 and 80% He atmosphere. In this atmosphere, the tri-layer system remains stable up to 350 °C. As mentioned before, the peak at around 30° is composed of the reflection from the (200) lattice planes of SmS and EuS. Above 350 °C, the diffraction intensities decrease—although the (111) reflection of SmS and EuS remains visible—and at 500 °C the observed diffraction positions match with those of Sm_2O_2S. For comparison, Figure 4b shows the stable character of 25 nm EuS thin film up to 450 °C in the ambient-like atmosphere, in line with previous investigations [28,29].

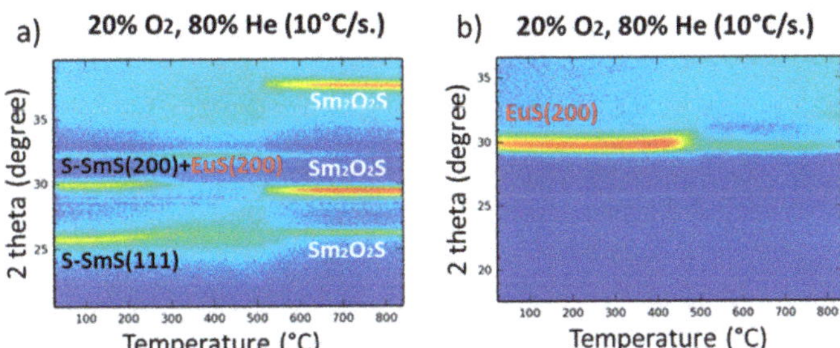

Figure 4. (**a**) In situ XRD patterns for increasing temperature of an as-deposited 35/30/35 nm SmS/EuS/SmS tri-layer in an ambient-like atmosphere of 20% O_2 and 80% He. (**b**) Same as in (**a**) for an as-deposited 25 nm EuS thin film on Si. Both thin films were deposited at 250 °C.

Then, we increased the substrate temperature to 400 °C to explore any changes in the switching properties and the degree of mixing in the tri-layer thin film stacks. In order to study the switching behavior, we rubbed the surface of a triple layer to induce the SMT, as shown earlier for the single SmS film (Figure 3). A clear SMT was still observed by following the shift of the (200) the diffraction peak (Figure 5a) from about 30.27° to 31.44° (corresponding to a change in d_{200} from 2.95 Å to 2.84 Å), although a smaller peak remained at the original position. While the former peak demonstrated a behavior typical for SmS, the latter peak is indicative of the presence of an EuS-like part in the tri-layer, which does not show switching. A similar picture is observed for the (111) peak. By annealing in vacuum at 400 °C, the semiconducting state can successfully be recovered (Figure 5a, blue curved), although the derived d_{200} lattice spacing of 2.92 Å is somewhat smaller than the original value.

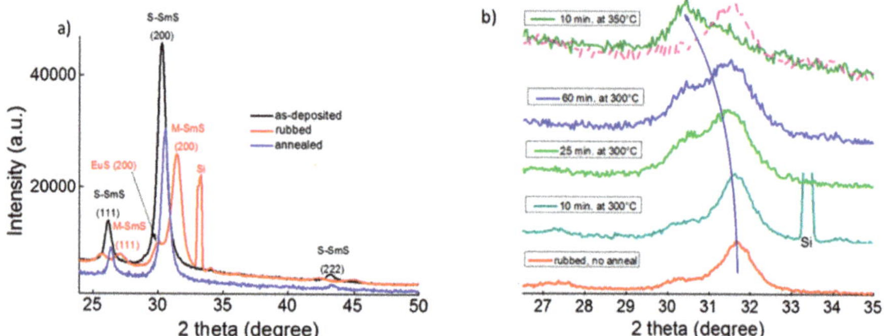

Figure 5. (**a**) XRD of an as-deposited SmS/EuS/SmS (35/30/35 nm) tri-layer at 400 °C (black curve), after rubbing (red curve) and after annealing in vacuum at 400 °C, for 10 min, and then cooling down (blue curve). (**b**) Cumulative annealing in ambient air of a similarly rubbed sample, as in (**a**), to provide the thermally triggered metal to semiconductor transition. For comparison, XRD pattern of a 100 nm SmS thin film, after 30 min annealing at 350 °C, is also depicted with the purple dashed line.

Subsequently, we investigated the thermally induced metal-semiconductor reverse transition (in ambient atmosphere) of the tri-layer system deposited at 400 °C. The red curve on the bottom of Figure 5b corresponds to the rubbed (without any annealing) tri-layer initially deposited at 400 °C. This sample was fabricated following the same process as in Figure 5a for the red curve. Then, 10 min of post-annealing at 300 °C in air was performed, which did not yield any significant structural change. Prolonging the accumulated annealing time to 25 or 60 min did not switch back the M-SmS part of the thin film. The back switching of the tri-layer to the semiconducting state (smaller 2θ value) was observed, however, after only 10 min of annealing at a slightly higher temperature of 350 °C. In contrast, a single SmS film is quite stable, without showing any back switching behavior, when it is annealed at 350 °C, for 30 min in air (purple dashed curve in Figure 5b).

3.3. Piezoresistive Behavior

Figure 6a shows the piezoresistance of the tri-layer system of SmS/EuS/SmS, deposited at 250 °C, with layer thicknesses of 35/30/35 nm, using Ir bottom and top electrodes. It is important to notice that the application of pressure with the indenter leads to a relatively small area (of the order of tens of μm^2) where the pressure is applied, as compared to the total top electrode area (1.8 mm^2). Hence, the measured resistance across the thin film is essentially determined by two parallel resistors, one with variable resistance (where the pressure is applied) and one with fixed resistance (outside the indenter area). The values given below are for the combined resistance, as it is difficult to estimate the pressed area, given that it is a function of the applied force and thus the contact area of the indenter. When the indenter just makes contact with the electrode, the resistance is 76 kΩ (HRS). Application of pressure first leads to a limited, gradual decrease in the resistance, until a first sudden drop to 5.2 kΩ appears at a force of about 0.5 N. We define this as an intermediate resistance state (IRS). Note that the actual change in resistance from the semiconducting to the metallic state is higher, taking into account the effect of the parallel resistance. The force threshold is similar to the behavior of single SmS thin films, leading to the conclusion that the SmS-like parts in the tri-layer switch first. Looking at higher force values, we see a second sharp resistance drop at around 1.55 N (to a resistance of 260 Ω), which is likely related to a partly mixed alloy of SmS and EuS. The resistance gradually drops further upon increasing force, to 160 Ω at 2 N, which can be related to the piezoresistive behavior of Eu-rich SmS, where the pressure leads to a narrowing of the *4f–5d* gap [16]. Upon gradual release of the pressure, the resistance increases slowly again, showing a major, rather discontinuous, increase between 0.8 and 0.5 N. This is likely related to the switching back of the alloyed part of the tri-layer, which had switched to an LRS around 1.5 N. Just before full release of the force (when the contact between the indenter and the top electrode is lost), the resistance is still an order of magnitude below the initial resistance of 76 kΩ. When contact is made again with the top electrode, the high resistance value is restored, indicating that the switching back of the relatively pure SmS part of the tri-layer (which had switched around 0.5 N during the loading) occurs very close to ambient pressure. As demonstrated in the inset of Figure 6a, the same piezoresistive behavior is found during 5 consecutive cycles of loading and unloading, showing that effectively a three state system, between an HRS, IRS and LRS, is obtained. It should be mentioned that the pressure needed for the second drop slightly changes upon cycling, with a variation between 1.3 and 1.6 N. A more integrated measurement approach where pressure is applied more uniformly is currently under development in order to further characterize the piezoresistive behavior.

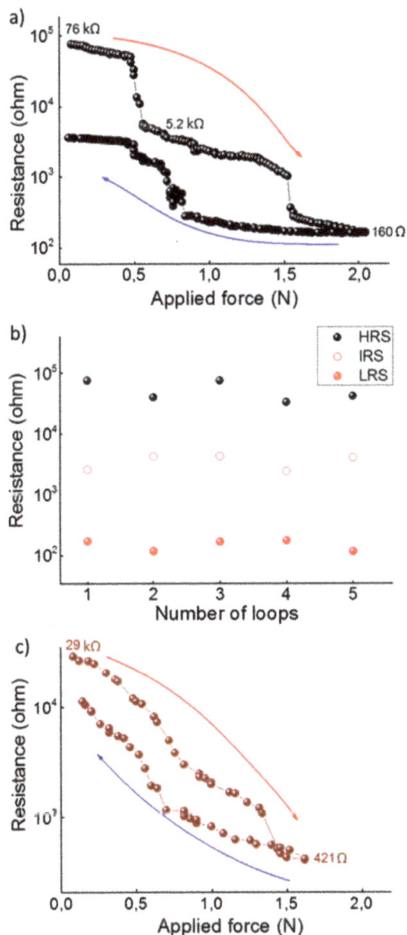

Figure 6. (**a**) Resistance across the SmS/EuS/SmS (35/30/35 nm, deposited at 250 °C) tri-layer, upon applying force up to 2 N. Red arrow for increasing force, the navy arrow when unloading. (**b**) Resistance value for five consecutive cycles of loading and unloading, at a force of 0.05, 0.9 and 1.6 N (corresponding to the high (HRS), intermediate (IRS) and low (LRS) resistance state) during the loading phase of the cycles. (**c**) Resistance of a thin film tri-layer with the same composition, deposited at 400 °C.

In Figure 6c, we demonstrate the piezoresistance of a sample with identical composition and electrode configuration, deposited at a substrate temperature of 400 °C. A significant difference with the previous case (sample deposited at 250 °C, Figure 6a) is that no sudden resistance drops are observed, at least not with large changes in the resistance. As will be demonstrated below by means of XPS depth profiling, the tri-layers deposited at 400 °C show a higher degree of mixing between the individual SmS and EuS layers. The high resistance state shows values of about 29 kΩ, smaller than the corresponding value (76 kΩ) for the tri-layer deposited at 250 °C, while at higher force values (~1.6 N) the resistance reaches a value of 421 Ω.

A pure EuS layer does not show significant changes in the resistance up to pressures similar to those where the SMT appears in single SmS films. This is due to a larger gap between the $4f$ and $5d$ bands for EuS compared to SmS. A structural change does occur, from fcc to bcc, at around 20 GPa [30], while 36 GPa is needed for the complete structural change, accompanied by a change in the valence state [31]. For a 100 nm EuS layer in our experiment, with bottom and top electrodes, the resistance across the EuS thin film was 2.2×10^7 Ω at close to ambient pressure (when the indenter just made contact with the top electrode), while it remained at a high value of 1×10^7 Ω, at 1.5 N. Measuring across the 100 nm EuS layer, the observed values were similar to those recorded for high quality EuS layers, via sheet resistance measurements, as mentioned above. The value of 2.2×10^7 Ω is slightly higher than expected from the sheet resistance for 25 nm EuS.

3.4. XPS Analysis

To correlate the piezoresistance properties of the studied tri-layer system to the extent of inter-diffusion between the three individual layers, XPS depth profiles were recorded for tri-layers with different compositions and annealing conditions. Photoelectrons from the four main elements (Sm, S, O, Eu) were recorded at the distinct spectral regions for each component: Sm $3d_{5/2}$, S $2p$, O $1s$, and Eu $3d_{5/2}$. The presented results cover the total thickness of the tri-layer system, while the Si substrate is not included.

An as-deposited composite thin film, at 250 °C, does not yield a fully alloyed system (see Figure 7a). The Sm concentration decreases from approximately 60% in the outer parts of the stack to 24% in the mixed region, where the Eu concentration reaches 40%, while the S concentration remains relatively constant throughout the sample. Care must be taken in the interpretation of these XPS results, since the concentration distribution can be influenced by knock-on sputtering, obscuring the interfaces between films. In addition, it was found by the authors of [32] that any exposed SmS surface oxidizes, even in UHV (ultrahigh vacuum) conditions. Nevertheless, the oxidation is surface limited, with the O remaining at very low concentration in the main volume of the tri-layer, even after 4 h of annealing at 400 °C (blue curve in Figure 7c). However, the main conclusion from Figure 7a remains, namely that there is incomplete mixing of the layers after deposition at a substrate temperature of 250 °C.

In this case, the tri-layer system is mainly consisting of the SmS-like parts that show a pressure-triggered resistance and color change at low applied pressure, and the partly mixed EuS/SmS intermediate thin film. Before the application of pressure, both materials (SmS, EuS) demonstrate semiconducting properties. For these materials, the $4f^6$ states of the Sm^{2+} and Eu $4f^7$ states lie between the valence band (formed by the $3p$ orbitals of S) and the CB (constituted of the $5d$ and $6s$ orbitals of the lanthanides). In the case of SmS, the gap between the $4f$ state and the bottom of the CB collapses at a pressure of about 0.65 GPa, providing metallic properties, or a low resistance state.

Based on previous investigation on bulk crystals [6], it can be concluded that a mixture between the layers leads to a shift of the pressure threshold value of resistance drop. This could be related either to the size effects or the electronic structure, with the occurrence of a Eu $4f^7$ level deep in the energy gap that is less important in the SMT of the mixed system [33]. Eu is among the rare-earth elements which do not induce the valence transition in SmS, so external pressure is needed to induce the SMT in those mixed systems. Initial studies have indicated that the size factor is the most significant reason for the SMT, especially for substituent elements with smaller ion sizes than Sm^{2+}. In that case, the bottom of the *5d* band lowers in energy, due to the local compression by the crystal lattice of the Sm ions. Nevertheless, other elements, such as Ca and Yb, though they show a smaller ion radius than Sm^{2+}, do not manage to induce the SMT under ambient conditions. This means that the ionic radius is not the only factor determining the pressure threshold, but also the electronic structure plays a significant role. In our case of Eu-substituted SmS, the Eu ions are all divalent, as the trivalent charge state is not stabilized [34]. In contrast, for the semiconducting SmS, there is a mixture of Sm^{2+} and Sm^{3+} ions in the thin films, as witnessed from the derived lattice constant and XPS analysis. Hence the introduction of Eu^{2+} in a SmS thin film forces the (alloyed) SmS to acquire a slightly larger lattice constant [35]. Indeed, in case of the 35/30/35 nm SmS/EuS/SmS tri-layer deposited at 400 °C, the (200) peak position is located at about 30.2°, slightly lower than the corresponding value (~30.4°) in a 100 nm SmS film. Consequently, the larger lattice constant leads to a larger energy spacing for the *4f-5d* gap [6]. The latter is likely the main physical reason for the second drop at higher force values (~1.5 N). Increasing the amount of Eu, we notice the change from a discontinuous to a continuous transition (see Figure 8), which is in accordance with previous results [6]. As a matter of fact, the three states system presented in Figure 6a is a consequence of the occurrence of more than one material system in the entire thin film stack.

As an increase in temperature can enhance the interdiffusion of SmS and EuS, the tri-layer thin film shows a better mixing upon deposition at 400 °C, with Eu more equally spread out towards the outer parts of the stack (Figure 7b). Additionally, the outer regions show a more stoichiometric composition (in terms of the relative concentration of Sm and S) in comparison with the deposition at 250 °C. Also, prolonged post-annealing of 4 h at 400 °C (Figure 7c) did not show any reliable progress on mixing. Besides, the oxygen relative concentration increases only close to the material's surface. Deposition at considerably higher substrate temperature than 400 °C would be needed for further investigate the mixing in the SmS/EuS/SmS system. Another option would also be the post-annealing at temperatures higher than 400 °C. Nevertheless, any attempt to provide further mixing should seriously take into consideration the corresponding HTXRD results (Figure 4a), which indicate the oxidation of the system at elevated temperature.

Figure 7d,e show representative Sm $3d_{5/2}$ and Eu$3d_{5/2}$ photoelectron spectra. The different traces correspond to measurements after 25 s (red curve), 150 s (navy curve) and 275 s (green curve) of argon ion sputtering. The XPS depth profiling process and the high surface reactivity of SmS can drastically change the ratio of the measured valence states of the lanthanide ions in the thin films [32]. Nevertheless, for the calculations of the element concentrations throughout the samples thickness, this is not an issue. For a further and detailed discussion of the valence state evaluation of Sm ions, as well as the surface oxidation of SmS thin films and the impact of depth profiling, we refer to the work done in [32].

Obviously, there is an intricate relation between the diffusion process (Figure 7) and the electrical changes (Figure 6) in the studied devices. For instance, in an inhomogeneous strongly correlated system, consisting of a semiconducting (EuS-like) and two metallic (M-SmS-like) phases, the measured property (e.g., resistance change) is an average value related to the entire system [36]. Also, the higher temperature deposition, at 400 °C, can provide an electronic state rearrangement, triggered by temperature [37]. This rearrangement can boost electrons either from the *4f* states of Sm or impurity levels within the energy gap to the CB. The measured electrical resistance, in the semiconducting state, can thus be lower upon deposition at 400 °C compared to films deposited at lower temperature, in accordance with our results (Figure 6a,c). Future work will focus on studying the influence of (measurement) temperature on the electrical behavior of these alloyed thin films, as this will yield insight in the dynamics of the metal-insulator transition [38].

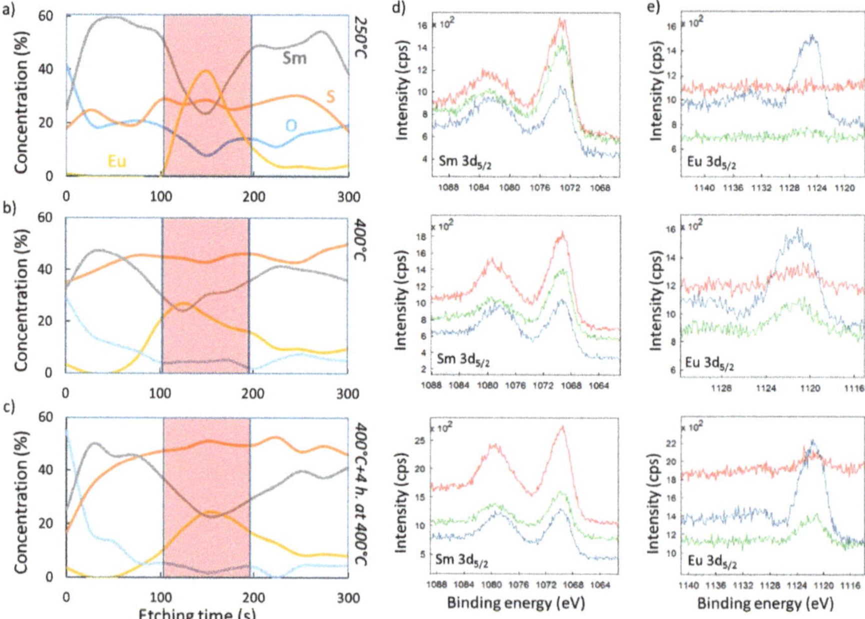

Figure 7. XPS depth profiling analysis of an as-deposited 35/30/35 nm SmS/EuS/SmS tri-layer system, deposited 250 °C (**a**), deposited at 400 °C (**b**), and after 4 h post-deposition annealing at 400 °C (**c**). (**d**) Sm 3d$_{5/2}$ photoelectron peaks for the fabricated samples in Figure 7a,c, from top to bottom, respectively. (**e**) Eu 3d$_{5/2}$ photoelectron peaks, as in (**d**). In both (**d**) and (**e**), the red curve corresponds to the photoelectron spectra after 25 s of sputtering, the navy one to 150 s, and the green curve to the sputtering time of 275 s.

In order to overcome the limited diffusion lengths, tri-layers consisting of thinner SmS and EuS layers were also deposited. Substrate temperature was chosen at 400 °C, in order to maximize the interdiffusion. Based on literature about Eu-doped SmS bulk crystals, we attempted to deposit two types of compositions, thus aiming at a different piezoresistive behavior. For a fully mixed tri-layer system that would show a discontinuous resistance change, the Eu concentration should be below the critical value of about 25% substitution [6]. For a second tri-layer, a higher Eu amount was chosen, in order to obtain a continuous resistance change. This is because only at lower Eu concentration is the resistance change due to the first-order valence change in the Sm ions.

For the first type, we deposited a tri-layer SmS/EuS/SmS with thicknesses of 18/4/18 nm, yielding an overall Eu fraction of 10%. Despite the small thickness, the thin film is well crystallized with both (111) and (200) reflections prominently visible (Figure 8a). For the corresponding piezoresistance behavior (Figure 8b), only one drop is observed. This is probably a result of the small thickness of the intermediate EuS layer, which is now more homogeneously diffused into the outer SmS layers. In the case of using 50% of EuS (6/12/6 nm SmS/EuS/SmS), only the (200) XRD peak appeared (Figure 8a), while the resistance response to the applied force is almost perfectly continuous (Figure 8c). Nevertheless, around 1.1 N, a sudden drop in the resistance can be observed (Figure 8c), which points to the mixing not being fully completed. In both cases there is a change of roughly one order of magnitude in the resistance, when increasing the applied force from about 0 N to 1.5 N. Taking into consideration that we used the same top electrode material (Ir) with the same area as in the devices in Figure 6, the moderate resistance change should be attributed to either the lower structural quality, the smaller thickness or the occurrence of EuS-rich films, which show a rather limited change in resistance.

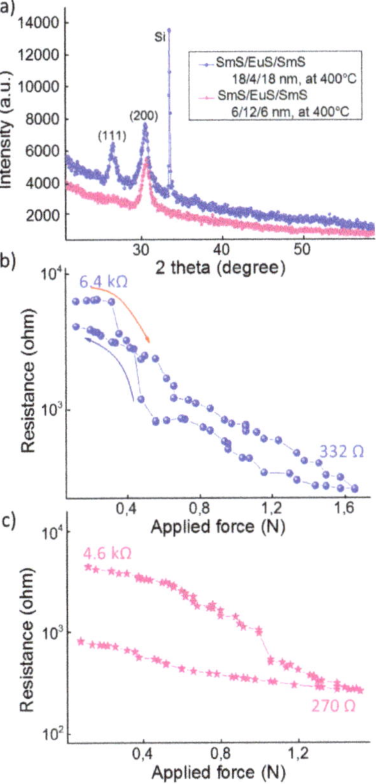

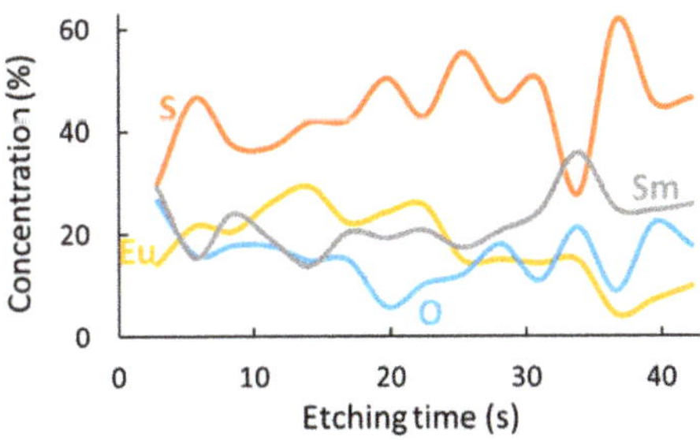

Figure 8. (**a**) XRD patterns of the thinner tri-layer systems (18/4/18 nm in blue dots and 6/12/6 nm in purple stars). Piezoresistance behavior for 18/4/18 nm tri-layer (**b**) and 6/12/6 nm tri-layer (**c**) deposited in between Ir electrodes. For clarity, in (**b**) red arrow represents the loading process, while the navy arrow shows the unloading process.

To shed light on the diffusion evolution, we used XPS depth profiling (with a sputter time of 3 s per step) for the case of 6/12/6 nm SmS/EuS/SmS (Figure 9). The two compounds homogeneously interdiffuse throughout the entire volume of the deposited tri-layer system, although some local variations could still occur. These results are in line with the previous analysis related to the resistive behavior.

Figure 9. XPS depth profiling for the 6/12/6 nm SmS/EuS/SmS tri-layer system deposited at 400 °C.

4. Conclusions

In this manuscript, we reported on high-quality SmS/EuS/SmS tri-layer thin films deposited by e-beam evaporation. The structural properties were determined, as well as measurements of their resistance response to the applied force. This work demonstrates a well-defined hysteretic pressure-triggered semiconductor to metal transition (SMT) in SmS/EuS/SmS thin films. Depending on the substrate temperature and thus the degree of interdiffusion, we were able to demonstrate either a three or a two state piezoresistive system. Three state piezoresistive systems are not common and can for instance be used in nano-sensors, where they can selectively operate at different regimes of force, providing either a continuous or discontinuous change of electrical properties. In case of a 35/30/35 nm stack deposited at a substrate temperature of 400 °C, the resistance change tends to become continuous. A post-annealing at 400 °C up to 4 h did not lead to significant additional diffusion between the layers. HTXRD results on the nanocomposite system demonstrated that an oxidation process begins at 500 °C. Thinner nanocomposite layers were also deposited to evaluate the influence of thickness to the diffusion, showing improved mixing between the layers. This resulted in a more continuous piezoresistive response. The change in resistance for the studied pressure range is more limited as in the case of pure SmS thin films, since no semiconductor to metal transition occurs. This work shows promising experimental results on the piezoresistance response of SmS/EuS/SmS tri-layer nanocomposites. On the one hand, this triggers the further exploration of this system, where the degree of interdiffusion could be further controlled to arrive at specific piezoresistive responses. On the other hand, these developments support the future application in contemporary integrated piezo-based electronic devices, such as piezo-electronic memories and RF switches. Last but not least, studying the behavior of SmS thin films, using other substituting lanthanides, like Gd or Y, could also be future avenues.

Author Contributions: Conceptualization, A.S., D.P. and P.F.S.; Investigation, A.S.; Methodology, A.S., D.P. and P.F.S.; Resources, D.P. and P.F.S.; Supervision, D.P. and P.F.S.; Writing—original draft, A.S.; Writing—review and editing, A.S., D.P. and P.F.S.

Funding: The authors are grateful to the European Union for the funding within Horizon 2020 research and innovation program under grant agreement No 688282. P.F.S. acknowledges the BOF GOA Project ENCLOSE at Ghent University for financial support.

Acknowledgments: A.S. would like to cordially thank Nico De Roo for the XPS depth profiling measurements, as well as Jonas Joos, for assistance with the fabrication of the EuS powder. Lastly, we would also like to thank the Draft and CoCoon research groups (both at Ghent University) for the set-up for measuring piezoresistance and HTXRD, respectively.

Conflicts of Interest: The authors declare no conflict of interest.

References

1. Solomon, P.M.; Bryce, B.A.; Kuroda, M.A.; Keech, R.; Shetty, S.; Shaw, T.M.; Copel, M.; Hung, L.W.; Schrott, A.G.; Armstrong, C.; et al. Pathway to the Piezoelectronic Transduction Logic Device. *Nano Lett.* **2015**, *15*, 2391–2395. [CrossRef] [PubMed]

2. Magdau, I.B.; Liu, X.H.; Kuroda, M.A.; Shaw, T.M.; Crain, J.; Solomon, P.M.; Newns, D.M.; Martyna, G.J. The piezoelectronic stress transduction switch for very large-scale integration, low voltage sensor computation, and radio frequency applications. *Appl. Phys. Lett.* **2015**, *107*, 073505. [CrossRef]

3. Josephine, B.C.; Hiroyuki, M.; Matthew, C.; Paul, M.S.; Xiao-Hu, L.; Thomas, M.S.; Alejandro, G.S.; Lynne, M.G.; Glenn, J.M.; Dennis, M.N. First realization of the piezoelectronic stress-based transduction device. *Nanotechnology* **2015**, *26*, 375201.

4. Jiang, Z.; Kuroda, M.A.; Tan, Y.; Newns, D.M.; Povolotskyi, M.; Boykin, T.B.; Kubis, T.; Klimeck, G.; Martyna, G.J. Electron transport in nano-scaled piezoelectronic devices. *Appl. Phys. Lett.* **2013**, *102*, 193501. [CrossRef]

5. Wachter, P.; Jung, A.; Steiner, P. Pressure-driven metal-insulator transition in La-doped SmS—Excitonic Condensation. *Phys. Rev. B* **1995**, *51*, 5542–5545. [CrossRef] [PubMed]

6. Jayarama, A.; Narayana, V.; Bucher, E.; Maines, R.G. Continuous and discontinuous semiconductor-metal transition in samarium monochalcogenides under pressure. *Phys. Rev. Lett.* **1970**, *25*, 1430. [CrossRef]

7. Domingo, N.; Lopez-Mir, L.; Paradinas, M.; Holy, V.; Zelezny, J.; Yi, D.; Suresha, S.J.; Liu, J.; Serrao, C.R.; Ramesh, R.; et al. Giant reversible nanoscale piezoresistance at room temperature in Sr2IrO4 thin films. *Nanoscale* **2015**, *7*, 3453–3459. [CrossRef]

8. Sousanis, A.; Smet, P.F.; Poelman, D. Samarium Monosulfide (SmS): Reviewing Properties and Applications. *Materials* **2017**, *10*, 953. [CrossRef]

9. Hickey, C.F.; Gibson, U.J. Optical response of switching SmS in thin films prepared by reactive evaporation. *J. Appl. Phys.* **1987**, *62*, 3912. [CrossRef]

10. Imura, K.; Matsubayashi, K.; Suzuki, H.S.; Deguchi, K.; Sato, N.K. Thermodynamic and transport properties of SmS under high pressure. *Phys. B* **2009**, *404*, 3028–3031. [CrossRef]

11. Sousanis, A.; Poelman, D.; Stewart, M.; Rungger, I.; Nielen, L.; Schmitz-Kempen, T.; Depla, D.; Smet, P.F. Controlling the piezoresistance hysteresis in SmS thin films. Unpublished work.

12. Sharenkova, N.V.; Kaminskii, V.V.; Golubkov, A.V.; Romanova, M.V.; Stepanov, N.N. Mechanism of stabilization of the $Sm_{1-x}Gd_xS$ metallic modification at the pressure-induced semiconductor-metal phase transition. *Phys. Solid State* **2009**, *51*, 1700–1702. [CrossRef]

13. Jayaraman, A.; Bucher, E.; Dernier, P.D.; Longinotti, L.D. Temperature induced explosive first order electronic phase transition in Gd-doped Sms. *Phys. Rev. Lett.* **1973**, *31*, 700–703. [CrossRef]

14. Jarrige, I.; Yamaoka, H.; Rueff, J.P.; Lin, J.F.; Taguchi, M.; Hiraoka, N.; Ishii, H.; Tsuei, K.D.; Imura, K.; Matsumura, T.; et al. Unified understanding of the valence transition in the rare-earth monochalcogenides under pressure. *Phys. Rev. B* **2013**, *87*, 115107. [CrossRef]

15. Antonov, V.N.; Harmon, B.N.; Yaresko, A.N. Electronic structure of mixed-valence semiconductors in the LSDA+U approximation. I. Sm monochalcogenides. *Phys. Rev. B* **2002**, *66*, 165208. [CrossRef]

16. Jayaraman, A.; Maines, R.G. Study of the valence transition in Eu-, Yb-, and Ca-substituted SmS under high pressure and some comments on other substitutions. *Phys. Rev. B* **1979**, *19*, 4154–4161. [CrossRef]

17. Robinson, J.M. Valence transitions and intermediate valence states in rare earth and actinide materials. *Phys. Rep.* **1978**, *51*, 1–62. [CrossRef]

18. Takenaka, K.; Asai, D.; Kaizu, R.; Mizuno, Y.; Yokoyama, Y.; Okamoto, Y.; Katayama, N.; Suzuki, H.S.; Imanaka, Y. Giant isotropic negative thermal expansion in Y-doped samarium monosulfides by intra-atomic charge transfer. *Sci. Rep.* **2019**, *9*, 122. [CrossRef]

19. Solomon, P.M.; Bryce, B.; Keech, R.; Shaw, T.M.; Copel, M.; Hung, L.W.; Schrott, A.; Theis, T.N.; Haensch, W.; Rossangel, S.M.; et al. The PiezoElectronic switch: A path to low energy electronics. In Proceedings of the 2013 Third Berkeley Symposium on Energy Efficient Electronic Systems (E3S), Berkeley, CA, USA, 28–29 October 2013; pp. 1–2.

20. Baek, S.H.; Park, J.; Kim, D.M.; Aksyuk, V.A.; Das, R.R.; Bu, S.D.; Felker, D.A.; Lettieri, J.; Vaithyanathan, V.; Bharadwaja, S.S.N.; et al. Giant Piezoelectricity on Si for Hyperactive MEMS. *Science* **2011**, *334*, 958–961. [CrossRef]

21. Chen, L.; Lee, H.; Guo, Z.J.; McGruer, N.E.; Gilbert, K.W.; Mall, S.; Leedy, K.D.; Adams, G.G. Contact resistance study of noble metals and alloy films using a scanning probe microscope test station. *J. Appl. Phys.* **2007**, *102*, 074910. [CrossRef]

22. Poulopoulos, P.; Goschew, A.; Kapaklis, V.; Wolff, M.; Delimitis, A.; Wilhelm, F.; Rogalev, A.; Pappas, S.D.; Straub, A.; Fumagalli, P. Induced spin-polarization of EuS at room temperature in Ni/EuS multilayers. *Appl. Phys. Lett.* **2014**, *104*, 112411. [CrossRef]

23. Poulopoulos, P.; Lewitz, B.; Straub, A.; Pappas, S.D.; Droulias, S.A.; Baskoutas, S.; Fumagalli, P. Band-gap tuning at the strong quantum confinement regime in magnetic semiconductor EuS thin films. *Appl. Phys. Lett.* **2012**, *100*, 211910. [CrossRef]

24. Rogers, E. Engineering the Electronic Structure of Lanthanide Based Materials. Ph.D. Thesis, TUDelft, Delft, The Netherlands, 2012.

25. Rogers, E.; Smet, P.F.; Dorenbos, P.; Poelman, D.; van der Kolk, E. The thermally induced metal-semiconducting phase transition of samarium monosulfide (SmS) thin films. *J. Phys. Condens Matter* **2010**, *22*, 015005. [CrossRef] [PubMed]

26. Sousanis, A.; Smet, P.F.; Detavernier, C.; Poelman, D. Stability of switchable SmS for piezoresistive applications. In Proceedings of the IEEE Nanotechnology Materials and Devices Conference (NMDC), Toulouse, France, 9–12 October 2016. [CrossRef]

27. Zenkevich, A.V.; Parfenov, O.E.; Storchak, V.G.; Teterin, P.E.; Lebedinskii, Y.Y. Highly oriented metallic SmS films on Si(100) grown by pulsed laser deposition. *Thin Solid Film.* **2011**, *519*, 6323–6325. [CrossRef]

28. Yang, Q.I.; Zhao, J.; Zhang, L.; Dolev, M.; Fried, A.D.; Marshall, A.F.; Risbud, S.H.; Kapitulnik, A. Pulsed laser deposition of high-quality thin films of the insulating ferromagnet EuS. *Appl. Phys. Lett.* **2014**, *104*, 082402. [CrossRef]

29. Ananth, K.P.; Gielisse, P.J.; Rockett, T.J. Synthesis and characterization of europium sulfide. *Mater. Res. Bull.* **1974**, *9*, 1167–1171. [CrossRef]

30. Gour, A.; Singh, S.; Singh, R.K. Theoretical analysis of pressure-induced phase transition of EuS and EuSe including the role of temperature. *J. Alloy. Compd.* **2009**, *468*, 438–442. [CrossRef]

31. Schmiester, G.; Wortmann, G.; Kaindl, G.; Bach, H.; Holtzberg, F. Pressure-induced valence changes in EuS and SmTe. *High Press. Res.* **1990**, *3*, 192–194. [CrossRef]

32. Sousanis, A.; Poelman, D.; Detavernier, C.; Smet, P.F. Switchable piezoresistive SmS thin films on large area. *Sensors* **2019**, *19*, 4390. [CrossRef]

33. Busch, G.; Guntherodt, G.; Wachter, P. Optical transitions and the energy level scheme of the europium chalcogenides. *J. Phys. Colloq.* **1971**, *32*, C1-928–C1-929. [CrossRef]

34. Kaminskii, V.V.; Stepanov, N.N.; Kazanin, M.M.; Molodykh, A.A.; Solov'ev, S.M. Electrical conductivity and band structure of thin polycrystalline EuS films. *Phys. Solid State* **2013**, *55*, 1074–1077. [CrossRef]

35. Kaminskii, V.V.; Solov'ev, S.M.; Khavrov, G.D.; Sharenkova, N.V.; Hirai, S. Structural features of $Sm_{1-x}Eu_xS$ thin polycrystalline films. *Semiconductors* **2017**, *51*, 828–830. [CrossRef]

36. Kim, H.-T.; Chae, B.-G.; Youn, D.-H.; Maeng, S.-L.; Kim, G.; Kang, K.-Y.; Lim, Y.-S. Mechanism and observation of Mott transition in VO_2-based two- and three-terminal devices. *New J. Phys.* **2004**, *6*, 52. [CrossRef]

37. Di Sante, D.; Fratini, S.; Dobrosavljević, V.; Ciuchi, S. Disorder-Driven Metal-Insulator Transitions in Deformable Lattices. *Phys. Rev. Lett.* **2017**, *118*, 036602. [CrossRef] [PubMed]

38. Möbius, A. The metal-insulator transition in disordered solids: How theoretical prejudices influence its characterization A critical review of analyses of experimental data. *Crit. Rev. Solid State Mater. Sci.* **2019**, *44*, 1–55. [CrossRef]

 *nanomaterials*

Article

Investigation of Pyrophosphates KYP_2O_7 Co-Doped with Lanthanide Ions Useful for Theranostics

Adam Watras [1,*], Marta Wujczyk [1], Michael Roecken [2], Katarzyna Kucharczyk [3,4], Krzysztof Marycz [3,4,5] and Rafal J. Wiglusz [1]

[1] Institute of Low Temperature and Structure Research PAS, Okolna 2 str. 50-422 Wroclaw, Poland; m.wujczyk@intibs.pl (M.W.); r.wiglusz@intibs.pl (R.J.W.)

[2] Faculty of Veterinary Medicine, Equine Clinic-Equine Surgery, Justus-Liebig-University, 35392 Giessen, Germany; Michael.Roecken@vetmed.uni-giessen.de

[3] International Institute of Translational Medicine, Jesionowa 11, Malin, 55-114 Wisznia Mala, Poland; kucharczyk.katarzyna@o2.pl (K.K.); krzysztof.marycz@upwr.edu.pl (K.M.)

[4] Department of Experimental Biology, Wroclaw University of Environmental and Life Sciences, 50-375 Wroclaw, Poland

[5] Collegium Medicum, Cardinal Stefan Wyszyński University (UKSW), Woycickiego 1/3, 01-938 Warsaw, Poland

* Correspondence: a.watras@intibs.pl

Received: 7 October 2019; Accepted: 7 November 2019; Published: 11 November 2019

Abstract: Diphosphate compounds (KYP_2O_7) co-doped with Yb^{3+} and Er^{3+} ions were obtained by one step urea assisted combustion synthesis. The experimental parameters of synthesis were optimized using an experimental design approach related to co-dopants concentration and heattreatment as well as annealing time. The obtained materials were studied with theinitial requirements showing appropriate morphological (X-Ray Diffraction (XRD), Scanning Electron Microscopy (SEM)) and spectroscopic properties (emission, luminescence kinetics). Moreover, the effect of Er^{3+} and Yb^{3+} ions doped KYP_2O_7 on morphology, proliferative and metabolic activity and apoptosis in MC3T3-E1 osteoblast cell line and 4B12osteoclasts cell line was investigated. Furthermore, the expression of the common pro-osteogenic markers in MC3T3-E1 osteoblast as well as osteoclastogenesis related markers in 4B12 osteoclasts was evaluated. The extensive in vitro studies showed that KYP_2O_7 doped with 1 mol% Er^{3+} and 20 mol% Yb^{3+} ions positively affected the MC3T3-E1 and 4B12 cells activity without triggering their apoptosis. Moreover, it was shown that an activation of mTOR and Pi3k signaling pathways with 1 mol% Er^{3+}, 20 mol% Yb^{3+}: KYP_2O_7 can promote the MC3T3-E1 cells expression of late osteogenic markers including RUNX and BMP-2. The obtained data shed a promising light for KYP_2O_7 doped with Er^{3+} and Yb^{3+} ions as a potential factors improving bone fracture healing as well as in bioimaging (so-called in theranostics).

Keywords: diphosphates; up-conversion; theranostics

---

1. Introduction

In recent years, much attention has been paid to rare earth phosphate phosphors due to their appealing features, such as chemical stability and diversity in crystallographic structure [1]. The phosphates could be used as a matrix for doping with optically active ions, such as the rare earth metals. Potential application of the rare earth phosphates could be related to such areas as: cell bioimaging [2–4], light-emitting diodes [5–8], solar cells [9–11] as well as regenerative medicine.

Potassium yttrium(III) diphosphate(V) KYP_2O_7 is a polymorphic compound. Depending on the annealing temperature so-called the low temperature phase (β-KYP_2O_7) or the high-temperature phase (α-KYP_2O_7) could be obtained. On the basis of ionic radius ratio ($r_K{}^+/r_Y{}^{3+} = 1.68$) value, polymorphism of KYP_2O_7 can be explained [12]. Three different synthesis routs were published for the KYP_2O_7: solid state reaction [13], one step urea-combustion synthesis [14] and boric acid flux method [15]. According to our knowledge the modern luminescent material KYP_2O_7 has never been employed as a matrix for investigation of up-conversion processes in biomedical applications.

Recently, tissue engineering together with regenerative medicine has become amore and more powerful tool in the field of bone regeneration as well as theranostics [16–19]. There are serious requirements for developing a strategy that could improve bone fracture regeneration, especially for elderly patients suffering from osteoporosis [20,21]. Bone fracture naturally involves two opposite processes, i.e., osteogenesis and osteoclastogenesis. The balance between these two processes ensures a new bone formation and finally bone regeneration. In the osteogenesis process the bone tissue formation is directly mediated by osteoblasts. This process is regulated on gene expression level by several transcripts including collagen type II, bone morphegenic protein 2 (BMP2), osteocalcin (OCL), osteopontin (OPN) and alkaline phospotase (ALP). The dynamic process thatis bone formation is mediated by several signaling pathways including mammalian target of rapamycin (mTOR) and phosphoinositide 3-kinase (Pi3k) regulating osteoblastogenesis and osteoclastogenesis. The activation of osteoclastis required for proper bone shaping and providing access to bone-stored minerals [22]. Thus, the induction of osteoblasts as well osteoclast activity and maintaining proper balance between them ensures proper bone remodeling and fracture regeneration, since over activity of osteoclast will lead to bone resorption. This phenomenon is well-known for several disorders including osteoporosis. The ability to improve osteoblasts viability with simultaneous inhibition of osteoclastogenesis seems to be a real challenge for novel materials. Moreover, the modern materials that serve additional functionality i.e., bioluminescence, which allows visualizing regenerative processes in a non-invasive way, are strongly required. The bioluminescent agents including Er^{3+} and Yb^{3+} besides their physical functions might additionally promote osteoblast activity, which can serve as their additional benefit.

In this paper, samples of KYP_2O_7 co-doped with Er^{3+} and Yb^{3+} ions, were obtained using the one step urea-combustion method. Moreover, the spectroscopic investigation of occurring up-conversion processes into KYP_2O_7 matrix doped with Er^{3+} and Yb^{3+} ions was presented. Furthermore, the effects of KYP_2O_7 doped with 1 mol% Er^{3+} and 20 mol% Yb^{3+} on MC3T3-E1 osteoblasts and 4B12 osteoclasts were investigated paying special attention to viability, apoptosis, mitochondrial activity as well as an expression of common osteogenic and osteoclastogenesis related markers on mRNA levels.

2. Materials and Methods

The x mol% Er^{3+}, y mol% Yb^{3+}:KYP_2O_7 (where $x = 0.25, 0.50, 0.75, 1, 2, 5$; $y = 1, 2, 5, 10, 15, 20$) powders were obtained by one step urea assisted combustion synthesis on the grounds of the synthesis route described elsewhere by R. Pązik et. al [14]. Reactants weight was calculated in stoichiometric manner with an exception to 10% excess for $K_2CO_3 \cdot 1.5H_2O$ as well as to 20% excess of CH_4N_2O in reference to metal cations. The raw materials used for the synthesis purpose are: Y_2O_3 (Alfa Aesar GmbH & Co KG, Karlsruhe, Germany, 99.99%), Er_2O_3 (Alfa Aesar GmbH & Co KG, 99.99%), Yb_2O_3 (Alfa Aesar GmbH & Co KG, 99.99%), $K_2CO_3 \cdot 1.5H_2O$ (Chempur, Piekary Slaskie, Poland, 99.0%), CH_4N_2O (PPH "POCh" S.A. Gliwice, Poland, 99.5%), $(NH_4)_2HPO_4$ (Carl Roth GmbH + Co. KG, Karlsruhe, Germany, 99.999%), HNO_3 (POCH S.A., Gliwice, Poland, 65%, ultrapure). Each of the final mixtures was dried for 24 h at 90 °C, later annealed at series of temperature ranging from 600 °C up to 800 °C for 4, 8 and 12 h.

The X-ray diffraction patterns were obtained by the use of X'Pert Pro PANalytical diffractometer (Cu, $K\alpha1$: 1.54060 Å) (Malvern Panalytical Ltd., Malvern, UK) in a 2θ range of 10°–50°, with a scan rate of 1.3°/min for 30 min at a room temperature. Investigation of morphology was performed using scanning microscope, specifically the FEI Nova NanoSEM 230 microscope (FEI Company, Hillsboro, OR, USA) equipped with the EDS spectrometer (EDAX PegasusXM4). Hydrodynamic size of the particles dispersed in water was determined by the use of dynamic light scattering technique supported by Zetasizer Nano-ZS (Malvern Panalytical Ltd., Malvern, UK) that is equipped with the He-Ne 633 nm laser (see Figure S1). Also, zeta potential was distinguished (see Figure S1). The emission spectra, as well as power dependence functions were recorded using the laser diode (λ_{exc} = 980 nm), with regulated power ranging from 0 to 4 W (Changchun New Industries Optoelectronics Tech. Co. Ltd., Jilin, China). For measurements the KG5 Schott filter was applied and as an optical detector the Hamamatsu PMA-12 photonic multichannel analyzer (Hamamatsu Photonics K.K., Hamamatsu City, Japan) was used. Furthermore, the obtained emission spectra are the result of averaged measurements, where the fixed parameters are the exposure time (200 ms) and the cumulative amount of measurements (15). Decay curves were collected using the tunable Ti:Sapphire laser (LOTIS TII, Minsk, Belarus) (λ_{exc} = 980 nm) pumped by the second harmonic of the YAG:Nd^{3+} pulse laser (f = 10 Hz, t < 10 ns).

Mice osteoblasts MC3T3-E1 and osteoclasts 4B12 were used in this study. The cells were cultured in Minimum Essential Medium (MEM) Alpha w/o ascorbic acid (Gibco A10490-01) supplemented with 10% of Fetal Bovine Serum (FBS) (SigmaAldrich, Lenexa, KS, USA) with addition of 1% Penicillin/Streptomycin (P/S) (Sigma Aldrich, USA). In turn 4B12 cells were cultured in EMEM Alpha (Sigma M0200) supplemented with 10% of FBS and 30% of calvaria-derived stromal cell conditioned media (CSCM) without addition of antibiotic. The MC3T3-E1 were cultivated at 80% of confluence and they were passaged every 5 days by enzymatic dissociation using Trypsin-EDTA solution (SigmaAldrich, Saint Louis, MO, USA). Cells were incubated at 37 °C in a humidified atmosphere with 5% CO_2.

Cell metabolic activity was measured by means of TOX-8 resazurin-based method using in vitro toxicology assay kit. MC3T3-E1 and 4B12 cells were plated into 96-well plates (3×10^3 cells per well) in 4 replicates. Next metabolic activity of MC3T3-E1 were measured when cells were exposed to x = 10, 15 and 20 mol% of KYP$_2$O$_7$:1 mol% Er^{3+}, x Yb^{3+}, which was incorporated into the culture medium. For examination, compound was diluted in phosphate-buffered saline (PBS). First 10 mg of compound was suspended in 1 mL of PBS. Next this solution was added directly to cell culture medium in the proper concentration. The MC3T3-E1 and 4B12 cells were cultured in the presence and absence of tested materials for 120 h. After 24 and 120 h, culture medium was replaced with 10% solution of resazurin in fresh complete medium and incubated at 37 °C for 2h in CO_2 cell culture incubator. Reduction of the dye was measured spectrophotometrically at a wavelength of 600 and 690 nm reference length (Epoch, Biotek, Bad Friedrichshall, Germany).

For analysis of genes expression, MC3T3-E1 and 4B12 cells were cultured 120 h onto KYP$_2$O$_7$:1 mol% Er^{3+}, 20 mol% Yb^{3+}. Then, cells were lysed in TRI Reagent and next total RNA was isolated using phenol-chloroform method described previously by Chomczynski and Sacchi [23]. To perform cDNA synthesis gDNA was digested with RNase-free (ThermoScientific^{TM}, Whaltam, MA, USA), DNase I and next cDNA was synthesized using Tetro cDNA Synthesis Kit (Bioline, London, UK). qRT-PCR was performed using CFX Connect^{TM} Real-Time PCR Detection System (Bio-Rad, Hercules, CA, USA) for gene expression analysis. Reaction mixture contained 1 µL of cDNA in a total volume of 10 µL using SensiFAST SYBR & Fluorescein Kit (Bioline, London, UK). The concentration of primers in each reaction was equal to 500 nM; primer sequences used in individual reactions are listed in Table 1. The algorithm used for quantitative expression of the investigated genes was performed using the $2^{-\Delta\Delta CT}$ method in relation to housekeeping gene (GAPDH).

To visualize theactin cytoskeleton and location of mitochondria the epifluorescent microscope (Olympus Fluoview FV1200, Tokyo, Japan) was used. Cells cultured onto KYP$_2$O$_7$:1 mol% Er^{3+}, 20 mol% Yb^{3+} were stained with PhalloidinAtto 488 staining for F-actin visualization. For this purpose, the cells were fixed in 4% paraformaldehyde (PFA) (Sigma Aldrich) for 45 min at RT, then washed with phosphate-buffered saline (PBS) (SigmaAldrich) three times and permeabilized using 0.3% Tween 20 (SigmaAldrich) in PBS for 15 min. For nuclei visualization PhalloidinAtto 488 in PBS (dilution 1:700) (SigmaAldrich) staining for 45 min was performed. Obtained pictures were analyzed using ImageJ software 1.51j version (NIH, Bethesda, MD, USA). For the visualization of the mitochondria network, staining with the MitoRed was performed. For this purpose the culture medium was removed and cells were washed twice with PBS. After that the culture medium with MitoRed (1:1000) was added to cells in an amount equivalent to 350 μL per well. Cells were incubated for 30 min, after that they were washed three times with PBS. Later 4% PFA was added in an amount equal to 300 μL per well for 45 min, then cells were washed three times with PBS and put on DAPI. To visualize the cells morphology the contrast phase photos was taken (Zeiss, Oberkochen, Germany).

Table 1. Sequences of primers used in qRT-PCR.

Gene	Primers (5′→3′)	Length of Amplicon	Accession No.
p53	F: AGTCACAGCACATGACGGAGG R: GGAGTCTTCCAGTGTGATGATGG	287	NM_001127233.1
BCL-2	F: GGATCCAGGATAACGGAGGC R: ATGCACCCAGAGTGATGCAG	141	NM_009741.5
BAX	F: AGGACGCATCCACCAAGAAGC R: GGTTCTGATCAGCTCGGGCA	251	NM_007527.3
p21	F: TGTTCCACACAGGAGCAAAG R: AACACGCTCCCAGACGTAGT	175	NM_001111099.2
Cas-9	F: CCGGTGGACATTGGTTCTGG R: GCCATCTCCATCAAAGCCGT	278	NM_001355176.1
GAPDH	F: TGCACCACCAACTGCTTAG R: GGATGCAGGGATGATGTTC	177	NM_001289726.1

F: forward; R: reverse; p53: tumor suppressor p53; BCl-2: B-cell lymphoma; BAX: Bcl-2 associated X protein; p21: cyclin dependent kinase inhibitor 1A; Cas-9: Caspase-9; GAPDH: Glyceraldehyde 3-phosphate dehydrogenase.

3. Results

3.1. Structural Analysis

The β-KYP$_2$O$_7$ crystallizes in monoclinic system that belongs to the *P2$_1$/c* space group and the α-phase crystallizes in orthorhombic system that can be assigned to the *Cmcm* space group. In the matrix Y^{3+} ions are substituted by the selected optically active ions RE^{3+}, herein meaning Er^{3+} and Yb^{3+} Figure 1.

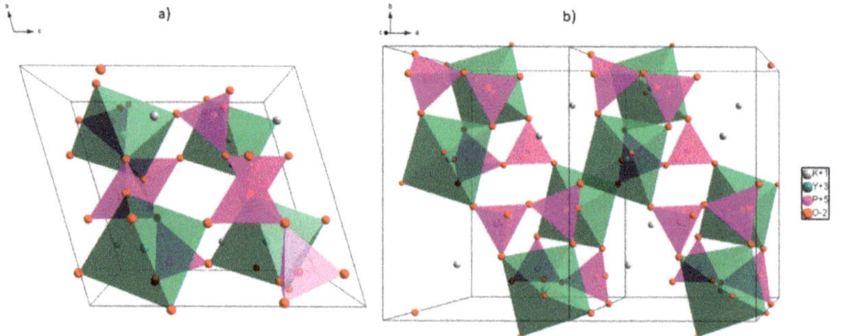

Figure 1. Projection of the β-KYP$_2$O$_7$ unit cell (**a**) and super cell (**b**) indicating the Y^{3+} and P^{5+} coordination polyhedra.

The X-ray diffraction patterns were collected for all of the samples (see Figure 2 and Figures S2 and S3). Independently of the dopant concentration and the annealing time, each collected X-ray diffraction pattern shows a match to the theoretical pattern no. 160190 from ICSD. Up to the annealing temperature of 700 °C the sample crystalizes in the low temperature phase β-KYP$_2$O$_7$ (see Figure 2b).

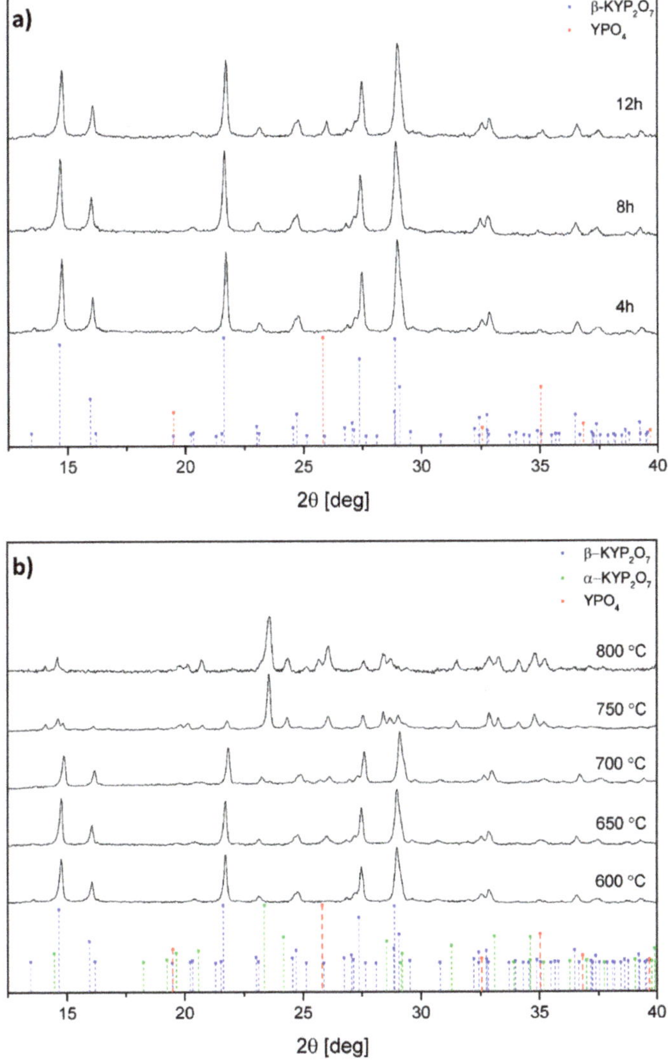

Figure 2. Representative XRD patterns of 1 mol% Er^{3+}, 1 mol% Yb^{3+}:KYP_2O_7 annealed at 600 °C for 4, 8 and 12 h (**a**) as well as annealed at 600–800 °C for 4 h (**b**).

Above the annealing temperature of 750 °C, the high-temperature phase α-KYP_2O_7 can be observed, matched with the pattern no. 75171 from ICSD. In the case of the annealing temperature from 750 to 800 °C, a decrease in the amount of β-KYP_2O_7 phase can be observed in favor of the high-temperature α-KYP_2O_7 phase. XRD patterns show presence of the YPO_4 phase. Although the peaks from the YPO_4 phase (ICSD no. 184543) overlap with β-KYP_2O_7, one could be noticed that this additional phase manifests itself in an increased intensity of certain peaks, when compared to the β-KYP_2O_7 theoretical pattern.

Obtained, representative SEM images of the 1 mol% Er^{3+}, 1 mol% Yb^{3+}: KYP_2O_7 material, annealed at 600 °C for 12 h have been shown in Figure 3 in two different magnifications.

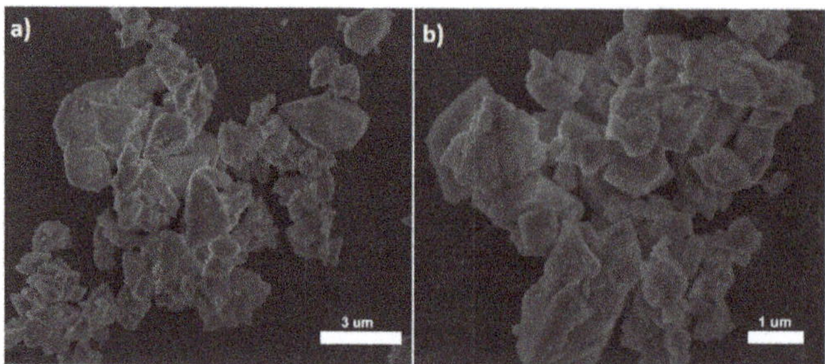

Figure 3. Representative SEM images of the 1 mol% Er^{3+}, 1 mol% Yb^{3+}:KYP_2O_7, annealed at 600 °C for 12 h with different magnifications (**a**) with 3 um scale bar and (**b**) with 1 um scale bar.

3.2. Luminescence Properties

The emission spectra were measured at room temperature (300 K) with excitation wavelength λ_{exc} = 980 nm of the continuous wave (CW) laser power of 1.56 W (see Figure 4 and Figures S4–S6). Measurements were carried out for the samples annealed at two temperatures, 600 and 650 °C for 12 h with varying content of the co-dopants. Each of the spectrum consists of five bands, three of them can be assigned as Er^{3+} transitions $^2H_{11/2}{\rightarrow}^4I_{15/2}$, $^4S_{3/2}{\rightarrow}^4I_{15/2}$, $^4F_{9/2}{\rightarrow}^4I_{15/2}$ observed respectively at 522, 540 and 650 nm. Samples annealed at 650 °C globally show more intense emission in comparison to those annealed at 600 °C. Within samples with varying content of Er^{3+} ions and fixed at 15 mol% concentration of Yb^{3+} ions, the highest emission intensity shows the one doped as follows: 1 mol% Er^{3+}, 15 mol% Yb^{3+}. Among all samples with varying content of Yb^{3+} and concentration of Er^{3+} fixed at 1 mol%, the most intense emission can be ascribed to the sample with 20 mol% of Yb^{3+}, regardless of the annealing temperature.

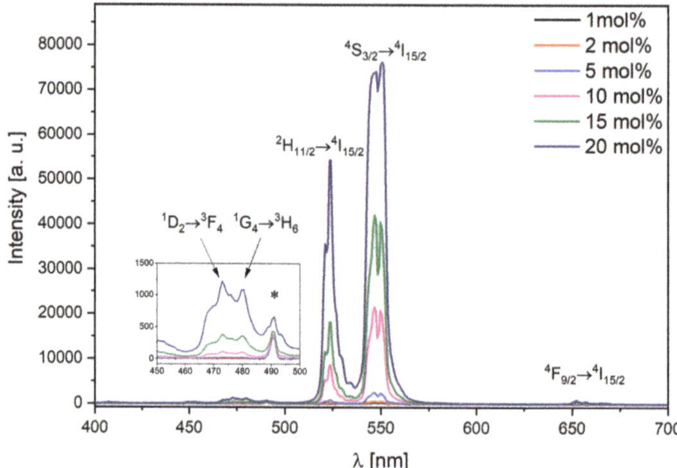

Figure 4. Representative emission spectra of KYP_2O_7 doped with x mol% Yb^{3+} ions and co-doped with 1 mol% Er^{3+} under the excitation wavelength λ = 980 nm, P = 1.56 W, heat-treated at 650 °C for 12 h.

In emission spectra, additional bands at 470 and 480 nm can be observed. These bands can be assigned to transitions occurring in Tm^{3+} ions, respectively $^1D_2{\rightarrow}^3F_4$ and $^1G_4{\rightarrow}^3H_6$.

In addition, in the emission spectra a band at the wavelength of 490 nm is noticed and marked with asterisks in Figure 4 and Figures S4–S6. The emission corresponds to the second harmonic generation (SHG) from the excitation source, which is a diode laser λ_{exc} = 980 nm.

Decay profiles were measured for the $^4S_{3/2} \rightarrow {}^4I_{15/2}$ transition prominent at 547 nm wavelength. Measurements were employed at room temperature (300 K) for the samples with varying concentration of the dopants, annealed 650 °C for 12 h. Each of decay curves was fitted with the double exponential function in Figure 5.

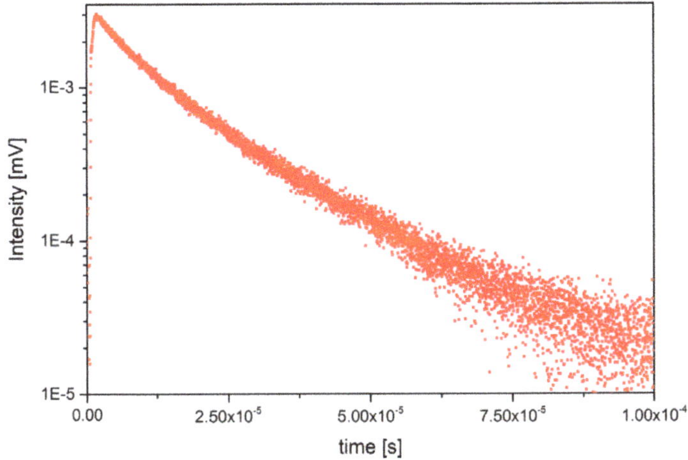

Figure 5. Decay time measured for 15 mol% Yb^{3+}, 1 mol% Er^{3+}:β-KYP_2O_7 annealed at 650 °C for 12 h.

Values of the two decay times, the fast component τ_1 and the slow component τ_2, are listed in Table 2. Due to the low emission intensity determination of the decay times for the samples with the lowest concentration of dopants were impossible. In case of the fixed value of erbium concentration, the increase in concentration of ytterbium is followed by the elongation of decay times. For higher concentrations of the dopants, reduction in the decay times can be observed. In addition, there is a correlation between the decay times and the intensity of the emission spectra. Those samples with the high intensity of emission can be assigned to the long decay times as well.

Table 2. Luminescence decay times for β-KYP_2O_7 samples annealed at 650 °C for 12 h.

Dopants Concentration		Decay Times	
Er^{3+} (mol%)	Yb^{3+} (mol%)	τ_1 (μs)	τ_2 (μs)
	5	1.13	8.59
	10	5.14	17.56
1	15	7.52	19.94
	20	6.96	18.85
0.50		0.91	6.73
0.75		5.85	19.90
1	15	7.52	19.94
2		5.92	12.06
5		3.26	6.25

3.3. Metabolic Activity, Morphology and Apoptosis of MC3T3-E1 Osteoblasts and 4B12 Osteoclast Cultured onto KYP_2O_7:1 mol% Er^{3+}, x mol% Yb^{3+}

The viability and proliferative rate analysis of osteoblasts cultured onto KYP_2O_7:1 mol% Er^{3+}, 20 mol% Yb^{3+} materials showed their beneficial effect on MC3T3-E1 number of cells (Figure 6). The highest proliferative activity of MC3T3-E1 cells was observed when they were exposed to KYP_2O_7: 1 mol% Er^{3+} doped with 20 mol% of Yb^{3+} in dose 500 μg/mL. Incorporation of 20 mol% Yb^{3+} in KYP_2O_7: 1 mol% Er^{3+} resulted in constant metabolic improvement through 120 h culture test. Similar effect was observed in 4B12 osteoclast cells, which reached the highest metabolic activity after 120 h, when cultured onto KYP_2O_7:1 mol% Er^{3+} doped with 20 mol% Yb^{3+} in dose 500 µg/mL. On the basis of mentioned results KYP_2O_7:1 mol% Er^{3+} doped with 20 mol% of Yb^{3+} in dose 500 µg/mL was used in further experiments.

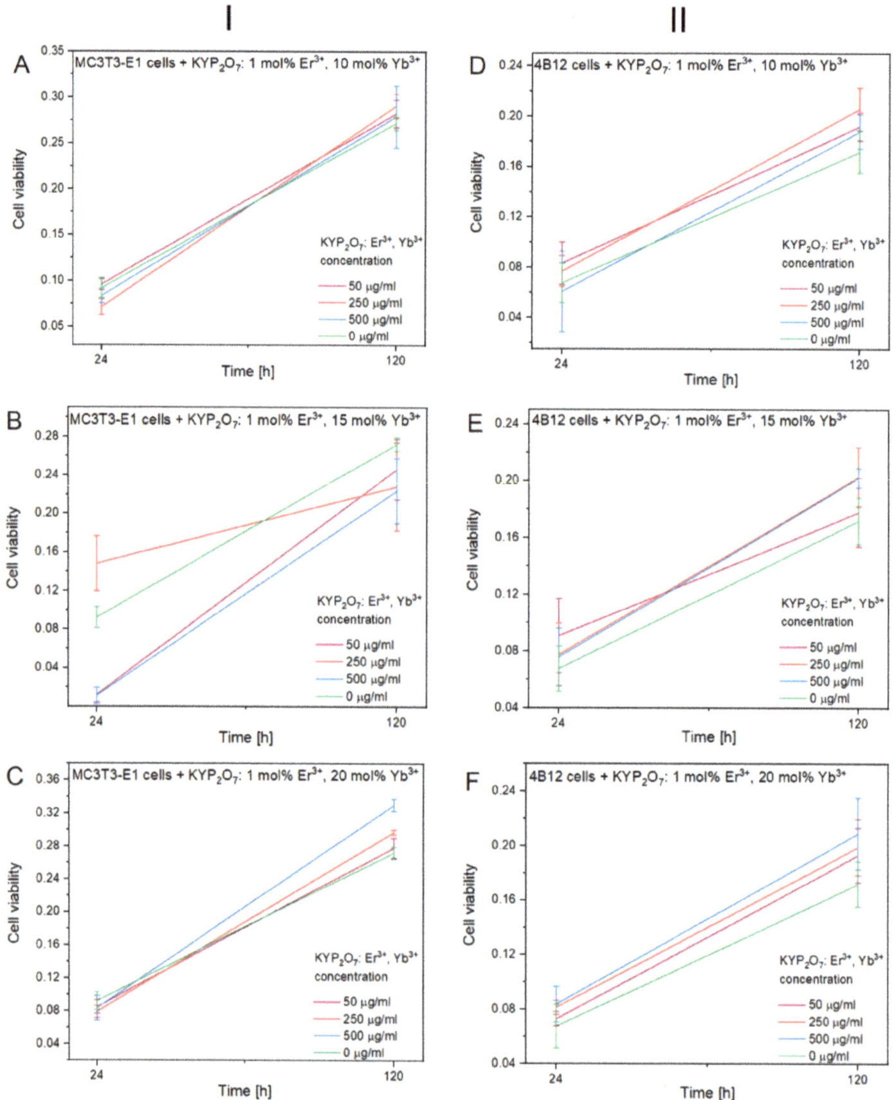

Figure 6. The viability and proliferative activity of MC3T3-E1 osteoblasts (I) and 4B12 osteoclast (II) cultured onto (**A,D**) 1 mol% Er^{3+}, 10 mol% Yb^{3+}:KYP_2O_7; (**B,E**) 1 mol% Er^{3+}, 15 mol% Yb^{3+}:KYP_2O_7 and (**C,F**) 1 mol% Er^{3+}, 20 mol% Yb^{3+}:KYP_2O_7 addition after 24 and 120 h.

For analysis of cells morphology the contrast phase pictures was taken (Figure 7). It was found that Yb^{3+} in the 500 ug/mL dosage positively affects morphology of both osteoblasts as well as osteoclasts. For analysis of the mitochondrial and actin network of MC3T3-E1 osteoblasts and 4B12 osteoclast, KYP_2O_7:1 mol% Er^{3+} co-doped with 20 mol% Yb^{3+} was proceeded. The creation of abundant actin network was observed in MC3T3-E1 osteoblasts when cultured onto KYP$_2$O$_7$:1 mol% Er^{3+} co-doped with 20 mol% Yb^{3+} when compared to the control group. The cells presented typical for osteoblast round-like shape morphology with a well visible nuclei. Moreover, cells communicated with each other and created a well-developed cell-to-cell network, as shown by the well-developed actin network (Figure 8). Actin network also testifies of increased adhesion of osteoblast when cultured onto KYP$_2$O$_7$:1 mol% Er^{3+}, 20 mol% Yb^{3+}. In the case of the osteoclasts, we also observed more developed cytoskeleton and actin staining showed well develop actin network when the cells were cultured with KYP$_2$O$_7$:1 mol% Er^{3+}, 20 mol% Yb^{3+}. Mitochondrial staining revealed that in both cells type i.e., osteoblasts and osteoclasts, dense network around nuclei was created, when the cells were cultured onto KYP$_2$O$_7$:1 mol% Er^{3+}, 20 mol% Yb^{3+}. It might suggest thata 500 µg/mL dose significantly promotes mitochondrial biogenesis, which resulted in creation of well-developed mitochondrial network (Figure 8). Moreover, it seems that this dose induces slight apoptosis in MC3T3-E1 osteoblasts while no prominent effect was observed in osteoclast cells. Incorporation of KYP$_2$O$_7$:1 mol% Er^{3+}, 20 mol% Yb^{3+} into osteoblasts culture resulted in significant up regulation of p21 and Cas-9 mRNA level since BAX transcript was significantly down regulated (Figure 9). It was found that p21 transcript was considerably down regulated in osteoclast cells when cultured onto KYP$_2$O$_7$:1 mol% Er^{3+}, 20 mol% Yb^{3+} in comparison for control culture.

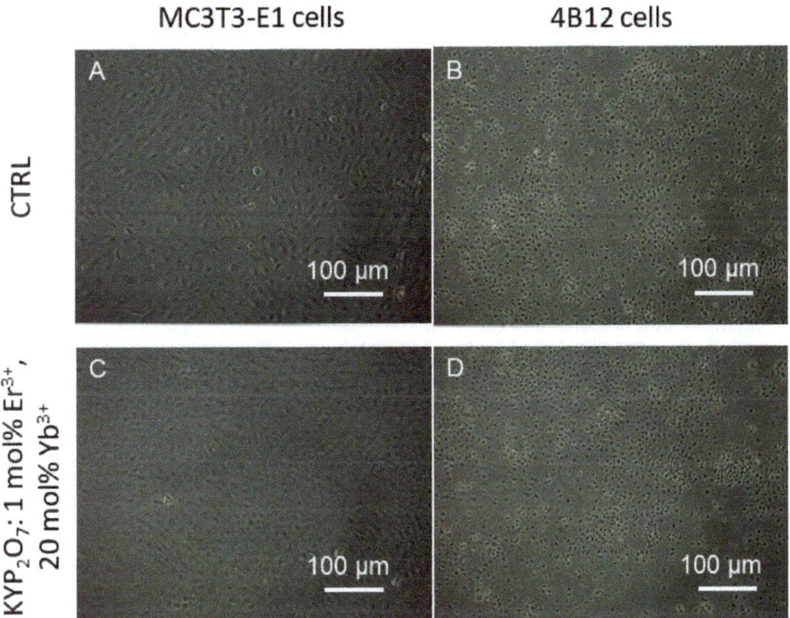

Figure 7. The MC3T3-E1 osteoblasts (**A**) and 4B12 osteoclasts (**B**) morphology visualized in control cells and in cells cultured with KYP$_2$O$_7$ doped with 1 mol% Er^{3+} ions co-doped with 20 mol% Yb^{3+} ions in dose 500 µg/mL in MC3T3 cells (**C**) and 4B12 cells (**D**) by contrast phase microscope. Magnification ×100, scale bars: 100 µm.

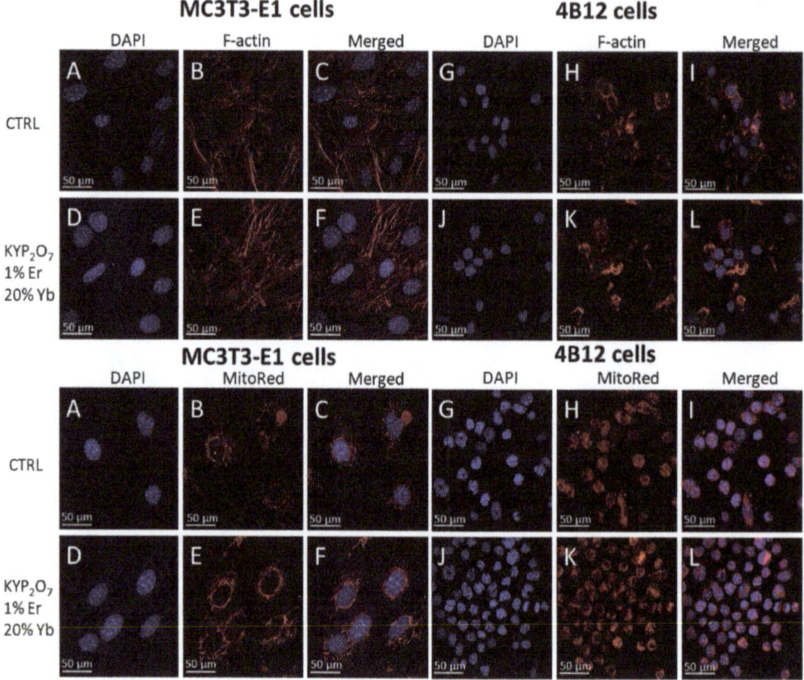

Figure 8. The F-actin, DAPI and MitoRed staining. (**A,B,C**) in upper graphs presented F-actin and DAPI staining of MC3T3-E1 cells; (**D,E,F**) showed F-actin and DAPI staining of MC3T3-E1 cells with investigated material KYP_2O_7 doped with 1 mol% Er^{3+} ions co-doped with 20 mol% Yb^{3+} ions in dose 500 μg/mL; (**G,H,I**) in upper graphs presented F-actin and DAPI staining off 4B12 cells; (**J,K,L**) showed F-actin and DAPI staining of 4B12 cells with investigated material KYP_2O_7 doped with 1 mol% Er^{3+} ions co-doped with 20 mol% Yb^{3+} ions in dose 500 μg/mL; (**A,B,C**) in lower graphs presented MitoRed and DAPI staining off MC3T3-E1cells; (**D,E,F**) showed MitoRed and DAPI staining of MC3T3-E1 cells with investigated material KYP_2O_7 doped with 1 mol% Er^{3+} ions co-doped with 20 mol% Yb^{3+} ions in dose 500 μg/mL; (**G,H,I**) in lower graphs MitoRed and DAPI staining off 4B12 cells; (**J,K,L**) showed MitoRed and DAPI staining of 4B12 cells with investigated material KYP_2O_7 doped with 1 mol% Er^{3+} ions co-doped with 20 mol% Yb^{3+} ions in dose 500 μg/mL. Scale bars presented in the images obtained using epifluroescent microscope were equal 50 μm.

MC3T3-E1 cells

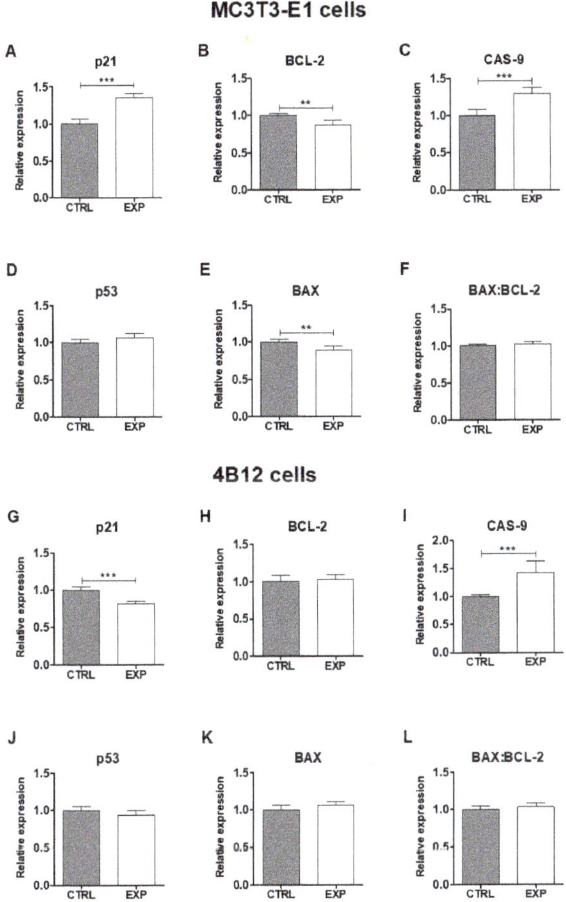

4B12 cells

Figure 9. Evaluation of apoptosis in MC3T3-E1 osteoblasts and 4B12 osteoclast. To evaluate apoptosis in cells, the expression of (**A,G**) p21, (**B,H**) Bcl-2, (**C,I**) CAS-9, (**D,J**) p53 (**E,K**) BAX was analyzed. The (**I,L**) BAX:BCL-2 ratio was calculated using relative expression values of both BCL-2 and BAX.

3.4. Expression of Osteogenic and Osteclastogenic Markers in MC3T3-E1 Osteoblasts and 4B12 Osteoclast Cultured onto KYP_2O_7: 1 mol% Er^{3+}, 20 mol% Yb^{3+} in Relation to mTOR and Pi3K Pathway

Evaluation of the expression of pro-osteogenic markers on mRNA level in MC3T3-E1 osteoblasts showed beneficial effect of KYP_2O_7:1 mol% Er^{3+}, 20 mol% Yb^{3+} material on osteogenesis process (Figure 10). It was found that KYP_2O_7:1 mol% Er^{3+}, 20 mol% Yb^{3+} promotes in MC3T3-E1 cells expression of RUNX-2 as well as BMP-2 mRNA level, since reduces expression of Coll-1 and ALP transcripts. In turn, it was observed that KYP_2O_7:1 mol% Er^{3+}, 20 mol% Yb^{3+} promotes in 4B12 osteoclast expression of PU. 1, which is involved in regulation of beta(3) integrin expression during osteoclast differentiation. Moreover, elevated expression of INTA5 in 4B12 osteoclasts was observed.

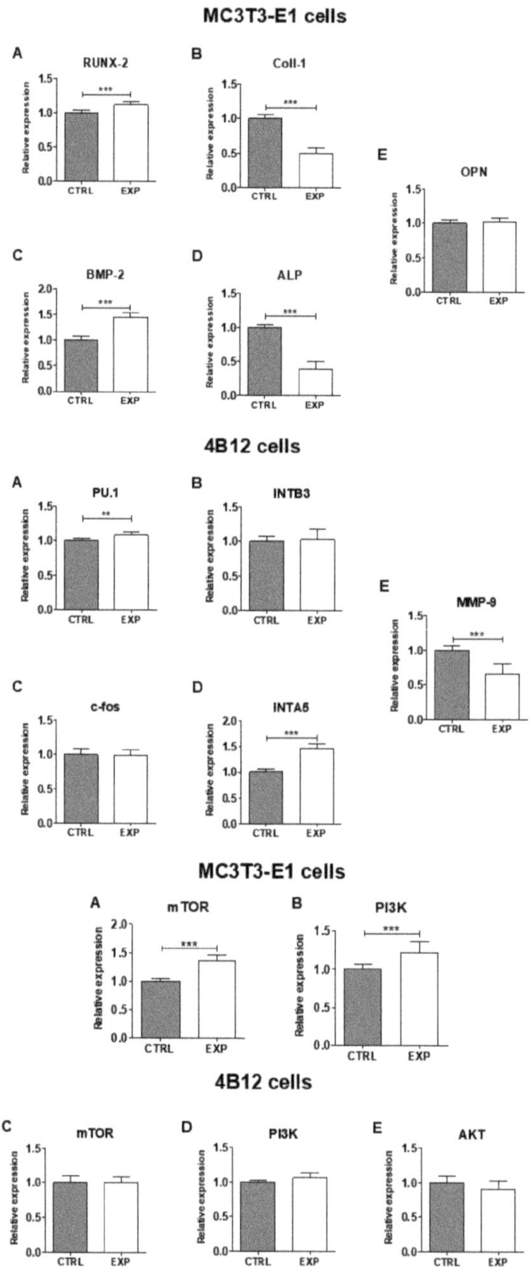

Figure 10. Comparison of the expression levels of osteogenesis-related genes using quantitative real-time PCR analysis. The expression of (**A**) RUNX-2, (**B**) Coll-1, (**C**) BMP2, (**D**) ALP, and (**E**) OPN in MC3T3-E1 osteoblasts and the expression of (**A**) PU.1, (**B**) INTB3, (**C**) c-fos, (**D**)INTA5 and (**E**) MMP-9 in 4B12 osteoclast cultured onto KYP_2O_7 doped with 1 mol% Er^{3+}, 20 mol% Yb^{3+} ions. In lower graphs the expression of (**A,C**) mTOR and (**B,D**) PI3K and (**E**) AKT in MC3T3-E1 cells and 4B12 cells was presented.

The elevated expression of both mTOR as well as Pi3K in MC3T3-E1 osteoblasts was observed when cultured onto KYP_2O_7:1 mol% Er^{3+}, 20 mol% Yb^{3+} material in comparison to the control group (Figure 10). Moreover, in 4B12 osteoclast a significant reduction of MMP-9 expression was found together with up regulation of INTA5 transcript. There were no significant differences between mTOR and Pi3K expression in 4B12 osteoclast exposed to KYP_2O_7:1 mol% Er^{3+}, 20 mol% Yb^{3+} and control.

4. Discussion

Obtained X-ray patterns for samples annealed at variety of temperature show decreasing presence of β-KYP_2O_7 phase in favor of α-KYP_2O_7 phase beginning at 750 °C. Optimal heat treatment parameters were set to be: 600 °C, 12 h and 650 °C, 12 h. Given parameters allow gratifying emission properties, shown in latter section, for up-conversion process characterization. Further spectroscopic and biological analysis were employed for the samples annealed with aforementioned parameters. Our observations are in accordance with literature data. It has already been proven, by us and others that α-KYP_2O_7 phase dominates over β-KYP_2O_7 one above 700 °C [14,15]. In reference to annealing time and doping level of β-KYP_2O_7 no reports were found. However, for up-conversion processes in different matrices doping level of Yb^{3+} and Er^{3+} ions similar to concentrations stated as optimal in this paper [24,25]. Size and morphology of the KYP_2O_7:Er^{3+}, Yb^{3+} powders were estimated using SEM microscopy, in Figure 3 shown are agglomerates ($\approx$ 3 μm) of elongated submicron particles with the shape of flat plates.

On the basis of the emission spectra, the highest intensity emission band can be assigned to the sample with concentration ratio of 1 mol% Er^{3+} and 20 mol% Yb^{3+}, when annealed at 650 °C for 12 h. For concentrations higher than 1 mol% Er^{3+} a decrease in emission band intensity can be observed, due to concentration quenching of activators' emission [26]. Hence, the optimal doping concentration was chosen to be 1 mol% Er^{3+}. The samples heavily doped with Yb^{3+} ions, where the above-mentioned phenomenon is not observed, show a monotonic increasement of intensity within analyzed concentration range (1–20 mol% Yb^{3+}). Therefore, the optimal doping concentration was chosen to be 20 mol% Yb^{3+}. Decay times of analyzed samples show direct correlation with emission spectra. Emission's intensity increasement is followed with decay time elongation. The samples exhibiting concentration quenching deviate from the mentioned trend and consequently reduction in decay time is being observed. Lengthening of the decay time might refer to an occurrence of energy transfer between upconverting ions [27].

Measurements of power dependence (PD) (see Figure S7), shown as a double-logarithmic function of emission intensity versus laser pump power, allow for estimating several absorbed photons vital for up-conversion process occurrence [28]. Results assert a two-photon nature of the $^2H_{11/2} \rightarrow {}^4I_{15/2}$ and the $^4S_{3/2} \rightarrow {}^4I_{15/2}$ transitions at λ = 522–540 nm with n values varying from 1.8 to 2.0. Therefore, the anti-Stokes emission may occur via two routes: Energy Transfer Up-conversion (ETU) or Excited State Absorption (ESA). ETU is the most efficient one out of all UC processes, as a resemblance to the full resonance is the closest [29]. These UC processes are not easy to distinguish by power dependence, owing to the fact that n value equals 2 for all cases. Short decay times for samples with minor content of co-dopants may indicate dominance of ESA, while highly doped samples might be favoring ETU, due to their longer decay times. Occurrence of the UC processes can be distinguished also with presence of arise and further prolongation of rise time in decay time function.

Weak emission intensity of the $^4F_{9/2} \rightarrow {}^4I_{15/2}$ transition at λ = 650 nm shows that metastable $^4F_{9/2}$ state is not being favorably populated. PD measurements, for aforementioned transition, show the n value equal to 1.0–1.3, letting us believe that the $^4F_{9/2} \rightarrow {}^4I_{15/2}$ transition is influenced by non-linear, nonradiative process, such as cross relaxation.

Presence of transitions from Tm^{3+} seen in emission spectra, may stem from contamination of reactants, herein especially erbium oxide. It is a well-known fact that Tm^{3+} ion can play a role of an activator in UC processes, similarly to Er^{3+} ions, if matrix is co-doped with Yb^{3+} ions. Hence, thulium ions compete as an activator with erbium ions.

The materials dedicated for bone fracture regeneration require specific characteristics including stimulation of bone formation processes as well as inducing matrix formation. The phosphates are well-known for their pro-osteogenic ability; however, KYP_2O_7 doped with rare earths elements including Er^{3+} and Yb^{3+} ions were not previously investigated. In this study, we showed that KYP_2O_7 doped with 1 mol% of Er^{3+} and 20 mol% Yb^{3+} in dose 500 μg/mL promotes osteoblasts metabolic activity and induces their highest proliferative potential. In previous research using nanometric hydroxyapatites doped with Er^{3+} we observed a similar effect; however on stem progenitor cells and olfactory ensheathing cells [30]. Moreover, we observed that KYP_2O_7:1 mol% Er^{3+}, 20 mol% Yb^{3+} promotes also proliferative and metabolic activity of osteoclast. Furthermore, the cytoskeleton development including actin formation was noted in MC3T3-E1 osteoblasts as well as 4B12 osteoclasts when they were exposed for KYP_2O_7:1 mol% Er^{3+}, 20 mol% Yb^{3+}. The observed arrangement of actin fibers indicates about fully stretched of cells and this allows us to evaluate the material as biocompatible [31]. Additionally, we observed improved cell-to-cell contact and creation of a well-developed network suggesting beneficial effect of KYP_2O_7:1 mol% Er^{3+}, 20 mol% Yb^{3+} on matrix formation. Interestingly, similar to osteoblasts, osteoclasts presented a well-developed cytoskeleton and actin network. The beneficial effect of KYP_2O_7:1 mol% Er^{3+}, 20 mol% Yb^{3+} on osteoblasts activity might be associated with improved mitochondrial biogenesis and creation of dense mitochondrial network. The mitochondria morphology and especially their fission and fusion is one of the elements of assessment cells viability, senescence and metabolism [32]. We indicated that examined material improved mitochondria network and not causes their fission what evidence about positive influence of KYP_2O_7:1 mol% Er^{3+}, 20 mol% Yb^3 on cells viability. Together with improved mitochondrial function, we observed that KYP_2O_7:1 mol% Er^{3+}, 20 mol% Yb^{3+} negatively affects expression of p21 and Cas-9 on mRNA level. Obtained data indicate on rather neutral role of KYP_2O_7:1 mol% Er^{3+}, 20 mol% Yb^{3+} on osteoblasts apoptosis although significant down regulation of BAX transcripts was observed. Moreover, the appearance of nuclei, after staining with DAPI showed that KYP2O7:1 mol% Er^{3+}, 20 mol% Yb^{3+} not implicates the chromatin condensation and DNA fragmentation, which indicates the lack of induce apoptosis by the examined material and well biocompatible of it [33]. What is more important, the beneficial effect of the material for pro-osteogenic genes expression including RUNX-2 as well as BMP-2 mRNA in MC3T3-E1 cells was observed. Interestingly, at the same time reduced expression of Coll-1 and ALP transcripts was noted. Obtained data clearly indicates on promotion of early markers of osteogenesis expression instead late markers expression. It suggests that KYP_2O_7:1 mol% Er^{3+}, 20 mol% Yb^{3+} might exert a beneficial effect on bone mineralization process and matrix formation. Observed pro-osteogenic effect of KYP_2O_7:1 mol% Er^{3+}, 20 mol% Yb^{3+} on MC3T3-E1 osteoblasts might be partially explained by the elevated expression of both mTOR as well as Pi3K signaling pathways. It was previously showed that both mTOR as well as Pi3K are positively associated with bone formation and bone remodeling effect [34]. What is more, we observed that KYP_2O_7:1 mol% Er^{3+}, 20 mol% Yb^{3+} enhanced the expression of BMP-2 and mTOR in MC3T3 osteoblasts. That fact indicates on pro-osteogenic properties of fabricated material as interplay between these two protein was shown to modulate and enhance osteogenesis [35]. Moreover, is worth adding that KYP_2O_7:1 mol% Er^{3+}, 20 mol% Yb^{3+} decreased expression of MMP-9. The increased level of this metalloproteinase is typical for osteoporotic bones. So the fact that modulation of the amount of transcripts MMP-9 by KYP_2O_7: 1 mol% Er^{3+}, 20 mol% Yb^{3+} affects the restoration of the balance between osteoblasts and osteoclasts in osteoporotic bones.

5. Conclusions

Optimization of the Potassium yttrium(III) diphosphate(V) synthesis parameters and a degree of doping was reached. Research shows a stable β-KYP_2O_7 crystallographic structure and gratifying spectroscopic properties, obtained by finding optimal synthesis conditions (such as annealing temperature, annealing time and degree of doping). The heating parameters of 600 and 650 °C as well as the heating time of 12 h were considered the best parameters of the synthesis process.

Globally the most intense emission was obtained for samples co-doped with 1 mol% Er^{3+} and 20 mol% Yb^{3+} ions. In addition, the studies were carried out to consider KYP_2O_7 co-doped with erbium and ytterbium ions, as a future material used in biomedical applications, especially theranostics. Additionally, phosphate KYP_2O_7 doped with 1 mol% of Er^{3+} and 20 mol% Yb^{3+} positively affects MC3T3-E1 osteoblasts morphology, proliferative as well as metabolic activity. Although no positive effect in relation to apoptosis was found, KYP_2O_7:1 mol% Er^{3+}, 20 mol% Yb^{3+} significantly promotes expression of early markers of osteogenesis via mTOR as well as Pi3K which sheds a promising light on that system as an agent promoting fracture bone regeneration. Moreover, observed inhibitory effect on osteoclastogenesis suggests the potential beneficial role of KYP_2O_7:1 mol% Er^{3+}, 20 mol% Yb^{3+} in treatment of osteoclast related disorders.

Supplementary Materials: The following are available online at http://www.mdpi.com/2079-4991/9/11/1597/s1, Figure S1: Results of the dynamic light scattering (DLS) expressed via z-average size parameter and zeta potential measurements for the representative sample KYP_2O_7:1 mol% Er^{3+}, 1 mol% Yb^{3+} heat-treated at 650 °C for 12 h; Figure S2: XRD patterns of β-KYP_2O_7 annealed at 600 °C for 12 h with varying content of Yb^{3+} ions and fixed 1 mol% Er^{3+} (a) as well as with varying content of Er^{3+} and fixed 15 mol% Yb^{3+} (b); Figure S3: XRD patterns of β-KYP_2O_7 annealed at 650 °C for 12 h with varying content of Yb^{3+} ions and fixed 1 mol% Er^{3+} (a) as well as with varying content of Er^{3+} and fixed 15 mol% Yb^{3+} (b); Figure S4: Emission spectra of KYP_2O_7 doped with x mol% Yb^{3+} ions and co-doped with 1 mol% Er^{3+} under the excitation wavelength λ = 980 nm, annealed at 600 °C for 12 h; Figure S5: Emission spectra of KYP_2O_7 doped with x mol% Er^{3+} ions and co-doped with 15 mol% Yb^{3+} under the excitation wavelength λ = 980 nm, annealed at 600 °C for 12 h.; Figure S6: Emission spectra of KYP_2O_7 doped with x mol% Er^{3+} ions and co-doped with 15 mol% Yb^{3+} under the excitation wavelength λ = 980 nm, annealed at 650 °C for 12 h; Figure S7: Power dependence measurements of the $^4F_{9/2} \rightarrow ^4I_{15/2}$ (a) and of the $^2H_{11/2}$, $^4S_{3/2} \rightarrow ^4I_{15/2}$ (b) for samples KYP_2O_7 annealed at 650 °C.

Author Contributions: A.W. and R.J.W. conceived and designed the experiments as well as analyzed all data; M.W. performed the experiments as well as analyzed all data; M.R. and K.K. contributed reagents/materials/analysis tools; K.M. designed the experiments as well as analyzed all data; all authors contributed to the writing of the paper.

Funding: This research was funded by the National Science Centre (NCN), grant number 'Preparation and characterization of biocomposites based on nanoapatites for theranostics' (No. UMO-2015/19/B/ST5/01330).

Acknowledgments: The authors would like to thank E. Bukowska for performing XRD measurements and D. Szymanski for SEM images.

Conflicts of Interest: The authors declare no conflict of interest.

References

1. Durif, A. *Crystal Chemistry of Condensed Phosphates*, 1st ed.; Springer Science + Business Media, LLC: New York, NY, USA, 1995.

2. Majeed, S.; Bashir, M.; Shivashankar, S.A. Dispersible crystalline nanobundles of YPO_4 and Ln (Eu, Tb)-doped YPO_4: Rapid synthesis, optical properties and bio-probe applications. *J. Nanopart. Res.* **2015**, *17*, 1–15. [CrossRef]

3. Mukhopadhyay, D.; Patra, C.R.; Bhattacharya, R.; Patra, S.; Basu, S.; Mukherjee, P. Lanthanide Phosphate Nanorods as Inorganic Fluorescent Labels in Cell Biology Research. *Clin. Chem.* **2007**, *53*, 2026–2029.

4. Di, W.; Li, J.; Shirahata, N.; Sakka, Y.; Willinger, M G.; Pinna, N. Photoluminescence, cytotoxicity and in vitro imaging of hexagonal terbium phosphate nanoparticles doped with europium. *Nanoscale* **2011**, *3*, 1263–1269. [CrossRef] [PubMed]

5. Mathur, A.; Halappa, P.; Shivakumara, C. Synthesis and characterization of Sm^{3+} activated $La_{1-x}Gd_xPO_4$ phosphors for white LEDs applications. *J. Mater. Sci. Mater. Electron.* **2018**, *29*, 19951–19964. [CrossRef]

6. Yahiaoui, Z.; Hassairi, M.A.; Dammak, M.; Cavalli, E.; Mezzadri, F. Tunable luminescence and energy transfer properties in YPO_4:Tb^{3+},Eu^{3+}/Tb^{3+} phosphors. *J. Lumin.* **2018**, *194*, 96–101. [CrossRef]

7. Li, H.; Gong, X.; Chen, Y.; Huang, J.; Lin, Y.; Luo, Z.; Huang, Y. Luminescence properties of phosphate phosphors $Ba_3Gd_{1-x}(PO_4)_3$:xSm^{3+}. *J. Rare Earths* **2018**, *36*, 456–460. [CrossRef]

8. Mbarek, A. Synthesis, structural and optical properties of Eu^{3+}-doped $ALnP_2O_7$ (A = Cs, Rb, Tl; Ln = Y, Lu, Tm) pyrophosphates phosphors for solid-state lighting. *J. Mol. Struct.* **2017**, *1138*, 149–154. [CrossRef]

9. Ferhi, M.; Rahmani, S.; Akrout, A.; Horchani-Naifer, K.; Charnay, C.; Durand, J.O.; Ferid, M. Hydrothermal synthesis and luminescence properties of nanospherical SiO_2@$LaPO_4$:Eu^{3+} (5%) composite. *J. Alloys Compd.* **2018**, *764*, 794–801. [CrossRef]

10. Pan, Y.; Li, L.; Chang, W.; Chen, W.; Li, C.; Chen, P.; Zeng, Z. Efficient near-infrared quantum cutting in Tb^{3+}, Yb^{3+} codoped $LuPO_4$ phosphors. *J. Rare Earths* **2017**, *35*, 235–240. [CrossRef]

11. Chen, Y.; Wang, J.; Zhang, M.; Zeng, Q. Light conversion material: $LiBaPO_4$:Eu^{2+},Pr^{3+}, suitablefor solar cell. *RSC Adv.* **2017**, *7*, 21221–21225. [CrossRef]

12. Yuan, J.L.; Zhang, H.; Chen, H.H.; Yang, X.X.; Zhao, J.T.; Gu, M. Synthesis, structure and X-ray excited luminescence of Ce^{3+}-doped $AREP_2O_7$-type alkali rare earth diphosphates (A = Na, K, Rb, Cs; RE = Y, Lu). *J. Solid State Chem.* **2007**, *180*, 3381–3387. [CrossRef]

13. Hamady, A.; Zid, M.F.; Jouini, T. Structure cristalline de KYP_2O_7. *J. Solid State Chem.* **1994**, *113*, 120–124. [CrossRef]

14. Pązik, R.; Watras, A.; Macalik, L.; Dereń, P.J. One step urea assisted synthesis of polycrystalline Eu^{3+} doped KYP_2O_7–luminescence and emission thermal quenching properties. *New J. Chem.* **2014**, *38*, 1129–1137. [CrossRef]

15. Jiang, Y.; Liu, W.; Cao, X.; Su, G.; Cao, L.; Gao, R. A new type of KYP_2O_7 synthesized by the boric acid flux method and its luminescence properties. *J. Alloys Compd.* **2016**, *657*, 697–702. [CrossRef]

16. Zhang, L.; Webster, T.J. Nanotechnology and nanomaterials: Promises for improved tissue regeneration. *Nano Today* **2009**, *4*, 66–80. [CrossRef]

17. Furth, M.; Atala, A.; van Dyke, M. Smart biomaterials design for tissue engineering and regenerative medicine. *Biomaterials* **2007**, *28*, 5068–5073. [CrossRef] [PubMed]

18. Szyszka, K.; Targonska, S.; Gazinska, M.; Szustakiewicz, K.; Wiglusz, R.J. The Comprehensive Approach to Preparation and Investigation of the Eu^{3+} Doped Hydroxyapatite/poly(L-lactide) Nanocomposites: Promising Materials for Theranostics Application. *Nanomaterials* **2019**, *9*, 1146. [CrossRef] [PubMed]

19. Yadid, M.; Feiner, R.; Dvir, T. Gold Nanoparticle-Integrated Scaffolds for Tissue Engineering and Regenerative Medicine. *Nano Lett.* **2019**, *19*, 2198–2206. [CrossRef] [PubMed]

20. Reid, I.R. Osteoporosis treatment: Focus on safety. *Eur. J. Intern. Med.* **2013**, *24*, 691–697. [CrossRef] [PubMed]

21. Lewiecki, E.M.; Watts, N.B. Assessing response to osteoporosis therapy. *Osteoporos. Int.* **2008**, *19*, 1363–1368. [CrossRef] [PubMed]

22. Henriksen, K.; Bollerslev, J.; Everts, V.; Karsdal, M.A. Osteoclast Activity and Subtypes as a Function of Physiology and Pathology—Implications for Future Treatments of Osteoporosis. *Endocr. Rev.* **2011**, *32*, 31–63. [CrossRef] [PubMed]

23. Chomczynski, P.; Sacchi, N. Single-step method of RNA isolation by acid guanidinium thiocyanate-phenol-chloroform extraction. *Anal. Biochem.* **1987**, *162*, 156–159. [CrossRef]

24. Szczeszak, A.; Runowski, M.; Wiglusz, R.J.; Grzyb, T.; Lis, S. Up-conversion green emission of Yb^{3+}/Er^{3+} ions doped YVO_4 nanocrystals obtained via modified Pechini's method. *Opt. Mater.* **2017**, *74*, 128–134. [CrossRef]

25. Yi, G.S.; Chow, G.M. Synthesis of Hexagonal-Phase $NaYF_4$:Yb,Er and $NaYF_4$:Yb,Tm Nanocrystals with Efficient Up-Conversion Fluorescence. *Adv. Funct. Mater.* **2006**, *16*, 2324–2329. [CrossRef]

26. Wiglusz, R.J.; Watras, A.; Malecka, M.; Deren, P.J.; Pazik, R. Structure Evolution and Up-Conversion Studies of ZnX_2O_4:Er^{3+}/Yb^{3+} (X = Al^{3+}, Ga^{3+}, In^{3+}) Nanoparticles. *Eur. J. Inorg. Chem.* **2014**, *2014*, 1090–1101. [CrossRef]

27. Vetrone, F.; Boyer, J.; Capobianco, J.A.; Speghini, A.; Bettinelli, M. Concentration-Dependent Near-Infrared to Visible Upconversion in Nanocrystalline and Bulk Y_2O_3:Er^{3+}. *Chem. Mater.* **2003**, *15*, 2737–2743. [CrossRef]

28. Pollnau, M.; Gamelin, D.R.; Lüthi, S.R.; Güdel, H.U.; Hehlen, M.P. Power dependence of upconversion luminescence in lanthanide and transition-metal-ion systems. *Phys. Rev. B* **2000**, *61*, 3337–3346. [CrossRef]

29. Auzel, F. Upconversion and Anti-Stokes Processes with f and d Ions in Solids. *Chem. Rev.* **2004**, *104*, 139–173. [CrossRef] [PubMed]

30. Marycz, K.; Sobierajska, P.; Smieszek, A.; Maredziak, M.; Wiglusz, K.; Wiglusz, R.J. Li^+ activated nanohydroxyapatite doped with Eu^{3+} ions enhances proliferative activity and viability of human stem progenitor cells of adipose tissue and olfactory ensheathing cells. Further perspective of nHAP: Li^+, Eu^{3+} application in theranostics. *Mater. Sci. Eng. C* **2017**, *78*, 151–162. [CrossRef] [PubMed]

31. Wang, H.; Zhao, B.J.; Yan, R.Z.; Wang, C.; Luo, C.C.; Hu, M. Evaluation of biocompatibility of Ti-6Al-4V scaffolds fabricated by electron beam melting. *Zhonghua Kou Qiang Yi XueZaZhi* **2016**, *51*, 667–672.

32. Schrepfer, E.; Scorrano, L. Mitofusins, from Mitochondria to Metabolism. *Mol. Cell* **2016**, *61*, 683–694. [CrossRef] [PubMed]

33. Sun, D.Y.; Jiang, S.; Zheng, L.M.; Ojcius, D.M.; Young, J.D. Separate metabolic pathways leading to DNA fragmentation and apoptotic chromatin condensation. *J. Exp. Med.* **1994**, *179*, 559–568. [CrossRef] [PubMed]

34. Pan, J.M.; Wu, L.G.; Cai, J.W.; Wu, L.T.; Liang, M. Dexamethasone suppresses osteogenesis of osteoblast via the PI3K/Akt signaling pathway in vitro and in vivo. *J. Recept. Signal Transduct. Res.* **2019**, *39*, 80–86. [CrossRef] [PubMed]

35. Karner, C.M.; Lee, S.-Y.; Long, F. Bmp Induces Osteoblast Differentiation through both Smad4 and mTORC1 Signaling. *Mol. Cell. Biol.* **2017**, *37*, 1–12. [CrossRef] [PubMed]

 *nanomaterials*

Article

Preparation and Characterization of Self-Assembled Poly(L-Lactide) on the Surface of β-Tricalcium Diphosphate(V) for Bone Tissue Theranostics

Jan A. Zienkiewicz [1], Adam Strzep [1], Dawid Jedrzkiewicz [2], Nicole Nowak [1], Justyna Rewak-Soroczynska [1], Adam Watras [1], Jolanta Ejfler [2] and Rafal J. Wiglusz [1,3,*]

[1] Institute of Low Temperature and Structure Research, Polish Academy of Sciences, Okolna 2 str., 50-422 Wroclaw, Poland; j.zienkiewicz@intibs.pl (J.A.Z.); a.strzep@intibs.pl (A.S.); n.nowak@intibs.pl (N.N.); j.rewak@intibs.pl (J.R.-S.); a.watras@intibs.pl (A.W.)

[2] Department of Chemistry, University of Wroclaw, F. Joliot-Curie 14 str., 50-383 Wroclaw, Poland; dawid.jedrzkiewicz@chem.uni.wroc.pl (D.J.); jolanta.ejfler@chem.uni.wroc.pl (J.E.)

[3] Centre for Advanced Materials and Smart Structures, Polish Academy of Sciences, Okolna 2 str., 50-950 Wroclaw, Poland

*   Correspondence: r.wiglusz@intibs.pl; Tel.: +48-071-395-42-74; Fax: +48-071-344-10-29

Received: 23 January 2020; Accepted: 11 February 2020; Published: 15 February 2020

Abstract: This work was aimed to obtain and characterize the well-defined biocomposites based on β-tricalcium diphosphate(V) (β-TCP) co-doped with Ce^{3+} and Pr^{3+} ions modified by poly(L-lactide) (PLLA) with precise tailored chain length and different phosphate to polymer ratio. The composites as well as β-tricalcium diphosphate(V) were spectroscopically characterized using emission spectroscopy and luminescence kinetics. Morphological and structural properties were studied using X-Ray Diffraction (XRD) and Scanning Electron Microscopy (SEM). The self-assembled poly(L-lactide) in a shape of rose flower has been successfully polymerized on the surface of the β-tricalcium diphosphate(V) nanocrystals. The studied materials were evaluated in vitro including cytotoxicity (MTT assay) and hemolysis tests. The obtained results suggested that the studied materials may find potential application in tissue engineering.

Keywords: β-tricalcium diphosphate(V); Ce^{3+} and Pr^{3+} ions co-doping; poly(L-lactide); theranostics

---

1. Introduction

Theranostics is a field of modern nanomedicine that is based on targeted therapy and medical diagnostic tests. This approach is involved with a dynamic development of chemistry, biology, biotechnology, physics and last, but not least—nanotechnology [1].

In the case of biomaterial, some requirements should be met to be treated as a theranostics agent, but one is obligatorily related to similarity to the original tissue [2]. Moreover, the bone minerals are mostly formed by calcium hydroxyapatites (herein CaHAp) [3]. There are a lot of materials related to the apatite family that could be used in the theranostics applications. Among them is β-tricalcium diphosphate (herein β-TCP) that could be a promising agent. This material crystallizes in rhombohedral structure in *R3C* space group [4]. Furthermore, the CaHAp and β-TCP are biocompatible, osteoconductive, and bioactive biomaterials. Comparing the Ca:P ratios in both materials, hydroxyapatite is more similar to mineralized bone, however resorption time of the β-TCP-based materials is much shorter due to its much greater solubility in water in 37 °C [5].

The simplest way to use this material as a bio-imaging agent is a structural modification in order to observe light emission in the range of one of biological optical windows. Our research has showed that β-TCP doped with Pr^{3+} ions exhibits an intense emission band near 650 nm. It has been shown that luminous efficiency of Pr^{3+} ions could be further enhanced by Ce^{3+} co-doping. The literature reports that significant increase of Pr^{3+} emission via energy transfer from excited Ce^{3+} ions is observed i.e., in YAG ($Y_3Al_5O_{12}$, Yttrium Aluminum Garnet) crystals [6]. Efficient luminescent properties are needed to precisely monitor temporal stability and spatial migration of introduced material in the bone tissue.

To insert inorganic material into the bone tissue, a carrier agent is needed. In human bone tissue, calcium hydroxyapatite crystals are distributed in the collagen fibrils forming a complex structure. Another way to obtain biomaterial similar to bone tissue is to use bioresorbable polymer. One of the most used ones is polylactide (PLA).

Biodegradable aliphatic polyesters, for example polylactide (PLA), are widely used polymers in a variety of bioapplications such as controlled drug release, gene therapy, regenerative medicine, or implants [7–11]. The most effective and controlled method for PLA synthesis is metal-catalyzed/-initiated ring-opening polymerization (ROP) of lactide [12–14]. Among the wide variety of catalytic systems, the most attractive so far are the single-site initiators based on structural motif L-M-OR, where L is the ancillary ligand, M is the metal center, and OR is the initiating group [15–23]. Alternative, similarly attractive in the controlled synthesis of PLA are binary catalytic systems based on the homoleptic complexes and external alcohol combination [24–26]. The most widely used system includes commercially available bis(2-ethylhexanoate)tin(II) ($Sn(Oct)_2$) catalyst, commonly applied in industry [12]. However, in the context of medical applications, biologically benign complexes (Zn, Mg, Ca) are the most searched due to their innocuous nature, ready availability, and their effectiveness for polymerization, both in terms of activity and stereoselectivity. Recently, polymer/inorganic nanocomposite materials have attracted considerable interest because of their excellent properties through synergism of polymer and inorganic nanoparticle (iNPs) components. The key to obtaining suitable in bioaplications polymer/iNPs composites is achieving the fine dispersion of inorganic nanoparticles in the polymer matrix. In terms of variety of synthetic strategies, some of them are more favorable for the formation of polymer/iNPs composites because of the simplicity in materials processing. Among others, it is worth mentioning about binding of polymer chains to iNPs by coating modification, dispersion of iNPs in polymer matrix, and formation of stabilizing polymer shell. PLA is suitable for surface coating of iNPs because of their biocompatibility and versatility while providing a platform for further biological modifications.

The proposed research has been intended to consider the well-characterized theranostics materials for bone damage tissue. Nanocrystalline β-tricalcium diphosphate (β-TCP) (so-called therapeutic part) has been co-doped with Ce^{3+} and Pr^{3+} ions (diagnostic part) and coated with poly(L-lactide) (so called a carrier). Moreover, they have been tested in vitro to examine their cytotoxicity using human chondrocyte cell line and mouse osteoblast cell line (MTT assay) and a standard hemolysis test.

2. Materials and Methods

2.1. Materials

All reactions and operations that required an inert atmosphere of N_2 (synthesis of zinc complex L_2Zn and polymers PLA) were performed using a glovebox (MBraun, Garching, Germany) or standard Schlenk-like apparatus (ILT&SR PAS, Wroclaw, Poland) and vacuum line techniques. The solvents for synthesis were purified by standard methods before use: n-hexane (VWR, Radnor, PA, USA) distilled from Na; MeOH (HPLC, VWR, Radnor, PA, USA) distilled from CaH_2; CH_2Cl_2 (99.8% VWR, Radnor, PA, USA) distilled from P_2O_5; deuterated solvents (C_6D_6), distilled from NaH. Unless otherwise stated, all chemicals were purchased from commercial sources and used without further purification:

Calcium carbonate (99.5%, Alfa Aesar, Haverhill, MA, USA), ammonium dihydrogen phosphate (99+%, for analysis, ACROS Organics, Geel, Belgium), cerium(III) nitrate hexahydrate (99.999%, Alfa Aesar, Haverhill, MA, USA), praseodymium(III, IV) oxide (99.999%, Sigma-Aldrich, Saint Louis, MO, USA), nitric acid (65%, Suprapure, Merck, Darmstadt, Germany), citric acid (99%+, Alfa Aesar, Haverhill,

MA, USA), ethylene glycol (ultrapure, Avantor Performance Materials Poland S.A., Gliwice, Poland) 2,4-di-*tert*-buthylphenol (99%, Sigma-Aldrich, Saint Louis, MO, USA), formaldehyde (37% solution in H_2O, Sigma-Aldrich, Saint Louis, MO, USA). *N*-methylcyclohexylamine (99%, Sigma-Aldrich, Saint Louis, MO, USA), ZnEt$_2$ (1.0 M solution in *n*-heptane, Sigma-Aldrich, Saint Louis, MO, USA). Proligand [L^{Cy}-*H*] *N*-[methyl(2-hydroxy-3,5-di-*tert*-butylphenyl)]-*N*-methyl-*N*-cyclohexylamine was synthesized according to a literature procedure [24].

2.2. Synthesis of β-tricalcium diphosphates (β-TCP) Doped with Pr^{3+} and Co-Doped with Ce^{3+} Ions

The nanocrystalline β-$Ca_3(PO_4)_2$ calcium phosphate co-doped with Ce^{3+} and Pr^{3+} ions were prepared by modified Pechini's method, using 18 mmol of $CaCO_3$, 12 mmol of $NH_4H_2PO_4$, 0.09 mmol of $Ce(NO_3)_3 \cdot 4H_2O$, and 0.015 mmol of Pr_6O_{11}. Intentional concentration of the Ce^{3+} and Pr^{3+} ions was set to 0.5 mol%, in replacement of overall molar content of Ca^{2+} ions.

In this method, stoichiometric amounts of $CaCO_3$ and Pr_6O_{11} were weighed and digested in excess of HNO_3 (Suprapure, Merck) in order to transform them into nitrates. Subsequently, cerium nitrate was dissolved in deionized water together with calcium and praseodymium (III) nitrates. Afterwards, the excess (12,5-fold relative to the total amount of cations) of citric acid as well as ethylene glycol were added under constant stirring at 60 °C, resulting in a viscous mixture. Finally, a suitable amount of ammonium hydrogen phosphate was added. The temperature was raised up to 120 °C. The heating was continued until a white voluminous foam was obtained. The mixture was further dried for 3 days at 90 °C. Afterwards, the resin thus obtained was calcinated in a temperature of 900 °C. As a result, a white powder (β-TCP) was obtained.

2.3. Synthesis of poly(ʟ-lactide) (PLLA)

Synthesis of PLLA was divided into two parts. At first, the initiator was synthetized. Afterwards, ring opening polymerization of L-lactide was carried out. Oligomeric chains based on L-lactide with methyl ending group were prepared in ring opening polymerization (ROP) reaction. Homoleptic zinc aminophenolate complex type ZnL$_2$ (L = *N*-methyl-(2-hydroxy-3,5-di-tert-buthylphenyl)]-*N*-cyclohexylamine) as initiator was used. High efficiency of this catalyst was described in the literature [24–26].

2.3.1. Synthesis of Initiator

[(L^{Cy})$_2$Zn] was synthesized following a modified procedure of our previous report [25].

To a stirred solution of 0.66 g (2.00 mmol) of proligand L^{Cy}-H in *n*-hexane (20 mL), 1.00 mL (1.00 mmol) of ZnEt$_2$ (1.0 M solution in *n*-heptane) was added dropwise at ambient temperature. Next, the solution was stirred for 12 h until a crude product precipitated. The white powder of [(L^{Cy})$_2$Zn] was collected by filtration, washed with cold n-hexane, and dried in vacuo. Yield: 92% (0.67 g, 0.92 mmol). Anal. Calcd (Found) for $C_{44}H_{72}N_2O_2Zn$ (726.41): C 72.75 (72.83), H 9.99 (9.86), N 1.96 (1.85)%. ^{1}H NMR for major form (500 MHz, C_6D_6, 298 K): δ = 7.72 (2H, d, ArH, J_{HH} = 2.6 Hz), 7.09 (2H, d, ArH, J_{HH} = 2.6 Hz), 4.16 (2H, d, N-CH$_2$-Ar, J_{HH} = 12.1 Hz), 3.47 (2H, d, N-CH$_2$-Ar, J_{HH} = 12.1 Hz), 3.20 (2H, m, C_6H_{11}), 2.48 (6H, s, NCH$_3$), 1.74 (18H, s, C(CH$_3$)$_3$), 1.66 (10H, m, C_6H_{11}), 1.59 (18H, s, C(CH$_3$)$_3$), 1.05 (10H, m, C_6H_{11}). ^{13}C NMR (500 MHz, C_6D_6, 298 K) δ = 164.3, 138.4, 135.5, 125.9, 124.4, 120.5 (12C, Ar), 65.3 (N-CH$_2$-Ar), 61.9 (2C, C_6H_{11}), 36.8 (C(CH$_3$)$_3$), 35.5 (C(CH$_3$)$_3$), 33.8 (NCH$_3$), 32.3 (C(CH$_3$)$_3$), 30.2 (C(CH)$_3$)$_3$), 26.9, 26.1, 24.4 (10C, C_6H_{11}).

2.3.2. Representative Procedure for Solution Polymerization

ROP polymerization of ʟ-Lactide while using binary catalytic system [(L^{Cy})$_2$Zn]/MeOH.

The solution of zinc complex [(L^{Cy})$_2$Zn] in CH$_2$Cl$_2$ (15 mL) was placed in a Schlenk flask, and MeOH and ʟ-lactide in molar ratio [(L^{Cy})$_2$Zn]/MeOH/ʟ-Lactide 1/1/15 was added. The obtained solution was stirred for 5 h. At certain time intervals, about 1 mL aliquots were removed, precipitated with hexanes, and dried in vacuo. The obtained precipitates were dissolved in C_6D_6 and used for the conversion monitoring, which was determined by ^1H NMR. After the reaction was completed, an excess of hexanes was added to the reaction mixture. The obtained crude polymer was next filtered off and dried in vacuo. The resulting powder was dissolved in dichloromethane and the PLA was precipitated with excess of cold hexanes. The PLA was collected by filtration, washed with hexanes, and dried in vacuo. The reaction mixture was prepared in glovebox, and the next subsequent operations for the isolation of pure PLLA were performed by using standard Schlenk apparatus and vacuum line techniques.

2.4. Representative Procedure for Preparation of β-TCP@PLLA Composites

TCP@PLA composites were obtained by using precipitation method.

To the sample of polymer, PLLA (50 mg) dissolved in 1 mL of CH_2Cl_2 nanocrystalline β-TCP (5 mg) was added and the mixture was stirred at room temperature for 0.5 h. The β-TCP@PLLA composites were precipitated with an excess of hexanes (50 mL), which was added dropwise over 10 min. Next the hexane was removed by decantation and composite material was purified by washing in hexane (3 × 50 mL) and isolated by drying the precipitate afterwards under reduced pressure during 72 h in the temperature 60 °C.

2.5. Physicochemical Characterization

^1H, ^{13}C NMR spectra were obtained using Bruker Avance 500 MHz spectrometer (Bruker, Billerica, MA, USA). The chemical shifts are given in ppm and referenced to the residual protons in the deuterated solvents. Microanalyses were conducted with an Elementar CHNS Vario EL III analyzer (Elementar, Langenselbold, Germany).

X-Ray Powder Diffraction (XRD) studies were carried out using PANalytical X'Pert Pro diffractometer (Malvern Panalytical Ltd, Malvern, UK) equipped with Ni-filtered Cu Kα radiation ($V = 40$ kV, $I = 30$ mA, $\lambda = 1.5406$ Å). The XRD patterns were collected during 3 h in the 2θ range of 10–60°.

The microstructure investigations and elemental analysis were carried out using a scanning electron microscope FEI Nova NanoSEM 230 (FEI Company, Hillsboro, OR, USA) equipped with EDS spectrometer (EDAX PegasusXM4) (EDAX Inc., Mahwah, NJ, USA). TCP sample was attached to the measuring table using graphite tape and imaged at an accelerated voltage 10 kV. In order to eliminate the effect of collecting charge on the surface of material, PLLA and composite samples were imaged at an accelerated voltage 5 kV and 10 kV.

High-resolution emission spectra and luminescence kinetics curves were recorded using excitation wavelength 445 nm. Opolette Nd:YAG Laser-OPO system (Opotek INC, CA, USA) was used as an excitation source. Emission spectra were recorded with use of DongWooOptron d750 monochromator (DongWoo Optron, Maesan-ri, South Korea). Light was collected by Hamamatsu R928 photomultiplier (Hamamatsu Photonics K.K., Hamamatsu, Japan). Signal from the photomultiplier was analyzed in parallel by SR250 Gated Integrator (Stanford Research System, Sunnyvale, CA, USA) for integration of signal and a Tektronix TDS 3050 digital oscilloscope (luminescence kinetics) (Tektronix Inc., Beaverton, OR, USA).

2.6. Evaluation of Biological Properties

Potential non cytotoxic properties of obtained materials were evaluated via MTT assay by using human chondrocyte cell line (TC28A2) and mouse osteoblast (7F2) cell lines. Hemolysis assay by using ram blood cells was performed to estimate possible hemolytic activity of our composites.

2.6.1. Human Chondrocytes Cell Line

TC28A2 human chondrocyte cell line was maintained in high glucose Dulbecco's Modyfied Eagle Medium (DMEM) with L-glutamine (Biowest, Nuaillé, France) and supplemented with 10%

Fetal Bovine Serum (FBS) South America Heat Inactivated (Biowest, Nuaillé, France), 200 U/mL penicillin, and 200 µg/mL streptomycin. Mouse osteoblast cell line (7F2) was cultured in Minimum Essential Medium Eagle–alpha modification (α-MEM) without nucleosides (Biowest, Nuaillé, France). To obtain a full cultured medium, α-MEM was supplemented with 10% FBS and 2mM stable glutamine (Biowest, Nuaillé, France). TC28A2 and 7F2 cell lines were incubated in standard conditions at 37 °C in humidified atmosphere of 5% CO_2 and 95% air. Passive cells were used three times in the experiments.

2.6.2. Influence of Obtained Composites on Chondrocytes Proliferation Rate

Proliferation capacity of human chondrocytes and mouse osteoblast were evaluated via performing MTT cytotoxicity assay Tc28A2 and 7F2 cells were seeded at density 10,000 cells per well in 96-well plates and allowed to attached and grow for 24 h. Then, cells were washed with sterile PBS (Biowest, Nuaillé, France), and fresh medium and adequate concentration of tested compounds, (50 µg/mL and 100 µg/mL of β-$Ca_3(PO_4)_2$, PLLA and composite) was added. MTT (BioReagent, >97.5%, Sigma-Aldrich, Saint Louis, MO, USA) assay was performed 24 h after cells treatment. Treatment medium was removed and sterile PBS containing 0.5 mg/mL MTT (tiazol blue tertazolium) was added, and cells were incubated 3 h at 37 °C. After incubation, medium containing MTT was removed without washing, and formed formazan crystals were dissolved in DMSO (99.5%, Sigma-Aldrich, Saint Louis, MO, USA). Absorbance was read with a Varioskan LUX plate reader (ThermoFisher Scientific, Waltham, MA, USA) at 560 nm with background reference at 670 nm. The experiment was performed three times. Percentage of cell viability was calculated using the following formula:

$$\text{Cells viability} \; = \frac{\text{sample absorbance}}{\text{control absorbance}} \times 100\%. \tag{1}$$

As a reference control (100% of cells viability), samples of non-treated cells were used in both cell lines.

2.6.3. Hemolysis Assay

Hemolysis assay was performed according to the protocol described elsewhere with the slight modification [27]. Ram blood (ProAnimali, Wroclaw, Poland) was centrifuged (3000 RPM, 10 min) in order to obtain erythrocyte fraction, which was washed with PBS (phosphate-buffered saline, pH 7.4) and mixed with fresh PBS (1:1 *v/v*). β-$Ca_3(PO_4)_2$; PLLA and the β-$Ca_3(PO_4)_2$/PLLA (1:10) composite were mixed with erythrocytes at final concentration of 50 and 100 µg/mL and incubated in 37 °C for 2 h. Then, samples were centrifuged in order to obtain supernatant (5000 RPM, 5 min) and the optical density was measured at 540 nm with a Varioskan LUX plate reader (ThermoFisher Scientific, Waltham, MA, USA). As a reference control (100% of hemolysis), the 1% solution of SDS (sodium dodecyl sulfate) was used and as negative control the solution of PBS (phosphate-buffered saline) was applied. Obtained results were compared with the absorbance of SDS sample and shown as a percentage of hemolysis and the hemolysis percentage was calculated as using following formula:

$$Hemolysis = \frac{sample\ absorbance\ -\ negative\ control\ absorbance}{positive\ control\ absorbance\ -\ negative\ control\ absorbance} \times 100 \tag{2}$$

3. Results and Discussion

3.1. Structural Properties

3.1.1. β-tricalcium diphosphate (β-TCP)

Structure and phase purity were checked with the powder XRD technique and were compared with the reference standard of the tetragonal β–$Ca_3(PO_4)_2$ lattice ascribed to the *R-3c* space group. Diffractograms of β-TCP co-doped with Ce^{3+} and Pr^{3+} and TCP@PLLA composite as well as the theoretical XRD pattern of β–$Ca_3(PO_4)_2$ (ICSD 97500) and the diffractogram of pure PLLA are shown in Figure 1.

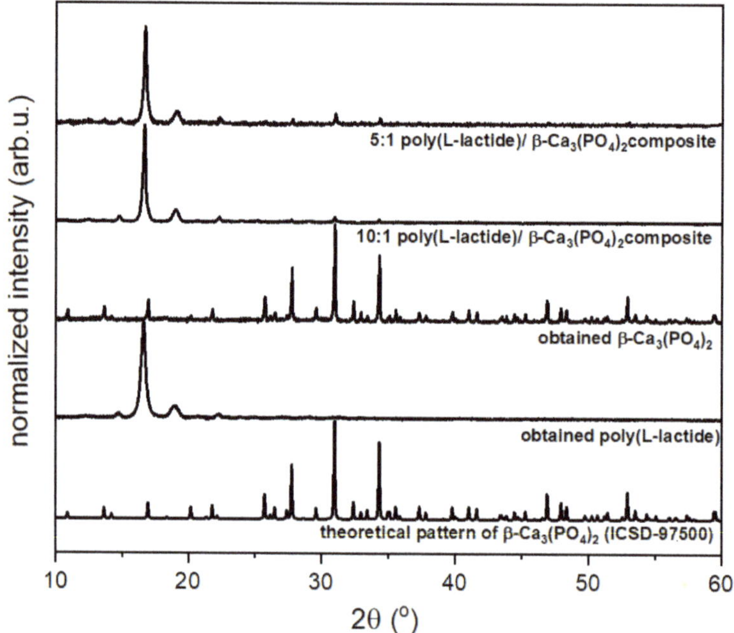

Figure 1. X-ray diffraction patters of poly(L-lactide) (PLLA), β-$Ca_3(PO_4)_2$ (β-TCP) co-doped with Ce^{3+} and Pr^{3+} ions and PLLA with β-$Ca_3(PO_4)_2$ composite compared to theoretical pattern of β-$Ca_3(PO_4)_2$ (ICSD-97500).

β-TCP crystallized in rhombohedral structure, in the space group *R3c*. Due to Shannon, Ca^{2+}, Ce^{3+}, and Pr^{3+} effective ionic radii were 1.00 Å, 1.01 Å, and 0.99 Å, respectively [28]. Due to similar ionic radiilanthanide, ions substituted Ca^{2+} sites in the β-TCP matrix lattice.

3.1.2. Poly(L-lactide) (PLLA)

The activity and effectiveness of binary catalytic system [$(L^{Cy})_2$Zn]/MeOH to synthesis of PLLA in living ROP of lactides from high and ultra-low molecular weight was recently published. That catalytic system is well suited especially for the synthesis of precisely defined oligolactides in comparison with the commercial $Sn(Oct)_2$, which is not as selective as [$(L^{Cy})_2$Zn]/ROH. Additionally, the most popular $Sn(Oct)_2$ used for the synthesis of low molecular PLLA produces the oligolactides with the fraction of alkyl-(S,S)-O-lactyllactate. Therefore, the precisely defined polymer matrix containing planned 15 lactide units has been obtained by using [$(L^{Cy})_2$Zn]/MeOH catalytic system. Under selected molar ratio of ROP components [$(L^{Cy})_2$Zn]/MeOH/L-LA = 1/1/15, the oligomer 15-PLLA-Me were obtained (Figure 2).

The end groups and the number of lactide units were detected by using NMR spectroscopy. The ^{1}H NMR spectra for 15-PLLA-Me oligomer showed expected resonances for both chains end, methyl ester and hydroxyl groups, and oligolactide backbone chain (Figure 3). The most intense signals A_3 and B_3 corresponded to methine and methyl groups of repetitive central open lactide units. The adequate resonances of open lactide units close to end groups were denoted as $A_{1\text{-}2}$ $B_{1\text{-}2}$ (a couple of signals coming from neighboring mer to hydroxy end) and $A_{4\text{-}5}$ $B_{4\text{-}5}$ (the first open lactide molecule close to ester chain end) (see Figure 2).

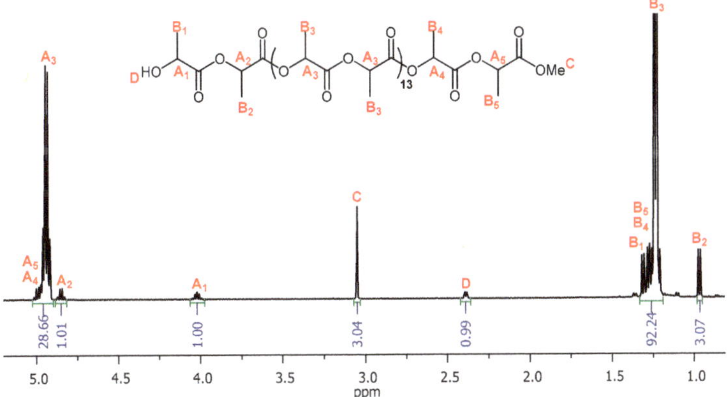

Figure 2. Synthesis of 15-PLLA-Me oligomer.

Figure 3. ^{1}H NMR spectrum of 15-PLLA-Me oligomer (C_6D_6).

3.2. Morfological Properties

SEM images of uncoated β-TCP (A) and poly(ʟ-lactide) (B) compared to β-TCP coated with PLLA in two β-TCP:PLLA ratios: 1:10 (C) and 1:20 (D) are presented in Figure 4.

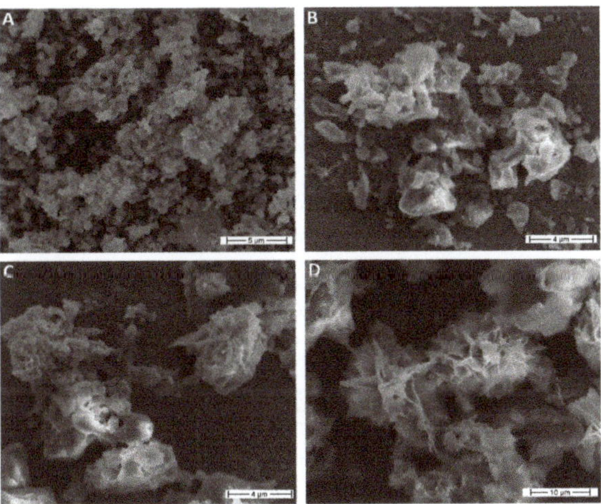

Figure 4. SEM images of (**A**) β-TCPco-doped with Ce^{3+} and Pr^{3+} ions, (**B**) PLLA, (**C**) PLLA with β-TCP composite with phosphate:polymer 1:10 ratio, and (**D**) PLLA with β-TCP composite with phosphate:polymer 1:20 ratio.

The surface of the sample coated with 1:20 ratio (D) clearly showed new morphology with the shape of rose flower, while surfaces of pure PLLA (B) and the sample coated with 1:10 ratio (C) were much smoother.

3.3. Spectroscopic Properties

3.3.1. Emission Spectra

Emission spectra were measured in the range of wavelength from 500 nm to 700 nm after excitation with 445 nm. Emission spectra of β-TCP:0.5%Ce; 0.5%Pr sintered at 900 °C for 3 h, pure and coated with poly(L-lactide) upon excitation at 445 nm, are presented in Figure 5.

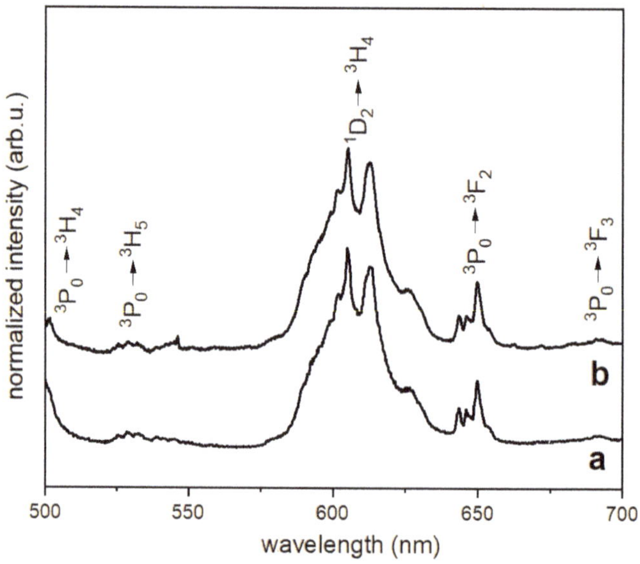

Figure 5. Emission spectra of PLLA with β-TCP co-doped with Ce^{3+} and Pr^{3+} ions (**a**) and PLLA with β-TCP composite with phosphate:polymer 1:20 ratio (**b**). Spectra were recorded after excitation with 445 nm.

The Ce^{3+} and Pr^{3+}-co-doped tricalcium β-diphosphate showed prominent $^1D_2 \rightarrow {}^3H_4$ and $^3P_0 \rightarrow {}^3F_2$ transitions. The $^3P_0 \rightarrow {}^3F_2$ transition was less intense than the $^1D_2 \rightarrow {}^3H_4$, but it was observed near 650 nm—in the range of first optical window for biological tissues. Spectrum of pure β-TCP co-doped with Ce^{3+} and Pr^{3+} ions was identical to the spectrum of coated with PLLA. This means that the presence of the coating layer did not change luminescent properties of β-TCP core. Furthermore, resemblance of both spectra suggests that Pr^{3+} ions remained in phosphate structure and did not diffuse to polymer body.

3.3.2. Luminescence Kinetics

Luminescence kinetics curves were recorded after excitation with 445 nm. Decays from the $^1D_2$ and $^3P_0$ state were observed at subsequently 605 and 650 nm.

Lifetimes of the $^1D_2$ and $^3P_0$ states of Pr^{3+} ions were not mono-exponential. Average lifetimes of all materials were calculated from the equation (Table 1):

$$\tau_{avg} = \frac{\int I(t)t\,dt}{\int I(t)\,dt} \tag{3}$$

where: *I*—intensity; *t*—time; τ_{avg}—average lifetime.

Table 1. Calculated lifetimes of β-TCP co-doped with Ce^{3+} and Pr^{3+} ions PLLA with β-$Ca_3(PO_4)_2$ composite with phosphate:polymer 1:10 ratio and PLLA with β-TCP composite with phosphate:polymer 1:20 ratio. Delays from the state $^3P_0$ were observed at 650 nm and delays from the state $^1D_2$ were observed at 605 nm.

Materials	State	Observed Wavelength [nm]	Lifetime [µs] (τ_{avg})
β-TCP:0.5%Ce/0.5% Pr, 900 °C	$^3P_0$	650	1.31
	$^1D_2$	605	84.39
β-TCP/PLLA composite 1:20	$^3P_0$	650	1.28
	$^1D_2$	605	84.50
β-TCP/PLLA composite 1:10	$^3P_0$	650	1.34
	$^1D_2$	605	77.28

$^1D_2$ lifetimes in pure TCP and 10% composite with PLLA were the same, but in the 5% composite, they were a little bit shorter; however, differences between samples were too scarce to consider them as a significant influence of polymer coating. Decay curves observed at 650 nm and 605 nm are presented in Figure 6.

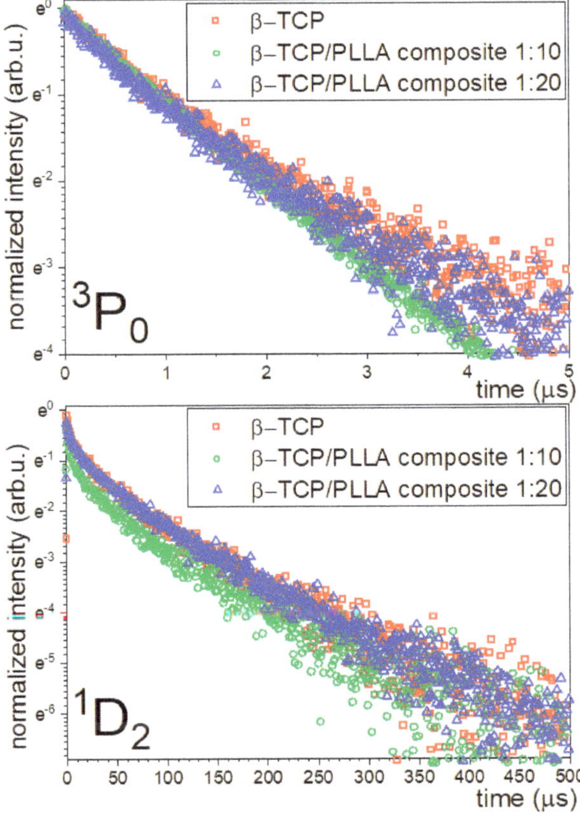

Figure 6. Luminescence kinetics curves of β-TCP co-doped with Ce^{3+} and Pr^{3+} ions PLLA with β-TCP composite with phosphate: polymer 1:10 ratio and PLLA with β-TCP composite with phosphate: polymer 1:20 ratio, recorded at 650 nm ($^3P_0$) and recorded at 605 nm ($^1D_2$).

Decay time of the luminescence from the state $^3P_0$ observed at 650 nm was about 1 µs in each studied case, which was much longer than the luminescence lifetime of emission exhibited by organic molecules from biological tissue [29,30]. This feature could be used to easily discern whether observed luminescence comes from tissue-related molecules or artificially introduced β-TCP. Time dependency and spectral shape of luminesce from introduced material could be used to precisely monitor temporal stability and spatial migration of β-TCP in the bone tissue. Location and stability monitoring are needed to fulfil diagnostic part of β-TCP/PLLA composite possible theranostics application.

3.4. Biological Features

3.4.1. Cytotoxicity Assay

Our results clearly showed that some of the tested materials increased cell proliferation rate in both concentrations 50 µg/mL and 100 µg/mL in the TC28A2 cell line, which may indicate enhancement of metabolic activity of treated cells. PLLA seems to be degraded in an extracellular environment to ʟ-lactide and then may be transported via MTCs (monocarboxylate transporters) and used by chondrocytes as an energetic fuel [31]. Additionally, chondrocytes proliferation rate was higher than 50% even in the 100 µg/mL concentration of all tested composites. Obtained results were collected in Figure 7. Intriguingly, in comparison with chondrocytes, viability of 7f2 cells treated with PLLA maintain edat 74% in concentration 50 µg/mL and seemed to gradually decrease in higher concentrations. The lower proliferation rate of 7F2 cells may be explained by increased lactate production in osteoblasts, which was caused by the aerobic glycolysis even in the presence of oxygen. This type of metabolism was similar to the Warburg effect, the major hallmark of cancer. It transpired that specific metabolic needs of osteoblasts demanded this particular type of metabolism. Elevated concentrations of lactic acid may additionally acidify extracellular environment, thus osteoblast cells do not use PLLA as an energy fuel like chondrocytes [32]. When compared to other compounds, the viability of mouse osteoblasts maintains around 75% when treated with β-TCP and the composite at a concentration of 50 µg/mL and around 60% at a concentration of 100 µg/mL. This may indicate that the osteoblast cell line is more sensitive to alterations in extracellular environment. However, what is the most important is that proliferation of TC28A2 and 7F2 cell lines was maintained above 50% in each sample and both concentrations.

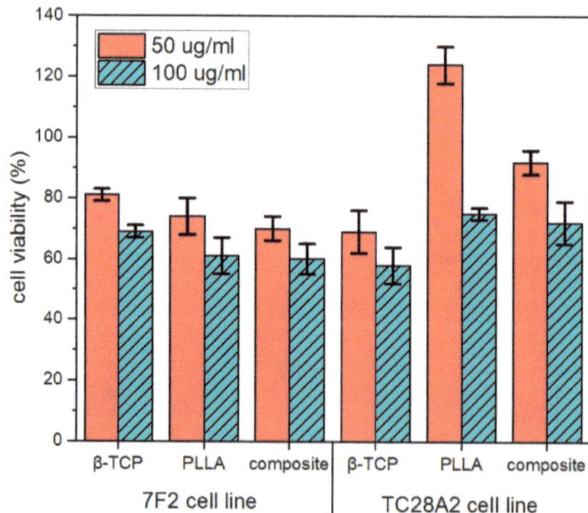

Figure 7. 7F2 and TC26A2 cells viability after 24 h of incubation and exposition on β-TCP co-doped with Ce^{3+} and Pr^{3+} ions, PLLA, and PLLA with β-TCP composite with phosphate:polymer 1:20 ratio, in concentrations of 50 µg/mL and 100 µg/mL.

3.4.2. Hemolysis

Tested compounds at concentrations of 50 and 100 µg/mL did not cause the hemolysis of ram erythrocytes. Obtained results were compared with the hemolysis caused by 1% of SDS solution (complete damage of the cellular membrane and hemoglobin release; data not shown) and are presented in Figure 8. Statistical analysis was also performed ($p < 0.05$) using one-way ANOVA test.

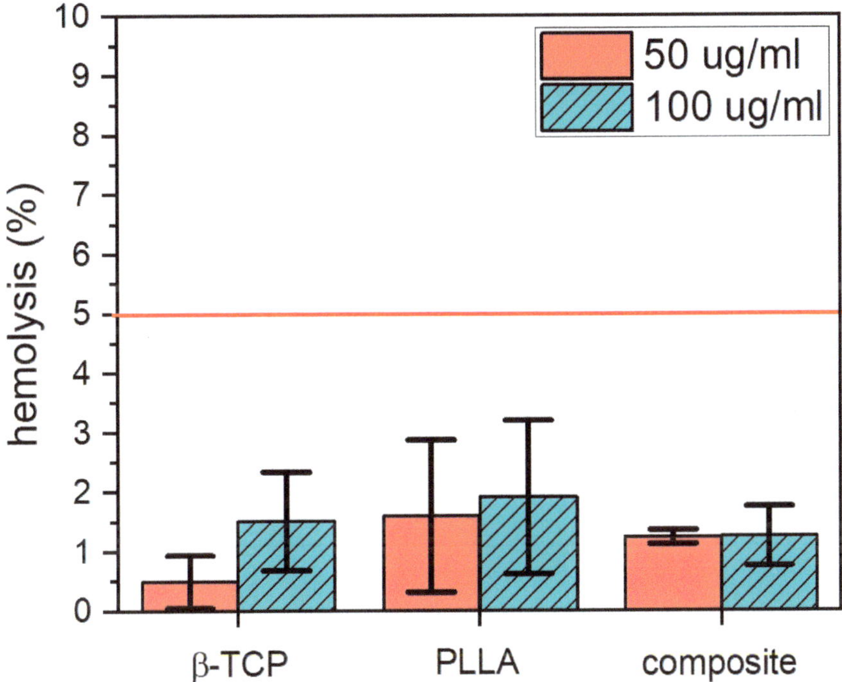

Figure 8. Activity of β-TCP co-doped with Ce^{3+} and Pr^{3+} ions, PLLA, and PLLA with β-TCP composite with phosphate:polymer 1:20 ratio, in concentrations 50 µg/mL and 100 µg/mL in reference to 1% of SDS solution (mean ± SD, $n = 3$; all results were statistically significant). The red line indicates the acceptable hemolysis level.

Performing such an experiment is essential for chemical compounds designed to be applied in direct contact with the human body and its fluids. The acceptable range of hemolysis is set below 5% [33], therefore tested substances could be regarded as erythrocyte safe. Similar results were obtained by Zibiao et al. who proved that selected polymers based on PLLA and PEG did not cause an hemolysis higher than 5%, even in higher concentrations [34]. The researchers proved that coating some commercial metal alloys with PEO/PLLA (plasma electrolytic oxidized/poly(L-lactide) composites can help reduce their hemolytic potential [35]. β-TCP was also tested before in order to check its hemolytic potential, but such effect was not observed [36].

4. Conclusions

In this work, the novel and self-assembled poly(L-lactide) (PLLA)/β-tricalcium diphosphate (β-TCP) composite has been prepared. In the first step, the well-β-TCP co-doped with Ce^{3+} and Pr^{3+} ions was obtained. Further, it was prepared precisely tailored to low-molecular mass PLLA on the β-TCP surface using biocatalysts from the group of zinc aminophenolates. The obtained materials may be potentially used to bone grafting and in vivo bioimaging in the near-infrared (NIR) window (also known as "optical window" or "therapeutic window"). PLLA has been used to enhance biocompatibility of the obtained composite for further application. Moreover, the polymer did not affect the luminescent properties of Ce^{3+} and Pr^{3+} ions co-doped β-TCP.

Furthermore, the hemolysis higher than 5% was not observed for all studied materials. It has been suggested that the obtained materials could be safely used in vivo because the cell proliferation and mitochondrial activity have not been disrupted. In conclusion, it might not interrupt the OXPHOS (Oxidative phosphorylation) and energy production in chondrocytes.

Author Contributions: R.J.W. conceived and designed the experiments as well as analyzed all data; J.A.Z., D.J., and J.E. contributed reagents/materials/analysis tools as well as analyzed data; A.W. and A.S. designed the experiments as well as analyzed luminescence data; N.N. and J.R.-S. contributed reagents/materials/analysis tools and analyzed biological data; all authors contributed to the writing of the paper. All authors have read and agreed to the published version of the manuscript.

Funding: Financial support of the National Science Centre in course of realization of the Projects "Preparation and characterization of biocomposites based on nanoapatites for theranostic (no. UMO-2015/19/B/ST5/01330) and "Biocompatible polyesters with accurately coded sequences: engineering, functionalization and applications" (no. UMO-2017/25/B/ST5/00597) and are gratefully acknowledge.

Acknowledgments: The authors would like to acknowledge Piotr Kuropka (Wrocław University of Environmental and Life Sciences) for the human chondrocyte cell line TC28A2 and Ewa Bukowskafor performing XRD measurements as well as Damian Szymański for SEM images (Institute of Low Temperature and Structure Research, Polish Academy of Sciences).

Conflicts of Interest: The authors declare no conflict of interest.

References

1. Zarrin, A.; Sadighian, S.; Rostamizadeh, K.; Firuzi, O.; Hamidi, M.; Mohammadi-Samani, S.; Miri, R. Design, preparation, and in vitro characterization of a trimodally-targeted nanomagnetic onco-theranostic system for cancer diagnosis and therapy. *Int. J. Pharm.* **2016**, *500*, 62–76. [CrossRef] [PubMed]

2. Wang, J.; Cui, H. Nanostructure-based theranostic systems. *Theranostics* **2016**, *6*, 1274–1276. [CrossRef]

3. Guyton, A.C.; Hall, J.E. *Textbook of Medical Physiology*, 11th ed.; Elsevier Saunders: Philadelphia, PA, USA, 2006; ISBN 0721602401.

4. Yashima, M.; Sakai, A.; Kamiyama, T.; Hoshikawa, A. Crystal structure analysis of β-tricalcium phosphate Ca3(PO4)2 by neutron powder diffraction. *J. Solid State Chem.* **2003**, *175*, 272–277. [CrossRef]

5. Wang, L.; Nancollas, G.H. Calcium Orthophosphates: Crystallization and Dissolution. *Chem. Rev.* **2008**, *108*, 4628–4669. [CrossRef] [PubMed]

6. Yang, H.; Kim, Y.-S. Energy transfer-based spectral properties of Tb-, Pr-, or Sm-codoped YAG:Ce nanocrystalline phosphors. *J. Lumin.* **2008**, *128*, 1570–1576. [CrossRef]

7. Murariu, M.; Doumbia, A.; Bonnaud, L.; Dechief, A.; Paint, Y.; Ferreira, M.; Campagne, C.; Devaux, E.; Dubois, P. High-Performance Polylactide/ZnO Nanocomposites Designed for Films and Fibers with Special End-Use Properties. *Biomacromolecules* **2011**, *12*, 1762–1771. [CrossRef] [PubMed]

8. Kim, H.; Park, H.; Lee, J.; Kim, T.H.; Lee, E.S.; Oh, K.T.; Lee, K.C.; Youn, Y.S. Highly porous large poly(lactic-co-glycolic acid) microspheres adsorbed with palmityl-acylated exendin-4 as a long-acting inhalation system for treating diabetes. *Biomaterials* **2011**, *32*, 1685–1693. [CrossRef]

9. Zhang, P.; Wu, H.; Wu, H.; Lù, Z.; Deng, C.; Hong, Z.; Jing, X.; Chen, X. RGD-Conjugated Copolymer Incorporated into Composite of Poly(lactide-co-glycotide) and Poly(l-lactide)-Grafted Nanohydroxyapatite for Bone Tissue Engineering. *Biomacromolecules* **2011**, *12*, 2667–2680. [CrossRef]

10. Bertrand, A.; Hillmyer, M.A. Nanoporous Poly(lactide) by Olefin Metathesis Degradation. *J. Am. Chem. Soc.* **2013**, *135*, 10918–10921. [CrossRef]

11. Dusselier, M.; van Wouwe, P.; Dewaele, A.; Jacobs, P.A.; Sels, B.F. Shape-selective zeolite catalysis for bioplastics production. *Science* **2015**, *349*, 78–80. [CrossRef]

12. Dechy-Cabaret, O.; Martin-Vaca, B.; Bourissou, D. Controlled Ring-Opening Polymerization of Lactide and Glycolide. *Chem. Rev.* **2004**, *104*, 6147–6176. [CrossRef] [PubMed]

13. Stanford, M.J.; Dove, A.P. Stereocontrolled ring-opening polymerisation of lactide. *Chem. Soc. Rev.* **2010**, *39*, 486–494. [CrossRef] [PubMed]

14. Sarazin, Y.; Carpentier, J.F. Discrete Cationic Complexes for Ring-Opening Polymerization Catalysis of Cyclic Esters and Epoxides. *Chem. Rev.* **2015**, *115*, 3564–3614. [CrossRef] [PubMed]

15. Chisholm, M.H.; Eilerts, N.W. Single-site metal alkoxide catalysts for ring-opening polymerizations. Poly(dilactide) synthesis employing {HB(3-ButPz)3}Mg(OEt). *Chem. Commun.* **1996**, 853–854. [CrossRef]

16. Chamberlain, B.M.; Cheng, M.; Moore, D.R.; Ovitt, T.M.; Lobkovsky, E.B.; Coates, G.W. Polymerization of lactide with zinc and magnesium β-diiminate complexes: Stereocontrol and mechanism. *J. Am. Chem. Soc.* **2001**, *123*, 3229–3238. [CrossRef]

17. Chisholm, M.H.; Gallucci, J.C.; Phomphrai, K. Well-defined calcium initiators for lactide polymerization. *Inorg. Chem.* **2004**, *43*, 6717–6725. [CrossRef]

18. Wu, J.-C.; Huang, B.-H.; Hsueh, M.-L.; Lai, S.-L.; Lin, C.-C. Ring-opening polymerization of lactide initiated by magnesium and zinc alkoxides. *Polymer* **2005**, *46*, 9784–9792. [CrossRef]

19. Chuang, H.J.; Weng, S.F.; Chang, C.C.; Lin, C.C.; Chen, H.Y. Synthesis, characterization and catalytic activity of magnesium and zinc aminophenoxide complexes: Catalysts for ring-opening polymerization of l-lactide. *Dalton Trans.* **2011**, *40*, 9601–9607. [CrossRef]

20. Huang, M.; Pan, C.; Ma, H. Ring-opening polymerization of rac-lactide and α-methyltrimethylene carbonate catalyzed by magnesium and zinc complexes derived from binaphthyl-based iminophenolate ligands. *Dalton Trans.* **2015**, *44*, 12420–12431. [CrossRef]

21. Williams, C.K.; Breyfogle, L.E.; Choi, S.K.; Nam, W.; Young, V.G.; Hillmyer, M.A.; Tolman, W.B. A Highly Active Zinc Catalyst for the Controlled Polymerization of Lactide. *J. Am. Chem. Soc.* **2003**, *125*, 11350–11359. [CrossRef]

22. Song, S.; Zhang, X.; Ma, H.; Yang, Y. Zinc complexes supported by claw-type aminophenolate ligands: Synthesis, characterization and catalysis in the ring-opening polymerization of rac-lactide. *Dalton Trans.* **2012**, *41*, 3266. [CrossRef]

23. Jędrzkiewicz, D.; Adamus, G.; Kwiecień, M.; John, Ł.; Ejfler, J. Lactide as the Playmaker of the ROP Game: Theoretical and Experimental Investigation of Ring-Opening Polymerization of Lactide Initiated by Aminonaphtholate Zinc Complexes. *Inorg. Chem.* **2017**, *56*, 1349–1365. [CrossRef]

24. Ejfler, J.; Kobyłka, M.; Jerzykiewicz, L.B.; Sobota, P. Highly efficient magnesium initiators for lactide polymerization. *Dalton Trans.* **2005**, 2047–2050. [CrossRef]

25. Wojtaszak, J.; Mierzwicki, K.; Szafert, S.; Gulia, N.; Ejfler, J. Homoleptic aminophenolates of Zn, Mg and Ca. Synthesis, structure, DFT studies and polymerization activity in ROP of lactides. *Dalton Trans.* **2014**, *43*, 2424–2436. [CrossRef]

26. Jędrzkiewicz, D.; Czeluśniak, I.; Wierzejewska, M.; Szafert, S.; Ejfler, J. Well-controlled, zinc-catalyzed synthesis of low molecular weight oligolactides by ring opening reaction. *J. Mol. Catal. A Chem.* **2015**, *396*, 155–163. [CrossRef]

27. Guan, R.; Johnson, I.; Cui, T.; Zhao, T.; Zhao, Z.; Li, X.; Liu, H. Electrodeposition of hydroxyapatite coating on Mg-4.0Zn-1.0Ca-0.6Zr alloy and in vitro evaluation of degradation, hemolysis, and cytotoxicity. *J. Biomed. Mater. Res. Part A* **2012**, *100*, 999–1015. [CrossRef] [PubMed]

28. Shannon, R.D. Revised Effective Ionic Radii and Systematic Studies of Interatomie Distances in Halides and Chaleogenides. *Acta Crystallogr. Sect. A Cryst. Phys. Diffr. Theor. Gen. Crystallogr.* **1976**, *32*, 751–767. [CrossRef]

29. Tadrous, P.J.; Siegel, J.; French, P.M.W.; Shousha, S.; Lalani, E.-N.; Stamp, G.W.H. Fluorescence lifetime imaging of unstained tissues: Early results in human breast cancer. *J. Pathol.* **2003**, *199*, 309–317. [CrossRef] [PubMed]

30. Chen, H.-M.; Chiang, C.-P.; You, C.; Hsiao, T.-C.; Wang, C.-Y. Time-resolved autofluorescence spectroscopy for classifying normal and premalignant oral tissues. *Lasers Surg. Med.* **2005**, *37*, 37–45. [CrossRef]

31. Halestrap, A.P.; Wilson, M.C. The monocarboxylate transporter family-Role and regulation. *IUBMB Life* **2012**, *64*, 109–119. [CrossRef]

32. Esen, E.; Long, F. Aerobic Glycolysis in Osteoblasts Aerobic Glycolysis in Osteoblasts. *Curr. Osteoporos. Rep.* **2015**, *12*, 433–438. [CrossRef]

33. Aditya, N.P.; Chimote, G.; Gunalan, K.; Banerjee, R.; Patankar, S.; Madhusudhan, B. Experimental Parasitology Curcuminoids-loaded liposomes in combination with arteether protects against Plasmodium berghei infection in mice. *Exp. Parasitol.* **2012**, *131*, 292–299. [CrossRef]

34. Li, Z.; Chee, P.L.; Owh, C.; Lakshminarayanan, R.; Loh, X.J. Safe and efficient membrane permeabilizing polymers based on PLLA for antibacterial applications. *RSC Adv.* **2016**, *6*, 28947–28955. [CrossRef]
35. Wei, Z.; Tian, P.; Liu, X.; Zhou, B. In vitro degradation, hemolysis, and cytocompatibility of PEO / PLLA composite coating on biodegradable AZ31 alloy. *J. Biomed. Mater. Res. Part B: Appl. Biomater.* **2015**, *103*, 342–354. [CrossRef] [PubMed]
36. Singh, R.K.; Kannan, S. Synthesis, Structural analysis, Mechanical, antibacterial and Hemolytic activity of Mg2+ and Cu2+ co-substitutions in β-Ca3(PO4)$_2$. *Mater. Sci. Eng. C* **2014**, *45*, 530–538. [CrossRef]

Article

The Influence of Ozonated Olive Oil-Loaded and Copper-Doped Nanohydroxyapatites on Planktonic Forms of Microorganisms

Wojciech Zakrzewski [1], Maciej Dobrzynski [2], Joanna Nowicka [3], Magdalena Pajaczkowska [3], Maria Szymonowicz [1], Sara Targonska [4], Paulina Sobierajska [4], Katarzyna Wiglusz [5], Wojciech Dobrzynski [6], Adam Lubojanski [1], Sebastian Fedorowicz [3], Zbigniew Rybak [1] and Rafal J. Wiglusz [4,*]

[1] Department of Experimental Surgery and Biomaterial Research, Wroclaw Medical University, Bujwida 44, 50-345 Wroclaw, Poland; wojciech.zakrzewski@student.umed.wroc.pl (W.Z.); maria.szymonowicz@umed.wroc.pl (M.S.); adam.lubojanski@student.umed.wroc.pl (A.L.); zbigniew.rybak@umed.wroc.pl (Z.R.)

[2] Department of Conservative Dentistry and Pedodontics, Wroclaw Medical University, Krakowska 26, 50-425 Wroclaw, Poland; maciej.dobrzynski@umed.wroc.pl

[3] Department of Microbiology, Wroclaw Medical University, Chalubinskiego 4, 50-368 Wroclaw, Poland; joanna.nowicka@umed.wroc.pl (J.N.); magdalena.pajaczkowska@umed.wroc.pl (M.P.); sebastian.fedorowicz@student.umed.wroc.pl (S.F.)

[4] Institute of Low Temperature and Structure Research, Polish Academy of Sciences, Okolna 2, 50-422 Wroclaw, Poland; s.targonska@intibs.pl (S.T.); p.sobierajska@intibs.pl (P.S.)

[5] Department of Analytical Chemistry, Wroclaw Medical University, Borowska 211 A, 50-566 Wroclaw, Poland; katarzyna.wiglusz@umed.wroc.pl

[6] Student Scientific Circle at the Department of Dental Materials, School of Medicine with the Division of Dentistry in Zabrze, Medical University of Silesia in Katowice, Akademicki Sq. 17, 41-902 Bytom, Poland; wojt.dobrzynski@wp.pl

* Correspondence: r.wiglusz@intibs.pl; Tel.: +48-71-3954-159; Fax: +48-71-344-10-29

Received: 15 July 2020; Accepted: 29 September 2020; Published: 10 October 2020

Abstract: The research has been carried out with a focus on the assessment of the antimicrobial efficacy of pure nanohydroxyapatite, Cu^{2+}-doped nanohydroxyapatite, ozonated olive oil-loaded nanohydroxyapatite, and Cu^{2+}-doped nanohydroxyapatite, respectively. Their potential antimicrobial activity was investigated against *Streptococcus mutans*, *Lactobacillus rhamnosus*, and *Candida albicans*. Among all tested materials, the highest efficacy was observed in terms of ozonated olive oil. The studies were performed using an Ultraviolet–Visible spectrophotometry (UV-Vis), electron microscopy, and statistical methods, by determining the value of Colony-Forming Units (CFU/mL) and Minimal Inhibitory Concentration (MIC).

Keywords: Cu^{2+} ions; ozonated olive oil; hydroxyapatite; antimicrobial activity; microorganisms

1. Introduction

Nowadays, the application of biomaterials is gaining popularity due to their high versatility. The development of modern medical science is based on the use of biomaterials, such as hydroxyapatite (HAp), to replace damaged hard tissue. Hydroxyapatite is the main inorganic component of bones and teeth, and it is related to the resorption and precipitation processes of calcium phosphates as well as the adsorption and formation of bones, dentine, and cementum [1]. Mainly, it crystallizes in the form of nanoplates or nanorods with an average size of approximately 50 nm × 25 nm × 2 nm [2,3]. The natural HAp is non-stoichiometric and poorly crystalline, and it contains numerous ionic substitutions, e.g., Mg^{2+}, Na^+, K^+, Sr^{2+}, Zn^{2+}, Mn^{2+}, Cu^{2+}, Co_3^{2-}, F^-, and Cl^- [1,4]. Its synthetic form is isostructural and chemically similar to bone apatite and possesses a strong affinity for ion exchange, which causes its high bioactivity [5]. Its biological properties are determined by such parameters as Ca/P molar ratio, the type of ionic dopants in the crystal lattice, or particle size and morphology. Stoichiometric HAp has a typical lattice structure described as $(A_{10}(BO_4)_6C_2)$, where A, B, and C are defined by Ca^{2+}, PO_4^{3-}, and OH^-, respectively, with a calcium-to-phosphate ratio of 1.67 [6]. The HAp is non-immunogenic and non-toxic due to the outstanding bioactivity and biocompatibility. Moreover, synthetic hydroxyapatite has been widely applied as a bone substitute for the reconstruction of bone defects in maxillofacial surgery as well as orthopedics [1,7]. Furthermore, it can be used as a filler for repairing cavities on the enamel surface [8].

In orthopedics, bacterial adhesion on implant surfaces is the most predominant problem of post-surgical infections. Several studies have reported that the Ag^+, Cu^{2+}, and Zn^{2+} ions are essential for preventing or minimizing initial microorganism adhesion [9–11]. Among them, the Cu^{2+} ion occupies a prominent position as an antibacterial agent, because it reveals the highest inhibition of bacteria growth with simultaneous tolerable cytotoxicity for tissue cells, as was reported by Heidenau et al. [10]. The antimicrobial activity of cooper ions can be ascribed by several mechanisms. Under aerobic conditions, the Cu^{2+} ion is proposed to be catalyzed producing hydroxyl radicals via the Fenton and Haber–Weiss reactions [12]. The possible mechanisms of action between the Cu^{2+} ion-containing compounds and the microorganism are based on the structural damage of the cell membrane causing its permeability and finally cell death, the deactivation of proteins by binding metal ions, and the interaction with microbial nucleic acids preventing microbial replication [13,14]. In the presented study, it has been decided to choose 1 mol% Cu^{2+} due to the fact that copper-doped nanohydroxyapatite (nHAp) retains an antimicrobial effect even at low Cu^{2+} content, while its cytotoxicity against normal cells remains low [15]. Chui Ping Ooi et al. showed that the survival ratio of osteoblasts decreased as the Cu^{2+} content increased [16], while Nam et al. [17] confirmed, that Cu^{2+} concentration and contact time do not affect to the phase composition, but affect the crystal size and morphology. Moreover, from the physicochemical point of view, the crystal structure of the hydroxyapatite is stable at this (1 mol%) concentration of dopant. The presence of secondary phases could be observed with an increase of copper ions content in nHAp crystal lattice, as was presented by Sumathi Shanmugam et al. [13].

Cu^{2+} ions have a strong activity against fungi and bacteria [18]. Moreover, the bactericidal effect of metal ions as well as nanoparticles has been attributed to their small size and a high surface-to-volume ratio, allowing close interaction with microbial membranes. Nanoparticles have coated surfaces and can be useful in various medical fields e.g., as cements or coatings in surgery, antimicrobial dressings, and actively targeted biomaterials [19,20].

The use of ozone in dentistry has increased in recent years due to its high oxidative power stimulating the immune response and blood circulation, together with its strong antimicrobial activity [21]. It has been demonstrated to be useful in controlling the physiology of microorganisms in dental plaque [22]. Ozone works synergistically—inducing the modification of intracellular contents and damaging the cytoplasmic membrane of cells [23]. Some medical products such as Ozonosept (see Section 2: Materials and Methods) contain ozone. It is fabricated during the process of ozonation of olive oil. Ozone is kept in the form of stable chemical compounds—ozonides. The ozonides show antibacterial, antifungal, and antiviral activity [24]. The antimicrobial activity of ozonated olive oil is related to the Criegee Mechanism i.e., a slow release of peroxides [25]. When an ozonide contacts with tissue, then carbonyl oxide reacts with water, and hydroxyhydroperoxide is produced. According to the Metrum Cryoflex leaflet, the Ozonosept has confirmed antimicrobial properties against *Staphylococcus aureus*, *Pseudomonas aeruginosa*, *Escherichia coli*, *Propionibacterium acnes*, and *Candida albicans*.

This study aimed to evaluate the selected materials against *Candida albicans*, *Streptococcus mutans*, and *Lactobacillus rhamnosus*. This set of microorganisms was used in our previous paper, Wiglusz et al. [26]. According to the literature, *C. albicans* has strong adhesive properties. Moreover, *S. mutans* and *L. rhamnosus* are related to formation of a subgingival plaque. Moreover, these strains are referential in the case of in vitro studies for biomaterials.

The co-administration of nanoparticles and ozonated olive oil has not been extensively studied against microbial species isolated from persistent endodontic infections. Recent studies [27,28] have shown that the combination of ozonated olive oil and chitosan nanoparticles has a more significant killing effect—it prevents biofilm formation and eradicates resistant endodontic pathogens from root canals. The novelty of this work is its evaluation and comparison of the antimicrobial activity of the proposed materials on their own and in various compositions. Such a comparative study gives the opportunity to reveal the specificity of these materials toward various microorganisms including bacterial strains and pathogen yeast as well as ultimately leading to the better utilization of nanoparticles and ozonated olive oil.

2. Materials and Methods

2.1. Synthesis of Nanocrystalline Hydroxyapatite

The studies were carried out on the following materials: (i) nHAp, (ii) nHAp doped with Cu^{2+} ions, (iii) nHAp with the addition of ozonated olive oil (Ozonosept, Metrum Cryoflex Sp. z o.o., Sp. K., Łomianki, Poland), and (iiii) nHAp doped with Cu^{2+} ions and loaded with ozonated olive oil. The amount of ozone in olive oil was 100 mg/mL, as proven by Magnetic Resonance Spectroscopy (600 Mhz/16 tesla) (Bruker Corporation, Billerica, MA, USA) by the manufacturer.

The nHAp nanocrystals of $Ca_{10}(PO_4)_6(OH)_2$ and $Ca_{9.9}Cu_{0.1}(PO_4)_6(OH)_2$ were synthesized by the wet chemistry method at the Institute of Low Temperature and Structure Research, Wroclaw, Poland. Analytical grade Ca $(NO_3)_2 \cdot 4H_2O$ (99.0%, Alfa Aesar, Haverhill, MA, USA), $NH_4H_2PO_4$ (99.0%, Fluka^{TM}, Honeywell Specialty Chemicals Seelze GmbH., Seelze, Germany), and Cu $(NO_3)_2 \cdot 2.5H_2O$ (98.0–102.0%, Alfa Aesar) were used as the starting materials. The pH was regulated by $NH_3 \cdot H_2O$ (99%, Avantor Performance Materials Poland S.A., Gliwice, Poland). The concentration of Cu^{2+} ions was 1 mol% to the overall molar content of calcium cations. All substrates were dissolved and mixed. The pH of the reaction mixture was adjusted to 10 with an ammonia solution. The reaction was performed at 100 °C for 60 min. The obtained product was washed several times with deionized water and dried at 70 °C for 24 h. The final product was heat-treated at 400 °C for 3 h.

2.2. Characterisation

The apatite crystal structure was confirmed by the X-ray diffraction technique (XRD). The XRD patterns were measured (five times for each samples) by using a PANalytical X'Pert Pro X-ray diffractometer (Malvern Panalytical Ltd., Malvern, UK) equipped with Ni-filtered Cu Kα1 radiation (Kα1 = 1.54060 Å, U = 40 kV, I = 30 mA). The measurements were done in the range of 3–70° (2θ). The Thermo Fisher Scientific Nicolet iS50 FT-IR spectrometer (Waltham, MA, USA) equipped with an Automated Beamsplitter exchange system (iS50 ABX containing DLaTGSKBr detector), which had a built-in all-reflective diamond Attenuated Total Reflectance (ATR) module (iS50 ATR), Thermo Scientific Polaris™ and HeNe laser, was used to record the FT-IR spectra (five times for each samples). The Fourier Transform Infrared (FT-IR) spectra in the mid-IR region (4000–400 cm^{-1}) were measured using the standard KBr pellet method, while in the case of the far-IR region (400–100 cm^{-1}), a Nujol suspension was used. Raman measurements (five times for each sample) were carried out with a Micro-Raman system Renishaw InVia Raman spectrometer equipped with a confocal DM 2500 Leica optical microscope (Wotton-under-Edge, Gloucestershire, UK), a thermoelectrically cooled Charge-Coupled Device CCD was used as a detector of the Raman spectra recorded. An argon laser operating at 831 nm was used. The chemical composition was performed by using an FEI Nova NanoSEM 230 scanning electron microscope (SEM, Hillsboro, OR, USA) with an energy-dispersive X-ray spectrometer (EDAX Genesis XM4). The Scanning Electron Microscopy coupled with Energy Dispersive X-ray Spectroscopy (SEM-EDS) was used for qualitative and quantitative analysis of materials. The spectra were recorded three times for each sample, and the calculated value is an average result.

The tests were carried out on reference strains *Streptococcus mutans* (ATCC 25175), *Lactobacillus rhamnosus* (ATCC 9595), and *Candida albicans* (ATCC 90028).

The aim of the study was to conduct preliminary studies associated with the activity of the tested compounds against different microorganisms in their planktonic forms. The next stage will be related to an evaluation of the activity of the selected compounds against a mature structure of microbial biofilms.

2.3. Spectrophotometric Examination

A suspension of 0.5 McFarland density (1.5×10^8 CFU/mL in case of bacteria and 1.5×10^6 CFU/mL in case of fungi) in liquid medium Sabouraud Broth (Biomaxima) Brain Heart Infusion (BHI) Broth (Biomaxima) and De Man, Rogosa and Sharpe MRS Broth (Biomaxima) were prepared from fresh culture of the analyzed strains for *Candida albicans*, *Streptococcus mutans*, and *Lactobacillus rhamnosus*, respectively. First, 1 mL of the suspension prepared in this way was incubated with nHAp at a concentration of 0.1% and 1% (both pure and with an admixture of Cu^{2+} and ozonated olive oil). According to previous studies including substituted hydroxyapatites with antibacterial properties it has been decided to choose two different concentrations of nHAp (0.1% and 1%) [15–17,28,29]. The samples were incubated at 37 °C (aerobic, anaerobic (GENbag anaer, Biomerieux), and microaerophilic (GENbag microaer, Biomerieux, Warsaw, Poland)) for 4 h and 24 h with shaking. After the incubation period, 100 µL of the suspension was transferred to the appropriate well of a 96-well plate according to the following scheme:

- Growth control (1 mL suspension of microorganisms);
- Sterility control (1 mL medium);
- Compound control (1 mL medium and 0.1% and 1% concentration);
- Test sample (1 mL of microorganism suspension and analyzed compound at a concentration of 0.1% and 1%).

The reading was made on a Biochrom Asys UVM 340 spectrophotometer at 595 nm (Biochrom Ltd., Holliston, MA, USA).

2.4. Determining the Value of the Colony-Forming Units, CFU/mL

Suspensions of 0.5 McFarland density (1.5×10^8 CFU/mL in case of bacteria and 1.5×10^6 CFU/mL in case of fungi) in liquid Sabouraud, BHI, and MRS medium were prepared from fresh culture of the analyzed strains for *C. albicans*, *S. mutans*, and *L. rhamnosus*, respectively. In this way, 1 mL of the prepared suspension was incubated with nHAp at concentrations of 0.1% and 1% (pure, as well as doped with Cu^{2+} ions and an addition of ozonated olive oil). The samples were incubated at 37 °C (aerobic, anaerobic (GENbag anaer, Biomerieux, Warsaw, Poland), and microaerophilic (GENbag microaer, Biomerieux)) for 24 h with shaking. After the incubation period, 100 µL of the suspension was withdrawn, and a series of dilutions were made in geometric progress (10^{-1}–10^{-6}). After plating on a solid medium (appropriate for the strain), the plates were incubated; then, the grown colonies were counted, and the value of the colony-forming units (CFU/mL) was determined. All tested samples were subjected to triplicate procedure.

Together with the test sample, a control test was done, which was a suspension of the analyzed strain. The antimicrobial properties of ozonated olive oil have also been evaluated. The proportions of the ozonated olive oil together with microbial strain were 1 mL of bacterial/fungal suspension and 1 mL of ozonated olive oil.

2.5. Determining the Value of the Minimal Inhibitory Concentration, MIC

First, 100 mL of the medium (appropriate for the strain) was applied to the wells of the 96-well plate. Then, 100 µL of the nHAp suspension (pure and doped with Cu^{2+} ions and an addition of ozonated olive oil) was applied to the appropriate plate rows and diluted geometrically to a concentration range of 9.7–5000 µg/mL. After adding 20 µL of the diluted microorganism culture, the plate was incubated (37 °C; aerobic, anaerobic and microaerophilic). After the incubation period, the minimum inhibitory concentration value was read visually.

2.6. Scanning Electron Microscopy

Suspensions of 0.5 McFarland density (1.5×10^8 CFU/mL in case of bacteria and 1.5×10^6 CFU/mL in case of fungi) in liquid Sabouraud, BHI, and MRS medium were prepared from a fresh culture of the analyzed strains for *C. albicans*, *S.mutans*, and *L. rhamnosus*, respectively. First, 1 mL of the suspension prepared in this way was incubated with nHAp at a concentration of 0.1% and 1% (pure HAp as well as doped with Cu^{2+} ions and the addition of ozonated olive oil). The samples were incubated at 37 °C (aerobic, anaerobic (GENbag anaer, Biomerieux), and microaerophilic (GENbag microaer, Biomerieux)) for 24 h with shaking. After the incubation period, 100 µL of the suspension was transferred to the appropriate well of a 12-well plate, fixed, sprayed with gold, and evaluated in a ZEISS scanning electron microscope model EVO LS15 (Carl Zeiss, Oberkochen, Germany).

2.7. Statistical Methods

For all quantitative features (number of colony-forming units, CFU/mL), their distribution was checked for compliance with the normal distribution. The conformity assessment was carried out with the Shapiro–Wilk test.

Qualitative variables (strains of microorganisms) are presented in the abundance tables (contingency) in the form of abundance (n) and proportion (%). The chi-squared test was used to assess the strength of the relationship between the two variables. In cases where the number expected in one of the tables (2×2) was less than 5, the Fisher's exact test was used. Mean values ($\pm M$) and standard deviations ($\pm SD$) were calculated for all measurable features. The homogeneity of variance was checked by the Bartlett and Levene test.

Analysis of variance (Anova) was used to compare the averages in several groups. Whether the analyzed feature in each of the examined groups had normal distribution and equal variances had been checked earlier. If the probability corresponding to the value of the Snedecor F distribution was lower than the assumed level of significance ($p < 0.05$), then multiple comparison tests (post hoc) were performed to determine which group significantly differs from the others. The Tukey test was used for this purpose.

The Statistica version 12.5 program (StatSoft, Tulsa, OK, USA) was applied for calculations and making charts.

3. Results

3.1. Structural Analysis

Structural analysis of the $Ca_{10}(PO_4)_6(OH)_2$ nanocrystals as well as nHAp doped with 1 mol% Cu^{2+} ions was performed by using the XRD technique. The X-ray diffraction patterns of pure and Cu^{2+}-doped nHAp are presented in Figure 1. As it can be seen, the observed XRD patterns are in good agreement with the reference hexagonal phase of nHAp (no. ICSD-26204) ascribed to the $P6_3/m$ space group [30]. The successful replacement of calcium ions by cooper ions in the crystal structure was confirmed—no additional peaks originating from other phases were observed. The efficient substitution of Ca^{2+} ions by Cu^{2+} ions has been clearly confirmed by shifting the positions of the diffraction peaks. Hydroxyapatite doped with Cu^{2+} ions exhibits a slight shift in the position of the (002) plane (c-plane) and (300) plane (a-plane), [13,28,31]. A shift toward higher 2θ angles was related to the decrease in the cell parameters induced by the substitution of the bigger Ca^{2+} cation (CN_9 = 1.18 Å, Ca^{2+} CN_7 = 1.06 Å, where CN is coordination number) by the smaller Cu^{2+} cation (CN_7 = 0.73 Å) [28,32,33].

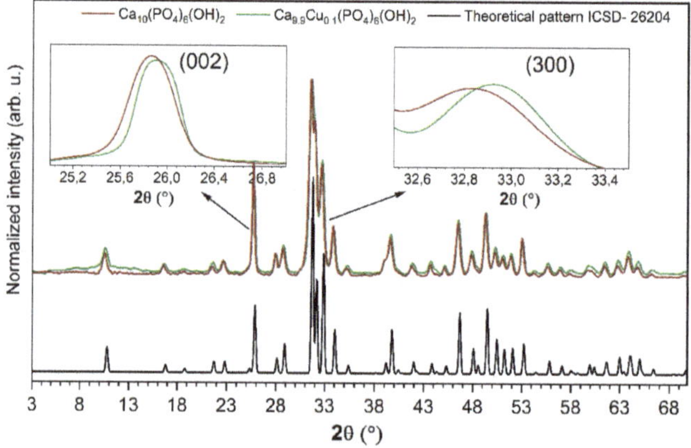

Figure 1. X-ray diffraction patterns of pure nanohydroxyapatite (nHAp) and nHAp doped with 1 mol% Cu^{2+} after heat treatment at 400 °C with the indication of (002) and (300) planes shift induced by doping with Cu^{2+} ions.

Moreover, the X-ray diffraction can be used to determine the presence of the OCP (octacalcium phosphate) phase. The OCP structure can be described as an alternative stacking of "apatite" layers and "hydrated" layers [34]. Most of the OCP reflections within the 2θ range of 10–60° overlapped with those belonging to the hydroxyapatite structure. However, three reflections are specific to OCP in the low 2θ range of 4–10°, at 4.7°, 9.5°, and 9.8° with the relative intensities equal to 100%, 8%, and 8%, respectively [35]. In the case of the studied materials, Bragg peaks at very low 2θ angles, especially the most intense (100) line, were not observed. Meanwhile, the most characteristic diffraction peaks belonging to the hydroxyapatite structure were found at 2θ equal to 31.8°, 32.2°, 32.9°, and 25.9°.

3.2. Infrared Spectra

The infrared spectra of investigated materials are shown in Figure 2. The absorption bands have been ascribed based on literature data [36–38]. The peaks at 1045 cm^{-1} and 1095 cm^{-1} correspond to the antisymmetric triply degenerate stretching vibrations of phosphate groups (PO$_4{}^{3-}$)v_3. The peak at 963 cm^{-1} belongs to the symmetric non-degenerate stretching vibrations of phosphate groups (PO$_4{}^{3-}v_1$), while the modes at 604 cm^{-1} and 570 cm^{-1} identify the triply degenerate vibration (PO$_4{}^{3-}$)v_4. The presence of the absorption band at 634 cm^{-1}, belonging to the librational mode of the –OH group, clearly indicates the nHAp structure. The peak observed at 3571 cm^{-1} is related to the stretching mode of the –OH group. The broad absorption band with a maximum at 3430 cm^{-1} corresponds to the typical vibrations of water molecules.

The ATR-IR absorption spectra (in the range of 4000–400 cm^{-1}) of the ozonated olive oil and nHAp with the addition of ozonated olive oil as well as nHAp doped with Cu^{2+} and loaded with ozonated olive oil are presented in Figure 3. The bands related to ozonated olive oil and hydroxyapatite have been marked with black and red dashed lines, respectively. According to the spectroscopic results, it has been proven that the ozonated olive oil was adsorbed on the obtained compounds. The most typical peak associated with ozonated olive oil, indicating the existence of an ozonide ion, is located at 1104 cm^{-1} and is correlated with the ozonide CO stretching mode [39]. The peak at 3006 cm^{-1} is associated with the C–H stretching vibration of the cis double bond. There are also two intense peaks at 2922 and 2853 cm^{-1}, which correspond to the C–H asymmetric stretching vibrations of both –CH$_2$ and –CH$_3$ groups. The peak at 1742 cm^{-1} demonstrates C=O vibrations. The modes around 1460 and

723 cm^{-1} are assigned to the bending C–H vibration, while the peak at 1160 cm^{-1} is attributed to the C–O bands [40].

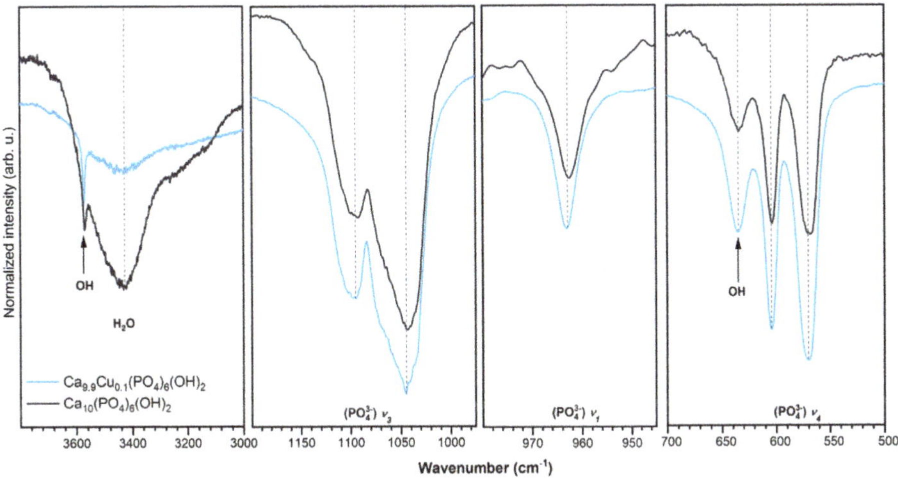

Figure 2. FT-IR spectra of nHAp and nHAp doped with 1 mol% Cu^{2+} with an indication of typical active vibrational bands.

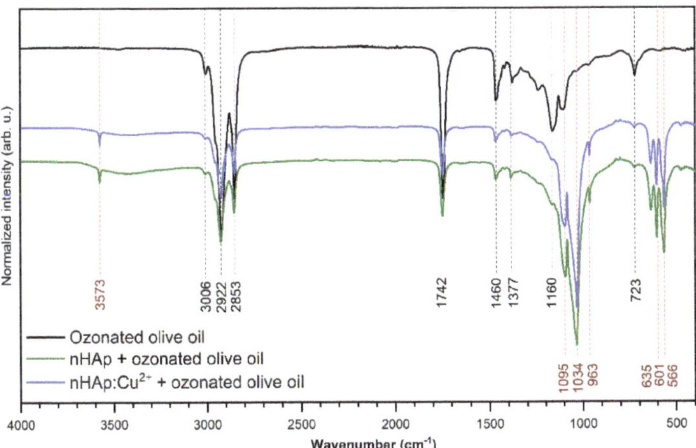

Figure 3. FT-IR spectra of ozonated olive oil as well as nHAp and nHAp doped with 1 mol% Cu^{2+} and loaded with ozonated olive oil.

3.3. Micro-Raman Spectra

The micro-Raman spectra of pure and Cu^{2+}-doped nHAp were recorded and presented in Figure 4. The spectra of the Cu^{2+}-doped nHAp contain four characteristic vibrational transitions of phosphate groups. The maximum of the most intense peak is located at 961 cm^{-1} and is correlated with the symmetric stretching mode of the phosphate groups (PO$_4^{3-}$)ν$_1$. The three overlapping vibration modes at 1075, 1046, and 1028 cm^{-1} are attributed to the asymmetric stretching of (PO$_4^{3-}$)ν$_3$. In the region of the (PO$_4^{3-}$)ν$_2$ bending mode, there are two peaks at 592 and 580 cm^{-1}. The positions of 452 and 430 cm^{-1} are associated with (PO$_4^{3-}$)ν$_4$ bending modes. The analysis of the Raman spectrum related to pure nHAp revealed one distinguishing peak at 961 cm^{-1} associated with (PO$_4^{3-}$)ν$_1$ vibration.

Moreover, the peaks correlated with the ν_2, ν_3, and ν_4 vibrational transitions of phosphate groups were not clearly detected [16,19]. The spectra of the samples loaded with ozonated olive oil reveal additional bands at 1660 cm^{-1} derived from aliphatic unsaturation, at 1442 cm^{-1} associated with the deformation of the CH$_2$ group, and at 1302 cm^{-1} correlated to the twisting of the CH$_2$ group [41].

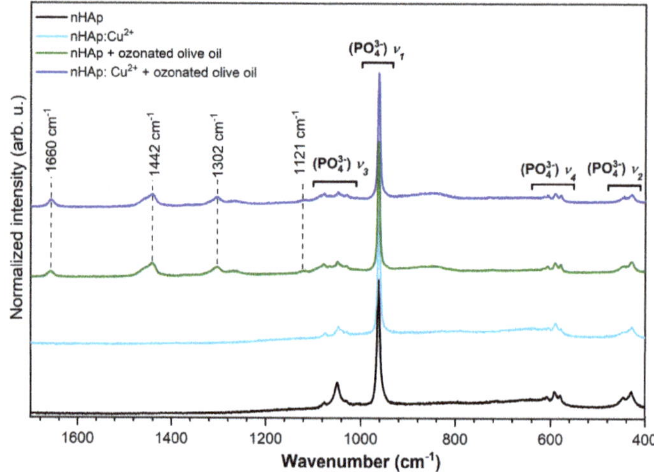

Figure 4. The micro-Raman spectra of nHAp and nHAp doped with 1 mol% Cu^{2+} as well as both materials loaded with ozonated olive oil.

3.4. EDS Analysis

The EDS spectra (Figure 5) recorded for the samples was applied to identify and quantify the elements in the nHAp:Cu^{2+}. The resulting contents of Ca, Cu, and P in the studied material were 28.2 at%, 0.39 at%, and 18.1 at%, respectively. The calculated value of the $(n_{Ca} + n_{Cu})/n_P$ ratio was equal to 1.5—close to the theoretical n_{Ca}/n_P ratio. The calculated concentration of Cu^{2+} ions was equal to 0.1 mol%, which stays in agreement with the theoretical value.

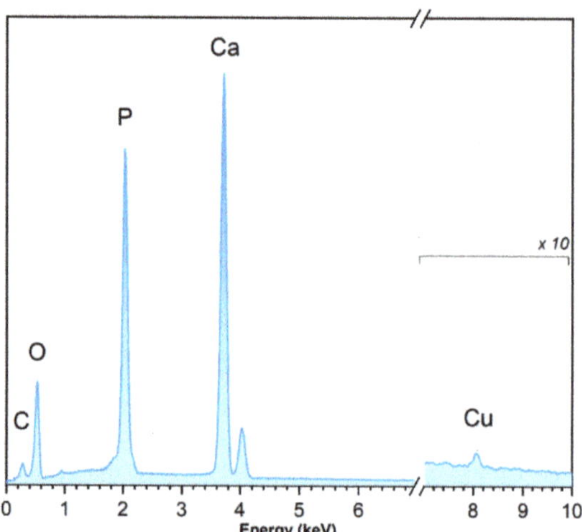

Figure 5. EDS spectrum of nHAp doped with 1 mol% Cu^{2+} after heat treatment at 400 °C.

3.5. Spectrophotometric Indication

The range of reduction of viable *C. albicans* cells after 4 h incubation under modified nHAp ranged from 46 to 87% for a concentration of 0.1% and 44 to 84% for a concentration of 1%. In the case of *S. mutans* and *L. rhamnosus*, it was respectively 47–97% and 77–100%, as well as 38–52% and 24–54% for concentrations of 0.1% and 1%.

After 24 h incubation, the degree of reduction was in the range of 21–89% and 50–52% in the case of *C. albicans*, and 41–75% and 19–81% for *S. mutans*. In the case of *L. rhamnosus*, these values ranged from 11 to 62% and 5 to 88% for 0.1% and 1%, respectively. The results are shown in Tables S1 and S2.

3.6. Determining the Value of the Minimal Inhibitory Concentration

The Minimum Inhibitory Concentration (MIC) value of undoped and Cu^{2+}-doped nHAp loaded with ozonated olive oil against *C. albicans* was >5000 and 5000 µg/mL, respectively. In the case of *S. mutans* MIC, undoped and Cu^{2+}-doped nHAp was >5000 µg/mL, and in the case of Cu^{2+}-doped nHAp loaded with ozonated olive oil, it reached 5000 µg/mL. In the case of *L. rhamnosus*, the value of the minimum growth inhibitory concentration for all analyzed compounds exceeded the value of 5000 µg/mL. The results are shown in Table 1.

Table 1. Values of Minimal Inhibitory Concentration (MIC) of nHAp for analyzed strains.

Species	MIC (µg/ mL)			
	I	II	III	IV
Candida albicans	>5000	5000	5000	5000
Streptococcus mutans	>5000	>5000	5000	5000
Lactobacillus rhamnosus	>5000	>5000	>5000	>5000

I—nHAp; II—Cu^{2+}-doped nHAp; III—nHAp with the addition of ozonated olive oil; IV—nHAp doped with Cu^{2+} and loaded with ozonated olive oil.

3.7. Determining the Value of Colony-Forming Units, CFU/mL

In the case of *C. albicans*, the range of Colony-Forming Units was 1×10^5–3.20×10^8 for 0.1% and 1.4×10^5–1.79×10^8 for 1%. In the case of *L. rhamnosus* and *S. mutans*, the range was 6.07×10^8–1.67×10^9, 1×10^3–3.36×10^9, 0–8.10×10^7, and 0–6.13×10^7, respectively. The results are presented in Tables S3–S5 and Figures 6–12. It has been shown that the HAp doped with Cu^{2+} ions and the ozonated olive oil-loaded HAp doped with Cu^{2+} ions have caused a readable reduction of CFU/mL for *Candida albicans* and *Lactobacillus rhamnosus*. Moreover, the highest value of CFU/mL has been observed when pure nHAp was used against the *Candida albicans* as well as Cu^{2+} ions-doped HAp against *Lactobacillus rhamnosus*. Furthermore, the highest CFU/mL value has also been observed in the case of pure nHAp, while the lowest occurred with both nHAp with the addition of ozonated olive oil and nHAp doped with Cu^{2+} ions and loaded with ozonated olive oil against the *Streptococcus mutans* strain. On the other hand, the CFU/mL had been evaluated by the lowest values in contact with 1 mol% Cu^{2+}-doped nHAp, 1 mol% Cu^{2+}-doped nHAp with the addition of ozonated olive oil, and 1 mol% Cu^{2+}-doped nHAp loaded with ozonated olive oil against *Streptococcus mutans* and *Candida albicans*.

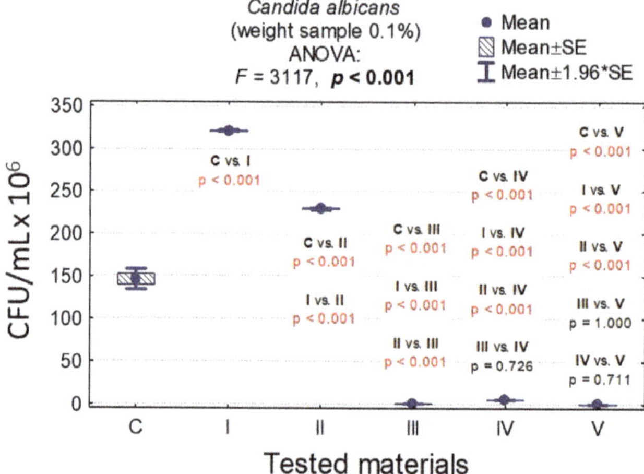

Figure 6. Average Colony-Forming Units (CFU)/mL values and standard deviation (M ± SD) for *Candida albicans* after contact with tested materials (C—control group, I—nHAp, II—Cu^{2+}-doped nHAp, III—nHAp with the addition of ozonated olive oil, IV—nHAp doped with Cu^{2+} and loaded with ozonated olive oil, V—ozonated olive oil) for weights of 0.1% and results of the analysis of variance (ANOVA) and multiple comparisons by a post-hoc test. Incubation took 24 h.

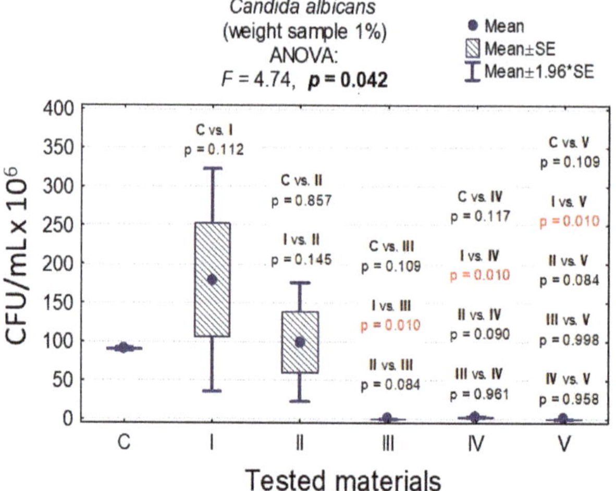

Figure 7. Average CFU/mL values and standard deviation (M ± SD) for *Candida albicans* after contact with the tested materials (C—growth control, I—nHAp, II—Cu^{2+}-doped nHAp, III—nHAp with the addition of ozonated olive oil, IV—nHAp doped with Cu^{2+} and loaded with ozonated olive oil, V—ozonated olive oil) for weights of 1% and results of the analysis of variance (ANOVA) and multiple comparisons by a post-hoc test. Incubation took 24 h.

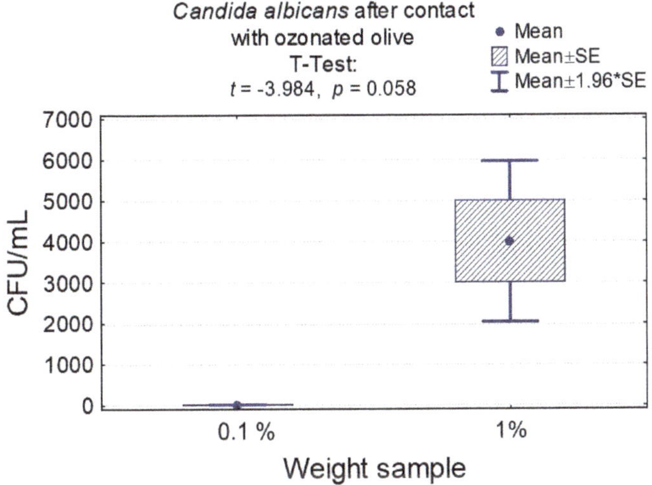

Figure 8. Average CFU/mL value and standard deviation (M ± SD) for *Candida albicans* after temporary contact with olive ozone in groups differing in weight and significance of test result. Incubation took 24 h.

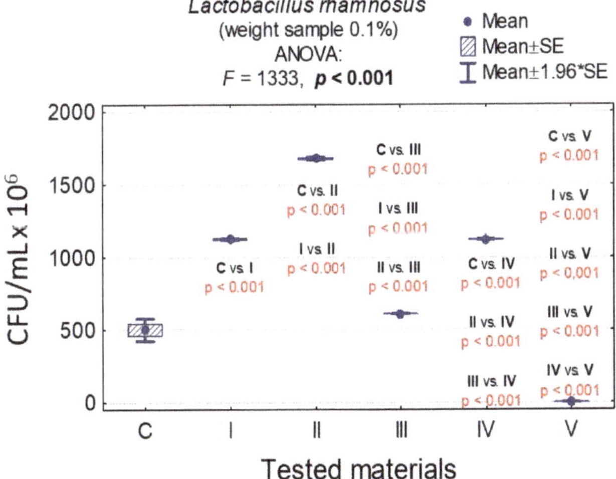

Figure 9. Average CFU/mL values and standard deviation (M ± SD) for *Lactobacillus rhamnosus* after temporary contact with the tested materials (C—growth control, I—nHAp, II—Cu^{2+}-doped nHAp, III—nHAp with the addition of ozonated olive oil, IV—nHAp doped with Cu^{2+} and loaded with ozonated olive oil, V—ozonated olive oil) for weights of 0.1% and results of analysis of variance (ANOVA) and multiple comparisons by a post-hoc test. Incubation took 24 h.

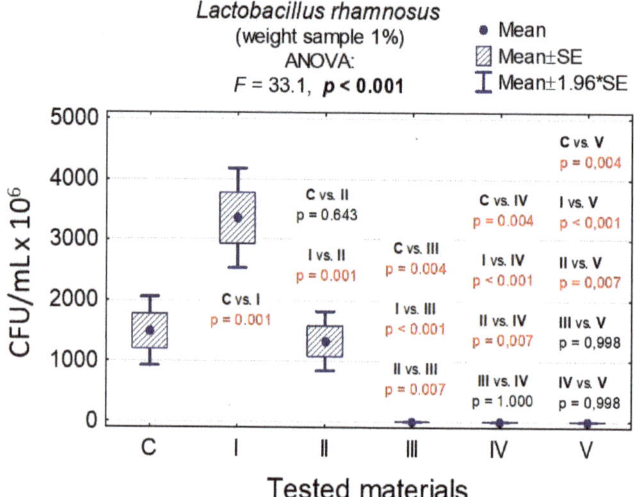

Figure 10. Average CFU/mL values and standard deviation (M ± SD) for *Lactobacillus rhamnosus* after temporary contact with the tested materials (C—growth control, I—nHAp, II—Cu^{2+}-doped nHAp, III—nHAp with the addition of ozonated olive oil, IV—nHAp doped with Cu^{2+} and loaded with ozonated olive oil, V—ozonated olive oil) for weights of 1% and results of analysis of variance (ANOVA) and multiple comparisons by a post-hoc test. Incubation took 24 h.

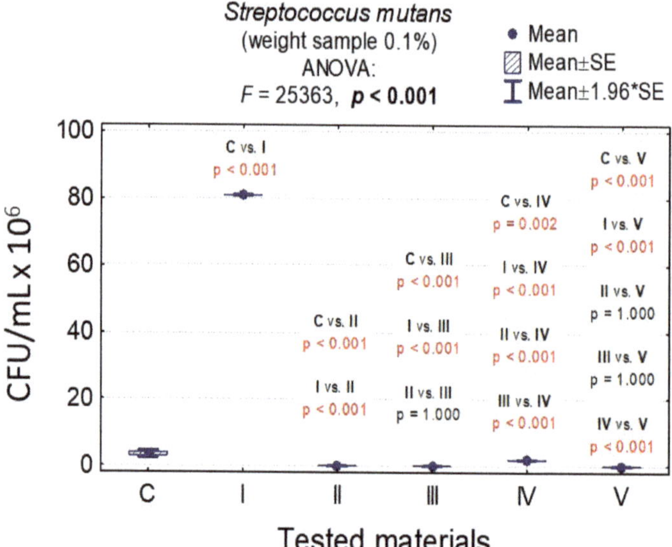

Figure 11. Average CFU/mL values and standard deviation (M ± SD) for *Streptococcus mutans* after temporary contact with the tested materials (C—growth control, I—nHAp, II—Cu^{2+}-doped nHAp, III—nHAp with the addition of ozonated olive oil, IV—nHAp doped with Cu^{2+} ions and loaded with ozonated olive oil, V—ozonated olive oil) for weights of 0.1% and results of the analysis of variance (ANOVA) and multiple comparisons by a post-hoc test. Incubation took 24 h.

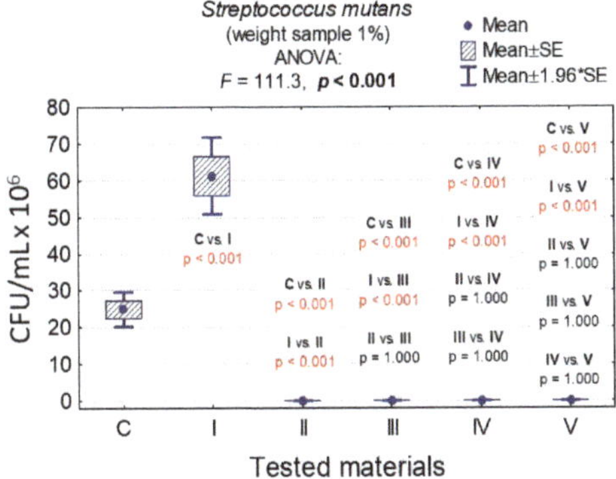

Figure 12. Average CFU/mL values and standard deviation (M ± SD) for *Streptococcus mutans* after temporary contact with the tested materials (C—growth control, I—nHAp, II—Cu^{2+}-doped nHAp, III—nHAp with the addition of ozonated olive oil, IV—nHAp doped with Cu^{2+} and loaded with ozonated olive oil, V—ozonated olive oil) for weights of 1% and results of the analysis of variance (ANOVA) and multiple comparisons by a post-hoc test. Incubation took 24 h.

In the case of a sample with 0.1% nHAp, we observed statistically significant differences in the growth of the *C. albicans* colony for the growth control and all other materials ($p < 0.001$), pure nHAp (material I) and other materials ($p < 0.001$), and nHAp doped with Cu^{2+} ions and other materials ($p < 0.001$). The differences between materials containing ozonated olive oil (materials III, IV and V) turned out to be insignificant ($p > 0.05$); see Table S3 and Figure 6.

In the case of a sample weight of 1% nHAp, it was observed that there were significant statistical differences in the growth of *C. albicans* for pure nHAp (material I) and materials containing ozonated olive oil (materials III, IV, V); see Table S4 and Figure 7.

Ozonated olive oil turned out to be the best material for *C. albicans* (smallest colony growth). After two measurements, the sample weight did not have a significant statistical impact on the colony concentration (Tables S3–S5 and Figure 8).

In the case of sample weights of 0.1% nHAp, it was shown that there were statistically significant differences in the growth of *L. rhamnosus* colonies for all materials except pure nHAp and nHAp doped with Cu^{2+} ions and loaded with ozonated olive oil ($p = 0.900$). It was proven that the ozone olive has been the best material (Table S3 and Figure 9).

In the case of sample weights of 1% nHAp, it was observed that there were statistically significant differences in the growth of *L. rhamnosus* colonies for all materials except those containing ozonated olive oil (III, IV, and V), confirming that the ozonated olive oil has been the best material (Table S4 and Figure 10).

In the case of a weight sample of 0.1% nHAp, the slight increase of *S. mutans* colonies under the influence of Cu^{2+}-doped nHAp, nHAp with the addition of ozonated olive oil and pure ozonated olive oil was observed, and the differences between them were insignificant ($p = 1.000$); see Table S3 and Figure 11.

In the case of a weight sample of 1% nHAp, the slight growth of *S. mutans* colonies under the influence of Cu^{2+}-doped nHAp, nHAp with the addition of ozonated olive oil, as well as nHAp doped with Cu^{2+} and loaded with the addition of ozonated olive oil and pure ozonated olive oil was observed, and differences between them were insignificant ($p = 1.000$); see Table S4 and Figure 12.

In all cases, the results of the variance analysis were statistically significant (for example, in Figure 12: $F = 111.3$ and $p < 0.001$. It means that the difference between at least one of the pairs of materials was significant). Moreover, the post-hoc tests were carried out Least Significant Difference (LSD) showing significant differences between the pairs of materials presented in Figure 12 ($p < 0.001$).

The difference between growth control and nHAp was significant ($p < 0.001$), and that was not the case in comparison with Cu^{2+}-doped nHAp and nHAp with the addition of ozonated olive oil ($p = 1.000$).

In the case of a sample with a concentration of 0.1% nHAp, statistically significant differences were observed in the growth of *L. rhamnosus* colonies for all materials, except for nHAp and nHAp doped with Cu^{2+} ions and loaded with ozonated olive oil ($p = 0.900$). Moreover, regarding the sample weights of 1.0% nHAp, it was observed that there were statistically significant differences in the growth of *L. rhamnosus* colonies for all materials, except for those containing ozonated olive oil (III, IV, and IV). Furthermore, for the weight sample of 0.1% nHAp, the slight increase of *S. mutans* colonies under influence of Cu^{2+}-doped nHAp, nHAp with the addition of the ozonated olive oil, and the pure ozonated olive oil was shown, and the differences between those materials were insignificant ($p = 1.000$).

Additionally, for the sample with a concentration of 1% nHAp, the growth of *S. mutans* colonies under the influence of materials II, III, IV, and V was the smallest, and the differences between those materials were insignificant ($p = 1.000$).

Taking into account the above, the result of the analysis of variance was statistically significant in all cases ($F = 111.2$ and $p < 0.001$; meaning that the difference was significant at least between one of the pairs of materials). Therefore, post-hoc tests were carried out (LSD) with the purpose of showing between which pairs of materials the differences were significant. For example, the difference between growth control and nHAp was significant ($p < 0.001$), but that was not the case in comparison with Cu^{2+}-doped nHAp and nHAp with the addition of ozonated olive oil ($p = 1.000$); see Tables S6–S8.

3.8. Scanning Electron Microscopy

In all cases, growth control is more abundant than when microorganisms are in contact with nHAp and Cu^{2+}-doped nHAp (Figure 13). The action of nHAp doped with Cu^{2+} ions definitely affects the reduction of the number of bacteria (*Lactobacillus rhamnosus*, *Streptococcus mutans*). Changes are visible on *Candida albicans* surface—it is not smooth and oval, but more angular instead. Additionally, it looks as if it was dehydrated.

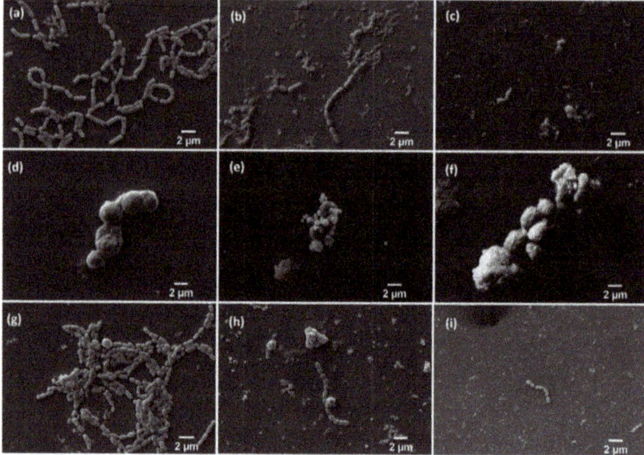

Figure 13. Scanning electron microscopy: (**a**) Growth control of *Lactobacillus rhamnosus*; (**b**) *Lactobacillus rhamnosus* with nHAp; (**c**) *Lactobacillus rhamnosus* with Cu^{2+}-doped nHAp; (**d**) Growth control of *Candida albicans*; (**e**) *Candida albicans* with nHAp; (**f**) *Candida albicans* with Cu^{2+}-doped nHAp; (**g**) Growth control of *Streptococcus mutans*; (**h**) *Streptococcus mutans* with nHAp; (**i**) *Streptococcus mutans* with Cu^{2+}-doped nHAp; Mag = 10,000×.

4. Discussion

Hydroxyapatite is an inorganic component of bones and teeth, acting as a scaffold and giving them mechanical properties. In addition, it stimulates bone development in small bone defects and can be used as a coating material for implants [42]. In a medium containing the nHAp particle, bacteria may adhere to the solid and co-aggregate. It has been reported that biofilms that could be formed at 15 min after inoculation on nHAp disks consist mainly of single, non-aggregated cells [43].

The relationship between the size of nHAp and bacterial adhesion is crucial because of an effect on slower plaque formation. The nHAp scale allows having enhanced physical and chemical properties, including increased wettability, roughness, and adsorption of proteins [44]. Non-aggregated and condensed nHAp particles adsorb on bacterial surfaces in vitro [45]. They interact with bacteria and thus reduce their adhesion. Severin AV et al. [46] investigated the interaction of the nHAp nanocrystals with *Staphylococcus aureus* bacteria. Moreover, the nHAp nanocrystallites adhere to the surface of bacteria, significantly reducing their ability to form colonies.

The activity of nHAp against the planktonic form of the chosne microorganisms has been evaluated in this study. The degree of reduction of viable cells by pure nHAp (% viable microorganisms) was minimal after using pure nHAp. However, it is worth noting that with increasing exposure time as well as a concentration of pure nHAp (material I), the reduction of *C. albicans* initially increased at a concentration of 0.1% nHAp from 1% after 4 h to 14% after 24 h and at a concentration of 1% nHAp from 9% after 4 h to 37% after 24 h. In the case of other microorganisms, no relationship was found. Whereas, after doping or mixing nHAp with other reagents, a significant difference in the reduction of cells viability was observed in groups of nHAp doped with Cu^{2+} ions, loaded with ozonated olive oil, or both. After 4 hours incubation, the reduction range of viable *C. albicans* cells for pure nHAp was 1% at a concentration of 0.1% and 9% for a concentration of 1%.

The nHAp appeared to enhance biofilm formation by increasing glucosyltransferase transcription, which resulted in an increase in the production of insoluble glucans. Since the demineralization of nHAp in enamel caused by acids, including those produced by bacteria in the plaque, is important in the development of dental caries, nHAp is used in toothpastes for the remineralization of enamel [47]. In the current study, we have examined the effect of nHAp on the growth of *S. mutans* in two different media and a nutrient-rich environment. While exploring the extinction value measured as the percentage of cell reduction, it is observed that the antimicrobial activity of nHAp combined with Cu^{2+} ions is higher than that of the pure nHAp, although it shows less efficacy than nHAp with the addition of ozonated olive oil.

The nHAp doped with Cu^{2+} ions is distinct from other materials such as nHAp, nHAp with ozonated olive oil, as well as nHAp with Cu^{2+} ions with ozonated olive oil and ozone olive alone, for the reason that it is more efficient after 4 hours than after 24 h in both concentrations of nHAp (0.1% and 1%), unlike the other materials. It significantly reduced microorganism growth.

With regard to nHAp containing Cu^{2+} ions and ozonated olive oil, which is nHAp doped with Cu^{2+} ions and loaded with olive oil, it is apparently more effective than other materials in reducing the number of microorganisms. The only exception is *L. rhamnosus* after 4 h incubation, compared to nHAp with ozonated olive oil. nHAp at a concentration of 0.1% is the most efficient in the reduction of *S. mutans* in nHAp doped with Cu^{2+} ions, similarly to nHAp with ozonated olive oil. A similar effect is obtained regarding nHAp doped with Cu^{2+} ions, where the lack of growth can be observed. Once again, the highest efficiency was detected in the case of *S. mutans*, which is the same in materials with ozonated olive oil. nHAp doped with Cu^{2+} ions caused a statistically insignificant reduction of growth in the *L. rhamnosus* colony and *C. albicans*. Other studies also prove the antimicrobial properties of Cu^{2+} ions [48,49].

Studies prove a wide spectrum of activity that increases the field of application of nHAp doped with Cu^{2+} ions. It is suggested that it would have a wide spectrum of future use, for instance, in orthopaedics and bone prosthesis or dentistry and teeth implants [13,50].

Virgin olive oil has an abundance of unsaturated fatty acids due to an especially high content of oleic acid, which is a monounsaturated omega−9 (*n*−9) fatty acid. The process of ozonization oxides unsaturated bonds with the simultaneous formation of peroxidic substances, which results in the higher antifungal and antibacterial potential of ozonated substances [51].

The ozonated olive oil oil and its antimicrobial properties had been tested in some clinical trials concerning different diseases, for instance, in the gynecological or dentistry field. In terms of vulvovaginal candidiasis—that is an inflammation of vagina often caused by *Candida* species such as *Candida albicans* or NCAC (*non-Candida albicans Candida* species) such as *Candida glabrata*, *Candida krusei*, *Candida parapsilosis*, *Candida tropicalis*—it was found that the ozonated olive oil is equally effective as clotrimazole in a significant reduction of symptoms. Moreover, it also led to a negative culture growth, and there were no significant differences in terms of reducing itching and leukorrhea, although the burning sensation was diminished more effectively by clotrimazole [52].

Other studies provide an evaluation of ozonated olive oil in the treatment of chronic periodontitis, which is a disease caused mainly by bacteria, resulting in an inflammatory process of gums and tissues of the oral cavity. Although the clinical study was limited in terms of patient numbers, it was found that ozonated olive oil was effective not only as an adjunct to scaling and root planning but also in monotherapy. Nevertheless, when used as an adjunctive therapy, the patients' dentinal hypersensitivity had significantly risen [53]. Those findings suggest a high level of efficacy of ozonated olive oil as an antimicrobial agent, which coincides with the presented results. Its antifungal and antibacterial potential had been tested in laboratory conditions as well as in the clinical trials [54].

According to the obtained results on the efficacy of the acquired materials containing ozonated olive oil (nHAp mixed with ozonated olive oil, mixed with olive oil and doped with Cu^{2+} ions, pure ozonated olive oil, respectively), their addition to microbiological samples causes the highest percentage reduction when compared to materials without ozonated olive oil (I and II). When analyzing the given data, there is an obvious conclusion that the highest efficacy of olive is presented in samples of *S. mutans*, especially in nHAp of a concentration of 0.1% for nHAp with the addition of ozonated olive oil, and also in 1% nHAp concentration for nHAp doped with Cu^{2+} ions and ozonated olive oil. The total growth reduction of *S. mutans* strains is visible in nHAp doped with Cu^{2+} ions and ozonated olive oil 4 h after application, as nearly 100% efficacy can be observed; however, in nHAp with ozonated olive oil, it occurs after 24 h. A complete reduction of *S. mutans* cells was observed after incubating this bacterial strain in the presence of ozonated olive oil. This study allows deducing that the *S. mutans* percentage reduction can reach the highest value among the measured microorganisms when it is mixed with nHAp doped with Cu^{2+} ions and loaded with ozonated olive oil. In the case of ozonated olive oil, the lowest percentage reduction value of *S. mutans* is present 24 h after the use of 0.1% nHAp with the addition of ozonated olive. Additionally, olive efficacy was also measured by the number of colonies concentration (thousand CFU/mL) concerning *S. mutans*. The value was the same—0.018 ± 0.008 when the species were mixed with selected materials at a weight of 1% (nHAp with ozonated olive oil, nHAp with Cu^{2+} and ozonated olive oil, ozonated olive oil). Interestingly, when the CFU/mL of *S. mutans* was compared after contact with selected materials at a weight of 0.1%, each material concerning olive lacked growth except for nHAp doped with Cu^{2+} ions and ozonated olive oil, which had $2 \times 10^6 \pm 0.01$.

The lowest percentage reduction value of ozonated olive oil inflicts *L. rhamnosus* when it is mixed with 0.1% nHAp with the addition of ozonated olive oil. Even when mixed with pure ozonated olive oil, *L. rhamnosus* represents the lowest percentage reduction among every tested microorganism strain. The CFU/mL of *L. rhamnosus* had the same value as *S. mutans* in most cases (0.018 ± 0.008), but this was only after contact with pure ozonated olive oil at 0.1% sample weight. On the other hand, for nHAp with ozonated olive oil as well as nHAp doped with Cu^{2+} ions and ozonated olive oil, the values were higher, ranging from 6.07×10^8 to 1.12×10^9.

The colony-forming unit of *L. rhamnosus* is less optimistic after contact with ozonated olive oil materials (nHAp with ozonated olive oil, nHAp with Cu^{2+} with ozonated olive oil, and ozonated olive oil alone). At a sample weight of 1%, values ranged between 1×10^3 and 1×10^6.

By analyzing data acquired by the authors in the study, it can be concluded that the percentage reduction of the measured microorganisms and CFU/mL. have the most positive results for materials containing ozonated olive oil (nHAp with the addition of ozonated olive oil, nHAp doped with Cu^{2+} ions and loaded with ozonated olive oil, ozonated olive oil). The percentage reduction in value after contact with pure ozonated olive oil is constantly high, ranging between 71 and 99%, and the highest results are obtained after 24 h of cultivation, ranging between 94 and 99% (see Tables S9 and S10). Similar results concerning a high efficacy of ozonated olive oil against these microorganisms were obtained by several other authors [54,55]. The CFU/mL of olive itself, with materials at a weight of 0.1%, has the best efficacy for each measured microorganism. On the other hand, at a sample weight of 1% nHAp, the same result is acquired only in the case of *S. mutans*, while the worst result is obtained for *L. rhamnosus*—$1 \times 10^6 \pm 10.14$.

Nanohydroxyapatite doped with Cu^{2+} ions or ozonated olive oil may limit the oral microbial activity. Moreover, it is successfully applied in dentine hypersensitivity treatment and maxillofacial bones regeneration. Doping it with Cu^{2+} ions or ozonated olive oil may enhance its antimicrobial characteristics and limit the postoperative complications.

5. Conclusions

Calcium hydroxyapatite and calcium hydroxyapatite doped with Cu^{2+} ions have been successfully synthetized by using the wet chemistry method. The obtained nanocrystals have been functionalized with ozonated olive oil, which resulted in the formation of a novel medical composition. In vitro screening of microorganism strains according to their activity in various experimental conditions may be a valuable method that could precede clinical efficacy treatments. The highest efficacy was observed for ozonated olive oil and nanocrystalline Cu^{2+}-doped nHAp followed by undoped nHAp. The obtained results indicate that 10 times higher concentrations of pure as well as doped nHAp have better antimicrobial activity. *Streptococcus mutans* had the highest sensitivity, while *Lactobacillus rhamnosus* had the lowest. Pure ozonated olive oil had the highest antimicrobial efficacy.

In the next stage of the study, the authors plan to evaluate the activity of the nHAp against the biofilms in order to complete a whole view of the bacteria cell properties as well as the effectiveness of the antimicrobial compounds.

Supplementary Materials: The following are available online at http://www.mdpi.com/2079-4991/10/10/1997/s1. Table S1: The effect of pure and doped nHAp on the analysed strains (optical density (OD) 595 nm); Table S2: Inhibition of the growth of microbial cells, %; Table S3: Average CFU/mL and standard deviation (M ± SD) for the tested strains with selected materials at a concentration of 0.1% nHAp; Table S4: Average CFU/mL and standard deviation (M ± SD) for the tested strains with selected materials with concentration of 1% nHAp; Table S5: Evaluation of the antimicrobial properties of ozonated olive oil (Inhibition % equals 99.99% for each strain); Table S6: Results of comparisons of *Candida albicans* colonies after contact with tested materials (C—growth control, I—nHAp, II—Cu^{2+}-doped nHAp, III—nHAp with the addition of ozonated olive oil, IV—nHAp doped with Cu^{2+} ions and loaded with ozonated olive oil, V—ozonated olive oil) for weight samples of 0.1 and 1% (M—average value); Table S7: Results of comparisons of *Lactobacillus rhamnosus* colonies after contact with tested materials (C—growth control, I—nHAp, II—Cu^{2+}-doped nHAp, III—nHAp with the addition of ozonated olive oil, IV—nHAp doped with Cu^{2+} ions and loaded with ozonated olive oil, V—ozonated olive oil) for weight samples of 0.1 and 1% (M—average value); Table S8: Results of comparisons of *Streptococcus mutans* colonies after contact with tested materials (C—growth control, I—nHAp, II—Cu^{2+}-doped nHAp, III—nHAp with the addition of ozonated olive oil, IV—nHAp doped with Cu^{2+} ions and loaded with ozonated olive oil, V—ozonated olive oil) for weight samples of 0.1 and 1% (M—average value); Table S9: Influence of 0.1% hydroxyapatite on microbial strains: *C. albicans*, *L. rhamnosus* i *S. mutans* (Inhibition growth%); Table S10: Influence of 1% hydroxyapatite on microbial strains: *C. albicans*, *L. rhamnosus* i *S. mutans* (Inhibition growth%).

Author Contributions: R.J.W. and M.D. conceived and designed the experiments in addition to analyzing all data; W.Z., S.T., P.S. and K.W. designed the experiments in addition to analyzing data; Z.R., W.D., A.L., S.F. and M.S. contributed reagents/materials/analysis tools and analyzing data; R.J.W., M.D., M.S. and Z.R. participated in funding acquisition; J.N. and M.P. designed and performed the microbiological experiments and analyzed data, and all authors contributed to the writing of the paper. All authors have read and agreed to the published version of the manuscript.

Funding: Financial support of the National Science Centre in the course of realization of the Projects "Preparation and characterization of biocomposites based on nanoapatites for theranostics" (no. UMO-2015/19/B/ST5/01330); "Elaboration and characteristics of biocomposites with anti-virulent and anti-bacterial properties against *Pseudomonas aeruginosa*" (no. UMO-2016/21/B/NZ6/01157) and "Preparation and investigation of multifunctional biomaterials based on nanoapatites for possible application in bone tumor treatment" (no. UMO-2017/27/N/ST5/02976).

Acknowledgments: The work was created as part of the cooperation of the Student Scientific Association of Microbiologists and the Scientific Association of Experimental Surgery and Biomaterials Research in Wroclaw Medical University, Wroclaw, Poland.

Conflicts of Interest: The authors declare no conflict of interest.

References

1. Liu, Q.; Huang, S.; Matinlinna, J.P.; Chen, Z.; Pan, H. Insight into biological apatite: Physiochemical properties and preparation approaches. *Biomed Res. Int.* **2013**, *2013*. [CrossRef] [PubMed]

2. Song, W.; Ge, S. Application of Antimicrobial Nanoparticles in Dentistry. *Molecules* **2019**, *24*, 1033. [CrossRef]

3. Sadat-Shojai, M.; Khorasani, M.T.; Dinpanah-Khoshdargi, E.; Jamshidi, A. Synthesis methods for nanosized hydroxyapatite with diverse structures. *Acta Biomater.* **2013**, *9*, 7591–7621. [CrossRef]

4. Boanini, E.; Gazzano, M.; Bigi, A. Ionic substitutions in calcium phosphates synthesized at low temperature. *Acta Biomater.* **2010**, *6*, 1882–1894. [CrossRef] [PubMed]

5. Al-Hazmi, F.; Alnowaiser, F.; Al-Ghamdi, A.A.; Al-Ghamdi, A.A.; Aly, M.M.; Al-Tuwirqi, R.M.; El-Tantawy, F. A new large—Scale synthesis of magnesium oxide nanowires: Structural and antibacterial properties. *Superlattices Microstruct.* **2012**, *52*, 200–209. [CrossRef]

6. Habibah, T.U.; Salisbury, H.G. *Hydroxyapatite Dental Material*; StatPearls Publishing: Treasure Island, FL, USA, 2020.

7. Pepla, E.; Besharat, L.K.; Palaia, G.; Tenore, G.; Migliau, G. Nano-hydroxyapatite and its applications in preventive, restorative and regenerative dentistry: A review of literature. *Ann. Stomatol.* **2014**, *5*, 108–114. [CrossRef]

8. Gupta, T.; Nagaraja, S.; Mathew, S.; Narayana, I.; Madhu, K.; Dinesh, K. Effect of desensitization using bioactive glass, hydroxyapatite, and diode laser on the shear bond strength of resin composites measured at different time intervals: An In vitro Study. *Contemp. Clin. Dent.* **2017**, *8*, 244.

9. Zawisza, K.; Sobierajska, P.; Nowak, N.; Kedziora, A.; Korzekwa, K.; Pozniak, B.; Tikhomirov, M.; Miller, J.; Mrowczynska, L.; Wiglusz, R.J. Preparation and preliminary evaluation of bio-nanocomposites based on hydroxyapatites with antibacterial properties against anaerobic bacteria. *Mater. Sci. Eng. C* **2020**, *106*, 110295. [CrossRef]

10. Heidenau, F.; Mittelmeier, W.; Detsch, R.; Haenle, M.; Stenzel, F.; Ziegler, G.; Gollwitzer, H. A novel antibacterial titania coating: Metal ion toxicity and in vitro surface colonization. *J. Mater. Sci. Mater. Med.* **2005**, *16*, 883–888. [CrossRef]

11. Chung, R.J.; Hsieh, M.F.; Huang, C.W.; Perng, L.H.; Wen, H.W.; Chin, T.S. Antimicrobial effects and human gingival biocompatibility of hydroxyapatite sol-gel coatings. *J. Biomed. Mater. Res.-Part B Appl. Biomater.* **2006**, *76*, 169–178. [CrossRef]

12. Liochev, S.I.; Fridovich, I. The Haber-Weiss cycle—70 years later: An alternative view. *Redox Rep.* **2002**, *7*, 55–57. [CrossRef] [PubMed]

13. Shanmugam, S.; Gopal, B. Copper substituted hydroxyapatite and fluorapatite: Synthesis, characterization and antimicrobial properties. *Ceram. Int.* **2014**, *40*, 15655–15662. [CrossRef]

14. Stanić, V.; Dimitrijević, S.; Antić-Stanković, J.; Mitrić, M.; Jokić, B.; Plećaš, I.B.; Raičević, S. Synthesis, characterization and antimicrobial activity of copper and zinc-doped hydroxyapatite nanopowders. *Appl. Surf. Sci.* **2010**, *256*, 6083–6089. [CrossRef]

15. Yang, L.; Perez-Amodio, S.; Barrère-de Groot, F.Y.F.; Everts, V.; van Blitterswijk, C.A.; Habibovic, P. The effects of inorganic additives to calcium phosphate on in vitro behavior of osteoblasts and osteoclasts. *Biomaterials* **2010**, *31*, 2976–2989. [CrossRef]

16. Li, Y.; Ho, J.; Ooi, C.P. Antibacterial efficacy and cytotoxicity studies of copper (II) and titanium (IV) substituted hydroxyapatite nanoparticles. *Mater. Sci. Eng. C* **2010**, *30*, 1137–1144. [CrossRef]

17. Nam, P.T.; Thom, N.T.; Phuong, N.T.; Xuyen, N.T.; Hai, N.S.; Anh, N.T.; Dung, P.T.; Thanh, D.T.M. Synthesis, characterization and antimicrobial activity of copper doped hydroxyapatite. *Vietnam J. Chem.* **2018**, *56*, 672–678. [CrossRef]

18. DeAlba-Montero, I.; Guajardo-Pacheco, J.; Morales-Sánchez, E.; Araujo-Martínez, R.; Loredo-Becerra, G.M.; Martínez-Castañón, G.-A.; Ruiz, F.; Compeán Jasso, M.E. Antimicrobial Properties of Copper Nanoparticles and Amino Acid Chelated Copper Nanoparticles Produced by Using a Soya Extract. *Bioinorg. Chem. Appl.* **2017**, *2017*, 1–6. [CrossRef]

19. Bobo, D.; Robinson, K.J.; Islam, J.; Thurecht, K.J.; Corrie, S.R. Nanoparticle-Based Medicines: A Review of FDA-Approved Materials and Clinical Trials to Date. *Pharm. Res.* **2016**, *33*, 2373–2387. [CrossRef]

20. Rajendran, N.K.; Kumar, S.S.D.; Houreld, N.N.; Abrahamse, H. A review on nanoparticle based treatment for wound healing. *J. Drug Deliv. Sci. Technol.* **2018**, *44*, 421–430. [CrossRef]

21. Bocci, V.A. Scientific and Medical Aspects of Ozone Therapy. State of the Art. *Arch. Med. Res.* **2006**, *37*, 425–435. [CrossRef] [PubMed]

22. Saini, R. Ozone therapy in dentistry: A strategic review. *J. Nat. Sci. Biol. Med.* **2011**, *2*, 151. [CrossRef] [PubMed]

23. Stübinger, S.; Sader, R.; Filippi, A. The use of ozone in dentistry and maxillofacial surgery: A review. *Quintessence Int.* **2006**, *37*, 353–359. [PubMed]

24. Ugazio, E.; Tullio, V.; Binello, A.; Tagliapietra, S.; Dosio, F. Ozonated oils as antimicrobial systems in topical applications. Their characterization, current applications, and advances in improved delivery techniques. *Molecules* **2020**, *25*, 334. [CrossRef]

25. Comparison of the Antibacterial Activity of an Ozonated Oil with Chlorhexidine Digluconate and Povidone-Iodine. A Disk Diffusion Test—PubMed. Available online: https://pubmed.ncbi.nlm.nih.gov/23912871/ (accessed on 23 September 2020).

26. Dobrzynski, M.; Pajaczkowska, M.; Nowicka, J.; Jaworski, A.; Kosior, P.; Szymonowicz, M.; Kuropka, P.; Rybak, Z.; Bogucki, Z.A.; Filipiak, J.; et al. Study of Surface Structure Changes for Selected Ceramics Used in the CAD/CAM System on the Degree of Microbial Colonization, in Vitro Tests. *Biomed. Res. Int.* **2019**, *2019*. [CrossRef] [PubMed]

27. Elshinawy, M.I.; Al-Madboly, L.A.; Ghoneim, W.M.; El-Deeb, N.M. Synergistic Effect of Newly Introduced Root Canal Medicaments; Ozonated Olive Oil and Chitosan Nanoparticles, Against Persistent Endodontic Pathogens. *Front. Microbiol.* **2018**, *9*, 1371. [CrossRef]

28. Huang, Y.; Zhang, X.; Mao, H.; Li, T.; Zhao, R.; Yan, Y.; Pang, X. Osteoblastic cell responses and antibacterial efficacy of Cu/Zn co-substituted hydroxyapatite coatings on pure titanium using electrodeposition method. *RSC Adv.* **2015**, *5*, 17076–17086. [CrossRef]

29. Kolmas, J.; Groszyk, E.; Kwiatkowska-Rózycka, D. Substituted hydroxyapatites with antibacterial properties. *Biomed Res. Int.* **2014**, *2014*. [CrossRef]

30. Sudarsanan, K.; Young, R.A. Significant precision in crystal structural details. Holly Springs hydroxyapatite. *Acta Crystallogr. Sect. B Struct. Crystallogr. Cryst. Chem.* **1969**, *25*, 1534–1543. [CrossRef]

31. Hiromoto, S.; Itoh, S.; Noda, N.; Yamazaki, T.; Katayama, H.; Akashi, T. Osteoclast and osteoblast responsive carbonate apatite coatings for biodegradable magnesium alloys. *Sci. Technol. Adv. Mater.* **2020**, *21*, 346–358. [CrossRef]

32. Shannon, R.D. Revised Effective Ionic Radii and Systematic Studies of Interatomie Distances in Halides and Chaleogenides. *Acta Cryst.* **1976**, *A32*, 751–767. [CrossRef]

33. Sobierajska, P.; Wiglusz, R.J. Influence of Li+ ions on the physicochemical properties of nanocrystalline calcium-strontium hydroxyapatite doped with Eu3+ ions. *New J. Chem.* **2019**, *43*, 14908–14916. [CrossRef]

34. Drouet, C. Apatite formation: Why it may not work as planned, and how to conclusively identify apatite compounds. *Biomed Res. Int.* **2013**, *2013*. [CrossRef] [PubMed]

35. Robin, M.; Von Euw, S.; Renaudin, G.; Gomes, S.; Krafft, J.M.; Nassif, N.; Azaïs, T.; Costentin, G. Insights into OCP identification and quantification in the context of apatite biomineralization. *CrystEngComm* **2020**, *22*, 2728–2742. [CrossRef]

36. Guerra-López, J.R.; Echeverría, G.A.; Güida, J.A.; Viña, R.; Punte, G. Synthetic hydroxyapatites doped with Zn(II) studied by X-ray diffraction, infrared, Raman and thermal analysis. *J. Phys. Chem. Solids* **2015**, *81*, 57–65. [CrossRef]

37. Zawisza, K.; Strzep, A.; Wiglusz, R.J. Influence of annealing temperature on the spectroscopic properties of hydroxyapatite analogues doped with Eu3+. *New J. Chem.* **2017**, *41*, 9990–9999. [CrossRef]

38. Targonska, S.; Szyszka, K.; Rewak-Soroczynska, J.; Wiglusz, R.J. A new approach to spectroscopic and structural studies of the nano-sized silicate-substituted hydroxyapatite doped with Eu3+ ions. *Dalt. Trans.* **2019**, *48*, 8303–8316. [CrossRef]

39. Tığlı Aydın, R.S.; Kazancı, F. Synthesis and Characterization of Ozonated Oil Nanoemulsions. *JAOCS J. Am. Oil Chem. Soc.* **2018**, *95*, 1385–1398. [CrossRef]

40. Penel, G.; Leroy, G.; Rey, C.; Sombret, B.; Huvenne, J.P.; Bres, E. Infrared and Raman microspectrometry study of fluor-fluor-hydroxy and hydroxy-apatite powders. *J. Mater. Sci. Mater. Med.* **1997**, *8*, 271–276. [CrossRef]

41. Gamberini, M.C.; Baraldi, C.; Freguglia, G.; Baraldi, P. Spectral analysis of pharmaceutical formulations prepared according to ancient recipes in comparison with old museum remains. *Anal. Bioanal. Chem.* **2011**, *401*, 1839–1846. [CrossRef]

42. Szymonowicz, M.; Korczynski, M.; Dobrzynski, M.; Zawisza, K.; Mikulewicz, M.; Karuga-Kuzniewska, E.; Zywicka, B.; Rybak, Z.; Wiglusz, R.J. Cytotoxicity evaluation of high-temperature annealed nanohydroxyapatite in contact with fibroblast cells. *Materials* **2017**, *10*, 590. [CrossRef]

43. Venegas, S.C.; Palacios, J.M.; Apella, M.C.; Morando, P.J.; Blesa, M.A. Calcium modulates interactions between bacteria and hydroxyapatite. *J. Dent. Res.* **2006**, *85*, 1124–1128. [CrossRef] [PubMed]

44. Zakrzewski, W.; Dobrzynski, M.; Rybak, Z.; Szymonowicz, M.; Wiglusz, R.J. Selected Nanomaterials' Application Enhanced with the Use of Stem Cells in Acceleration of Alveolar Bone Regeneration during Augmentation Process. *Nanomaterials* **2020**, *10*, 1216. [CrossRef] [PubMed]

45. Hannig, C.; Hannig, M. Natural enamel wear—A physiological source of hydroxylapatite nanoparticles for biofilm management and tooth repair? *Med. Hypotheses* **2010**, *74*, 670–672. [CrossRef] [PubMed]

46. Severin, A.V.; Mazina, S.E.; Melikhov, I.V. Physicochemical aspects of the antiseptic action of nanohydroxyapatite. *Biophysics* **2009**, *54*, 701–705. [CrossRef]

47. Huang, S.; Gao, S.; Cheng, L.; Yu, H. Remineralization Potential of Nano-Hydroxyapatite on Initial Enamel Lesions: An in vitro Study. *Caries Res.* **2011**, *45*, 460–468. [CrossRef]

48. Grass, G.; Rensing, C.; Solioz, M. Metallic copper as an antimicrobial surface. *Appl. Environ. Microbiol.* **2011**, *77*, 1541–1547. [CrossRef]

49. Gross, T.M.; Lahiri, J.; Golas, A.; Luo, J.; Verrier, F.; Kurzejewski, J.L.; Baker, D.E.; Wang, J.; Novak, P.F.; Snyder, M.J. Copper-containing glass ceramic with high antimicrobial efficacy. *Nat. Commun.* **2019**, *10*, 1–8. [CrossRef]

50. Vincent, M.; Duval, R.E.; Hartemann, P.; Engels-Deutsch, M. Contact killing and antimicrobial properties of copper. *J. Appl. Microbiol.* **2018**, *124*, 1032–1046. [CrossRef]

51. Díaz, M.F.; Hernández, R.; Martínez, G.; Vidal, G.; Gómez, M.; Fernández, H.; Garcés, R. Comparative study of ozonized olive oil and ozonized sunflower oil. *J. Braz. Chem. Soc.* **2006**, *17*, 403–407. [CrossRef]

52. Tara, F.; Zand-Kargar, Z.; Rajabi, O.; Berenji, F.; Akhlaghi, F.; Shakeri, M.T.; Azizi, H. The Effects of Ozonated Olive Oil and Clotrimazole Cream for Treatment of Vulvovaginal Candidiasis. *Altern. Ther. Health Med.* **2016**, *22*, 44–49. [CrossRef]

53. Patel, P.V.; Patel, A.; Kumar, S.; Holmes, J.C. Effect of subgingival application of topical ozonated olive oil in the treatment of chronic periodontitis: A randomized, controlled, double blind, clinical and microbiological study. *Minerva Stomatol.* **2012**, *61*, 381–398. [PubMed]

54. Pietrocola, G.; Ceci, M.; Preda, F.; Poggio, C.; Colombo, M. Evaluation of the antibacterial activity of a new ozonized olive oil against oral and periodontal pathogens. *J. Clin. Exp. Dent.* **2018**, *10*, e1103. [CrossRef] [PubMed]

55. Tara, F.; Zand-Kargar, Z.; Rajabi, O.; Berenji, F.; Azizi, H. Comparing the effect of ozonated olive oil to clotrimazole cream in the treatment of vulvovaginal candidiasis. *BMC Complement. Altern. Med.* **2012**, *12*, P196. [CrossRef]

 *nanomaterials*

Article

Investigation of Physicochemical Properties of the Structurally Modified Nanosized Silicate-Substituted Hydroxyapatite Co-Doped with Eu^{3+} and Sr^{2+} Ions

Sara Targonska * and Rafal J. Wiglusz *

Institute of Low Temperature and Structure Research, Polish Academy of Sciences, Okolna 2, 50-422 Wroclaw, Poland
* Correspondence: s.targonska@intibs.pl (S.T.); r.wiglusz@intibs.pl (R.J.W.); Tel.: +48-071-3954-159 (R.J.W.)

Abstract: In this paper, a series of structurally modified silicate-substituted apatite co-doped with Sr^{2+} and Eu^{3+} ions were synthesized by a microwave-assisted hydrothermal method. The concentration of Sr^{2+} ions was set at 2 mol% and Eu^{3+} ions were established in the range of 0.5–2 mol% in a molar ratio of calcium ion amount. The XRD (X-ray powder diffraction) technique and infrared (FT-IR) spectroscopy were used to characterize the obtained materials. The Kröger–Vink notation was used to explain the possible charge compensation mechanism. Moreover, the study of the spectroscopic properties (emission, emission excitation and emission kinetics) of the obtained materials as a function of optically active ions and annealing temperature was carried out. The luminescence behavior of Eu^{3+} ions in the apatite matrix was verified by the Judd–Ofelt (J-O) theory and discussed in detail. The temperature-dependent emission spectra were recorded for the representative materials. Furthermore, the International Commission on Illumination (CIE) chromaticity coordinates and correlated color temperature were determined by the obtained results.

Keywords: spectroscopy; nanocrystallites; silicate-substituted hydroxyapatite; Eu^{3+} and Sr^{2+} ion co-doping; microwave-assisted hydrothermal method

Citation: Targonska, S.; Wiglusz, R.J. Investigation of Physicochemical Properties of the Structurally Modified Nanosized Silicate-Substituted Hydroxyapatite Co-Doped with Eu^{3+} and Sr^{2+} Ions. *Nanomaterials* **2021**, *11*, 27. https://doi.org/10.3390/nano11010027

Received: 18 November 2020
Accepted: 16 December 2020
Published: 24 December 2020

Publisher's Note: MDPI stays neutral with regard to jurisdictional clai-ms in published maps and institutio-nal affiliations.

1. Introduction

Nanotechnology has exerted a considerable impact on a vast number of scientific fields in the last decade. Studies on the preparation and characterization of apatite materials have been influenced by this. Nanoapatites are the focus of great research interest due to their biocompatible and nontoxic properties to encourage bone and tissue filling. The structure of the apatite allows for various modifications, providing the opportunity to create a material with intentional and targeted properties. Moreover, two unequal calcium positions can be substituted by ions with +1, +2 or +3 charges, such as Sr^{2+}, Ba^{2+}, K^+, Na^+, Mn^{2+}, Li^+, Mg^{2+} as well as lanthanide ions, etc. [1–3].

There is particular interest in the characterization of nanoapatite materials doped with optically active ions. Consequently, it is possible to successfully replace Ca^{2+} ion by Eu^{3+} [1,4,5], Ce^{3+} [6], Tb^{3+} [7], Dy^{3+} [8,9], Nd^{3+} [9,10], Sm^{3+} [8,9] ions to obtain materials with characteristic emission in red [1,4], green [11,12], violet [6] as well as blue [6] spectral regions. Recently, the apatite matrix has been the focus of attention and investigated in terms of white light emission. Promising materials are co-doped with Dy^{3+}, Li^+ and Eu^{3+} [13] or La^{3+}, Dy^{3+} and Sr^{3+} ions [8]. The vast number of possibilities and favorable results encourage the detailed investigation of a variety of apatite modifications.

Apatite-based structures can tolerate numerous ionic substitutions in order to improve their properties for medical application. Recent research has shown that the combination of silica and strontium co-doping may improve their properties for medical application [14]. In vivo observations as well as in vitro studies have exposed the beneficial effects of using silica-based materials for bone treatment. It has been shown that silica promotes prolyl

171

hydroxylase, stimulates the enzyme involved in collagen synthesis and participates in the proliferation and differentiation of bone mesenchymal stem cells and osteoblasts [15,16]. The synergy of great luminescence properties with bioactivity may result in obtaining a new group of specific materials dedicated to bioimaging and regeneration.

In this paper, we show the physicochemical characterization of silicate-substituted hydroxyapatite co-doped with Sr^{2+} and Eu^{3+} ions. Considerable attention has been paid to the luminescence properties, including emission and excitation spectra depending on Eu^{3+} ion concentration and heat-treating temperature as well as the influence of ambient temperature.

2. Materials and Methods

2.1. Synthesis of the Co-Doped Materials

Synthesizing of silicate-substituted hydroxyapatite co-doped with Eu^{3+} and Sr^{2+} ions involved a hydrothermal process. As substrates, the following were used: $Ca(NO_3)_2 \cdot 4H_2O$ (99.0–103.0% Alfa Aesar, Haverhill, MA, USA), $(NH_4)_2HPO_4$ (>99.0% Acros Organics, Schwerte, Germany), Eu_2O_3 (99.99% Alfa Aesar, Haverhill, MA, USA), $Sr(NO_3)_2$ (99.0% min Alfa Aesar, Haverhill, MA, USA) and tetraethyl orthosilicate TEOS (>99% Alfa Aesar, Haverhill, MA, USA). The concentration of strontium ions for all obtained materials was fixed to 2 mol% in a ratio of calcium ion molar content. Moreover, the concentration of optical active Eu^{3+} ions was set to 0.5; 1.0 and 2.0 mol% in a ratio to the calcium ion molar content. A stoichiometric number of substrates were dissolved separately in deionization water (see Table S1). A stoichiometric amount of Eu_2O_3 was digested in an excess of HNO_3 (65% suprapure Merck KGaA, Darmstadt, Germany) to generate water-soluble $Eu(NO_3)_3 \cdot xH_2O$. Afterwards, all starting substrates were added and mixed into a Teflon vessel. The ammonia solution ($NH_3 \cdot H_2O$ 25% Avantor, Poland) was used to obtain a pH level of around 10. The hydrothermal process was conducted in a microwave reactor (ERTEC MV 02-02, Wrocław, Poland) for 90 min at elevated temperature (250 °C) and under autogenous pressure (42–45 bar). The achieved materials were centrifuged, cleaned by deionization water several times and dried for 24 h. Then, the obtained materials were heat-treated in the temperature range of 400–600 °C for 3 h, increased in steps of 3.3 °C/min.

2.2. Physical–Chemical Characterization

X-ray powder diffraction studies were carried out using a PANalytical X'Pert Pro X-ray diffractometer (Malvern Panalytical Ltd., Malvern, UK) equipped with Ni-filtered Cu $K\alpha_1$ radiation ($K\alpha_1$ = 1.54060 Å, U = 40 kV, I = 30 mA) in the 2θ range of 10°–70°. XRD patterns were analyzed by Match! software version 3.7.0.124.

The surface morphology and the element mapping were assessed by a FEI Nova NanoSEM 230 scanning electron microscope (SEM, Hillsboro, OR, USA) equipped with EDS spectrometer (EDAX GenesisXM4) and operating at an acceleration voltage in the range 3.0–15.0 kV and spots at 2.5–3.0 were observed. EDX analysis was carried out to confirm the chemical formula.

The Thermo Scientific Nicolet iS50 FT-IR spectrometer (Waltham, MA, USA) equipped with an Automated Beamsplitter exchange system (iS50 ABX containing DLaTGS KBr detector), built-in all-reflective diamond ATR module (iS50 ATR), Thermo Scientific Polaris™, was used to record the Fourier-transformed infrared spectra. As an infrared radiation source, we used the HeNe laser. FT-IR spectra of the powders were recorded in KBr pellets at 295 K temperature in the middle infrared range, from 4000 to 500 cm^{-1}, with a spectral resolution of 2 cm^{-1}.

2.3. Spectroscopy Properties

The emission, excitation emission spectra and luminescence kinetics were recorded by an FLS980 Fluorescence Spectrometer (Edinburgh Instruments, Kirkton Campus, UK) from Edinburgh Instruments equipped with a 450 W Xenon lamp and a Hamamatsu

R928P photomultiplier. The hydroxyapatite powders were placed into a quartz tube. The excitation of 300 mm focal length monochromator was in Czerny–Turner configuration. All spectra were corrected during measurement according to the characteristics of the intensity of the excitation source. All spectra were recorded at room temperature. The spectral resolution of the excitation and emission spectra was 0.1 nm. Excitation spectra were recorded, monitoring the maximum of the emission at 618 nm that relates to the $^5D_0 \rightarrow {}^7F_2$ transition, and emission spectra were recorded upon excitation wavelength at 394 nm. Before analysis, the emission spectra were normalized to the $^5D_0 \rightarrow {}^7F_1$ magnetic transition. The luminescence kinetics profiles were recorded according to the $^5D_0 \rightarrow {}^7F_2$ electric dipole transition.

Temperature-dependent emission spectra were recorded using the laser diode ($\lambda_{exc} = 375$ nm), and as an optical detector, we used the Hamamatsu PMA-12 photonic multichannel analyzer (Hamamatsu Photonics K.K., Hamamatsu City, Japan). The presented emission spectra are the average result of 15 measurements with an exposure time of 500 ms.

3. Results and Discussion

3.1. X-ray Diffraction

The formation of the silicate-substituted hydroxyapatite crystalline nanopowders was investigated by the XRD measurements and is shown in Figure 1 as a function of the concentration of optically active ions as well as heat-treating process. All samples prepared via the hydrothermal method have shown detectable crystallinity for the entire range of sintering temperatures (400–600 °C for 3 h). The presence of the single phase of the final products was confirmed by the reference standard of hexagonal strontium-substituted hydroxyapatite ICSD-75518 [17]. No other phase was detected in the studied powders, indicating that dopant ions were completely dissolved in the silicate-substituted host lattice.

Structural refinement was carried out using the Maud software version 2.93 [18,19] and was based on apatite crystals with a hexagonal structure using better approximations, as well as indexing of the crystallographic information file (CIF). The formation of the hexagonal phase, as well as the successful incorporation of Eu^{3+} and Sr^{2+} ions into the apatite lattice, was confirmed by the results. The average grain sizes of silicate-substituted hydroxyapatite nanopowders were in the range of 16 to 56 nm (see Figure S2 and Table S2). The representative SEM image is presented in Figure S1a. The effect of dopant ion substitution is confirmed by EDS measurements (see Figure S1b).

The most intense diffraction peaks corresponding to hydroxyapatite structures are located at 25.9° (002), 31.7° (211), 32.2° (112), 32.9° (300) and 34.0° (202), assigned to the crystallographic planes in brackets. In the obtained materials, Eu^{3+} and Sr^{2+} ions replaced Ca^{2+} ions. In the apatite host lattice, Ca^{2+} ions are located in two different sites with various chemical and structural environments, Ca(1) and Ca(2) sites with C_3 and C_S symmetry, respectively. The Ca(1) site is surrounded by nine oxygen atoms coming from $PO_4{}^{3-}$ groups, which formed a tricapped trigonal prism with formula CaO_9. The Ca(2) site is an irregular polyhedron with formula CaO_6OH formed by six oxygen atoms from $PO_4{}^{3-}$ and one hydroxyl group [3,20]. The difference between the ionic radii of trivalent europium and divalent strontium ions permit the occupation of two possible crystallographic positions of Ca^{2+} ions in the apatite host lattice (Ca^{2+} (CN_9) = 1.18 Å, Eu^{3+} (CN_9) = 1.12 Å, Sr^{2+} (CN_9)$-$1.31 Å (Ca(1) site); Ca^{2+} (CN_7) = 1.06 Å and Eu^{3+} (CN_7) = 1.01 Å, Sr^{2+} (CN_7)$-$1.21 Å (Ca(2) site)) [5,21].

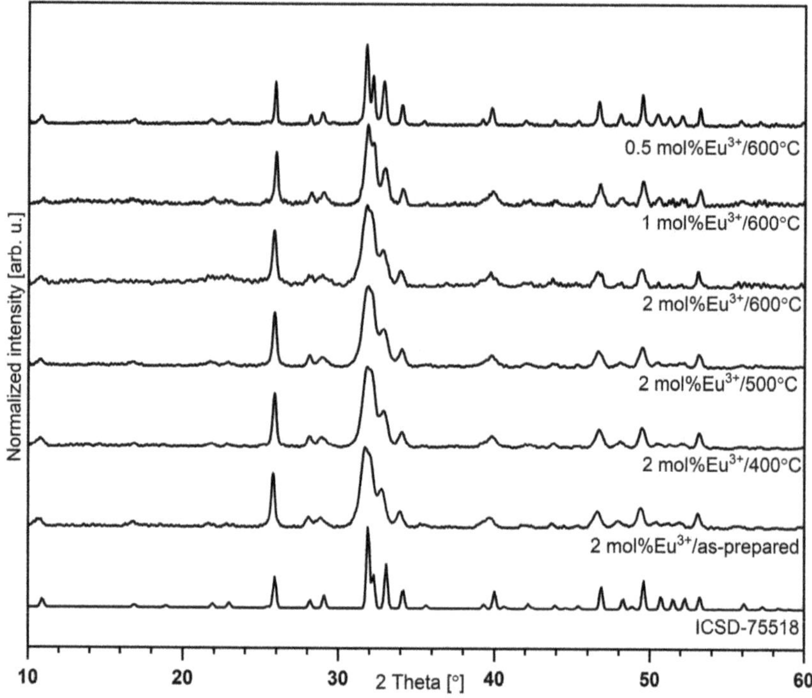

Figure 1. X-ray diffraction pattern of silicate-substituted hydroxyapatite co-doped with 2 mol% Sr^{2+} and x mol% Eu^{3+} ions.

3.2. Kröger–Vink Notation

The cationic vacancies formed by replacing Ca^{2+} ions with Eu^{3+} ions with higher charge could be balanced by SiO_4^{4-} ions substituted into the PO_4^{3-} position in the silicate-substituted apatite matrix. The occupation of divalent calcium ion sites by trivalent europium ions could be described according to the Kröger–Vink notation by the charge compensation phenomenon. According to this theory, the total charge in the material should be compensated by the creation of relatively positive or negative charge. The following processes may be observed:

A double negative vacancy on the Ca^{2+} position (V''_{Ca}) is created by the substitution of divalent calcium ions by trivalent europium ions (Equation (1)):

$$Eu_2O_3 + 3Ca^*_{Ca} \rightarrow 2Eu^{\cdot}_{Ca} + 3CaO + V''_{Ca} \tag{1}$$

The substitution of divalent calcium ions by trivalent rare earth ions could be explained by the creation of interstitial oxygen O''_i with double relative negative charge. The mechanism could be described as follows (Equation (2)):

$$Eu_2O_3 + 2Ca^*_{Ca} \rightarrow 2Eu^{\cdot}_{Ca} + 2CaO + O''_i \tag{2}$$

Eu^{3+} ions first replace into the Ca(1) site and this preference changes with increasing Eu^{3+} concentration in favor of the Ca(2) site. In case of substitution into the Ca(2) site, where calcium(II) ions are surrounded with one hydroxyl group and six oxygen atoms from PO_4^{3-}, these hydroxyl groups could participate in the charge compensation mechanism, expressed as (Equation (3)):

$$Eu_2O_3 + 2Ca^*_{Ca(2)} + 2OH^*_{OH} \rightarrow 2Eu^{\cdot}_{Ca(2)} + 2O'_{OH} + 2CaO + H_2O \tag{3}$$

In the case of the obtained materials, the substitution of the PO_4^{3-} group by the more negative SiO_4^{4-} group could create a negative charge on the PO_4^{3-} position and two positive charge vacancies (V^-_{Ca}) on the hydroxyl group position. The mechanism can be expressed as follows (Equation (4)):

$$2SiO_2 + 2PO_4^*{}_{PO4} + 2OH^*{}_{OH} \rightarrow 2SiO_{4'PO4} + 2V\cdot_{OH} + P_2O_5 + H_2O \tag{4}$$

In the present study, the charge compensation mechanism could be described as a combination of Equations (1)–(4). Equation (5) combines the creation of negative and positive vacancy, because of Ca^{2+} substitution by Eu^{3+} and PO_4^{3-} substitution by SiO_4^{4-}, respectively.

$$2SiO_2 + Eu_2O_3 + 2Ca^*{}_{Ca} + 2PO_4^*{}_{PO4} \rightarrow 2Eu\cdot_{Ca} + 2SiO_{4'PO4} + 2\,CaO + P_2O_5 \tag{5}$$

3.3. Infrared Spectra

To confirm the presence of phosphate, silicate and hydroxyl groups, the infrared spectra were measured and are presented in Figure 2. According to previous reports, characteristic peaks are ascribed to the compound of hydroxyapatite [3,22,23]. The triply degenerated antisymmetric stretching vibration $v_3(PO_4^{3-})$ of phosphate groups is observed at 1101.1 and 1049.1 cm^{-1}. At 966.4 cm^{-1}, lines are detected which can be described as non-degenerated symmetric stretching bands $v_1(PO_4^{3-})$ vibrations. Strong absorption bands associated with the $v_4(PO_4^{3-})$ triply degenerated vibrations are located at 566.2 and 604.8 cm^{-1}. Two bands related to the stretching and bending modes of OH^- groups are observed at 3571.3 and at 604.5 cm^{-1}, respectively. These bands clearly confirm the presence of hydroxyl groups in the crystal structure. The broad peak between 3600 and 3200 cm^{-1} belongs to H_2O vibration. Peaks assigned to $(SiO_4)^{4-}$ vibrational modes and Si–O–Si stretch modes are observed at 890 and 478 cm^{-1}, respectively. It should be pointed out that there are similarly located vibrational modes of the $(PO_4)^{3-}$ and the silicate groups in the hydroxyapatite matrix, causing some interpretation problems. The Si–O symmetric stretching mode is located at 945 cm^{-1} and the weak peak corresponding to the P–O symmetric stretching mode is located at 962 cm^{-1} [23].

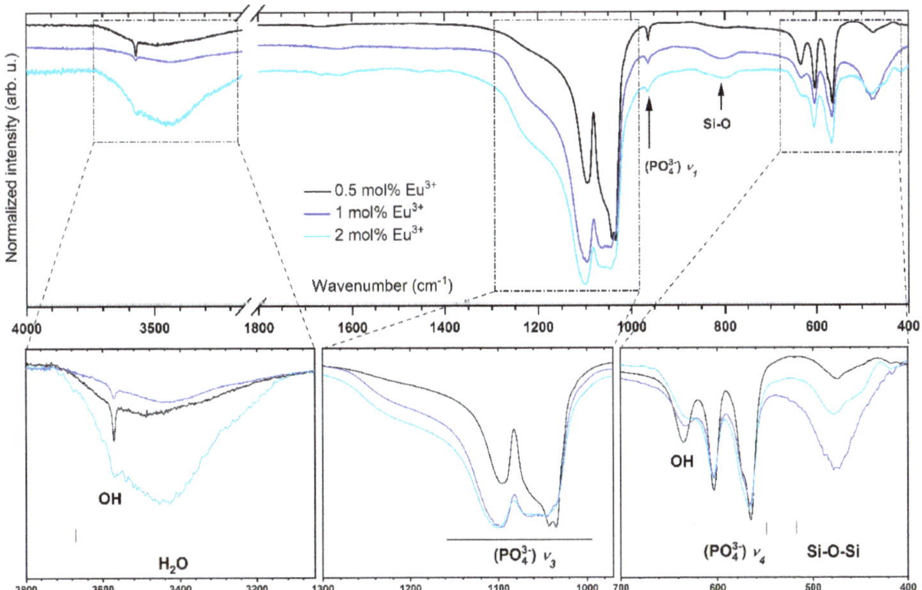

Figure 2. FT-IR spectra of the silicate-substituted hydroxyapatite co-doped with Sr^{2+} (2.0 mol%) and Eu^{3+} (0.5, 1.0 and 2.0 mol%) ions.

3.4. Spectroscopy Properties

The emission excitation spectra of the silicate-substituted appetites were recorded as a function of the europium ion concentration and heat-treated temperature in Figure 3a,b, respectively. The spectra were recorded at 300 K at an observation wavelength of 616 nm (16,233 cm^{-1}). The presented spectra were normalized to the most intensity bands. In relation to the most intense band of the $^5D_0 \rightarrow {}^7F_2$ transition, the spectra were recorded at an observation wavelength of 618 nm. In the UV range was observed a broad, intense band ascribed to the $O^{2-} \rightarrow Eu^{3+}$ charge transfer (CT) transition with a maximum located around 205 nm (48,780 cm^{-1}). Increasing the dopant concentration and heat-treating temperature does not have an influence on the CT maximum position. In the composition of the excitation spectra were recorded sharp, narrow bands, attributed to the 4f–4f transitions of Eu^{3+} ions, at: $^7F_0 \rightarrow {}^5F_{(4,1,3,2)}$, $^3P_0$ at 299.3 nm (33,411 cm^{-1}), $^7F_0 \rightarrow {}^5H_{(6,5,4,7,3)}$ at 319.8 nm (31,269 cm^{-1}), $^7F_0 \rightarrow {}^5D_4$, $^5L_8$ at 363.7 nm (27,495 cm^{-1}), $^7F_0 \rightarrow {}^5G_2$, $^5L_7$, $^5G_3$ at 383.7 nm (26,062 cm^{-1}), $^7F_0 \rightarrow {}^5L_6$ at 394.4 nm (25,354 cm^{-1}), $^7F_0 \rightarrow {}^5D_3$ at 413.8 nm (24,166 cm^{-1}) and $^7F_0 \rightarrow {}^5D_2$ at 465.1 nm (21,500 cm^{-1}). In lanthanide ions, strongly isolating the 4f orbitals by the external 5s, 5p and 5d shells causes only slight changes in the positions of the electronic transition bands.

Figure 4a represents the emission spectra for the 2 mol% Eu^{3+}-doped sample as a function of sintering temperature. Figure 4b shows the emission spectra as a function of optically active ion concentration for samples sintered at 600 °C. The emission spectra were detected at an excitation wavelength of 394 nm to directly excite the f electrons of Eu^{3+} ions. All spectra were normalized according to the $^5D_0 \rightarrow {}^7F_1$ magnetic transition. As should have been expected, the emission transitions of the Eu^{3+} ions are forbidden by selection rules but do not consider the subtle influence of atom vibrations which consequently change the dipole moment, causing the occurrence of forbidden transitions on the spectrum. In the spectra are presented emission lines due to the $^5D_0 \rightarrow {}^7F_J$ transitions for J = 0, 1, 2, 3 and 4, typical of Eu^{3+} ions, which are ascribed in reference to previous reports [5,24–27]. The bands located at the listed wavelength were attributed to the following transition: $^5D_0 \rightarrow {}^7F_0$ at 577 nm (17,331 cm^{-1}); $^5D_0 \rightarrow {}^7F_1$ at 588 nm (17,006 cm^{-1}); $^5D_0 \rightarrow {}^7F_2$ at 616 nm (16,233 cm^{-1}); $^5D_0 \rightarrow {}^7F_3$ at 653 nm (15,313 cm^{-1}) and $^5D_0 \rightarrow {}^7F_0$ at 700 nm (14,285 cm^{-1}). In general, the spectroscopic properties can provide information about the local chemical environment of Eu^{3+} ions. The essential change in emission spectra can be dependent on the quantity of possible crystallographic positions and therefore the amount of potential sites of substitution, as well as the presence of defects, additional phases or impurities [3,27].

The typical emission spectrum of 4f–4f electrons of Eu^{3+} ions is recorded in the red range of the electromagnetic radiation spectrum. In this range, the $^5D_0 \rightarrow {}^7F_{0,1,2}$ transitions are observed and the shape of the lines in this region is correlated with the structural properties of the material. Due to these transitions, the trivalent europium ion is called an optical probe. If the Eu^{3+} ion is located in a centrosymmetric crystal lattice, the $^5D_0 \rightarrow {}^7F_0$ transition is not observed. In the contrary situation, if the Eu^{3+} ion is placed in a non-centrosymmetric lattice, the $^5D_0 \rightarrow {}^7F_0$ transition is detected on the emission spectra. Moreover, this is possible only in the low-symmetry crystal position as in C_n, C_{nv} as well as C_S symmetry. In the hydroxyapatite matrix, calcium ions are located at two different crystal positions: Ca(1) and Ca(2) with local symmetry at C_3 and C_S, respectively. Both calcium positions could be occupied by Eu^{3+} ions. The additional crystallographic position appears as a result of reverse cis and trans symmetry of the Ca(2) site with the same point symmetry. In the environment of the Ca(1) site, nine oxygen atoms from phosphate groups are present. The calcium ion located on the Ca(2) site is in sevenfold coordination with six oxygen atoms from the phosphate group and one from the hydroxyl group [28]. The number and ratio of the intensity of the lines in the range between 570 and 580 nm give information about Eu^{3+} ion site-occupied preference. The coordination polyhedra of the Ca(1) and Ca(2) cations are presented in Figure 5a,b, respectively.

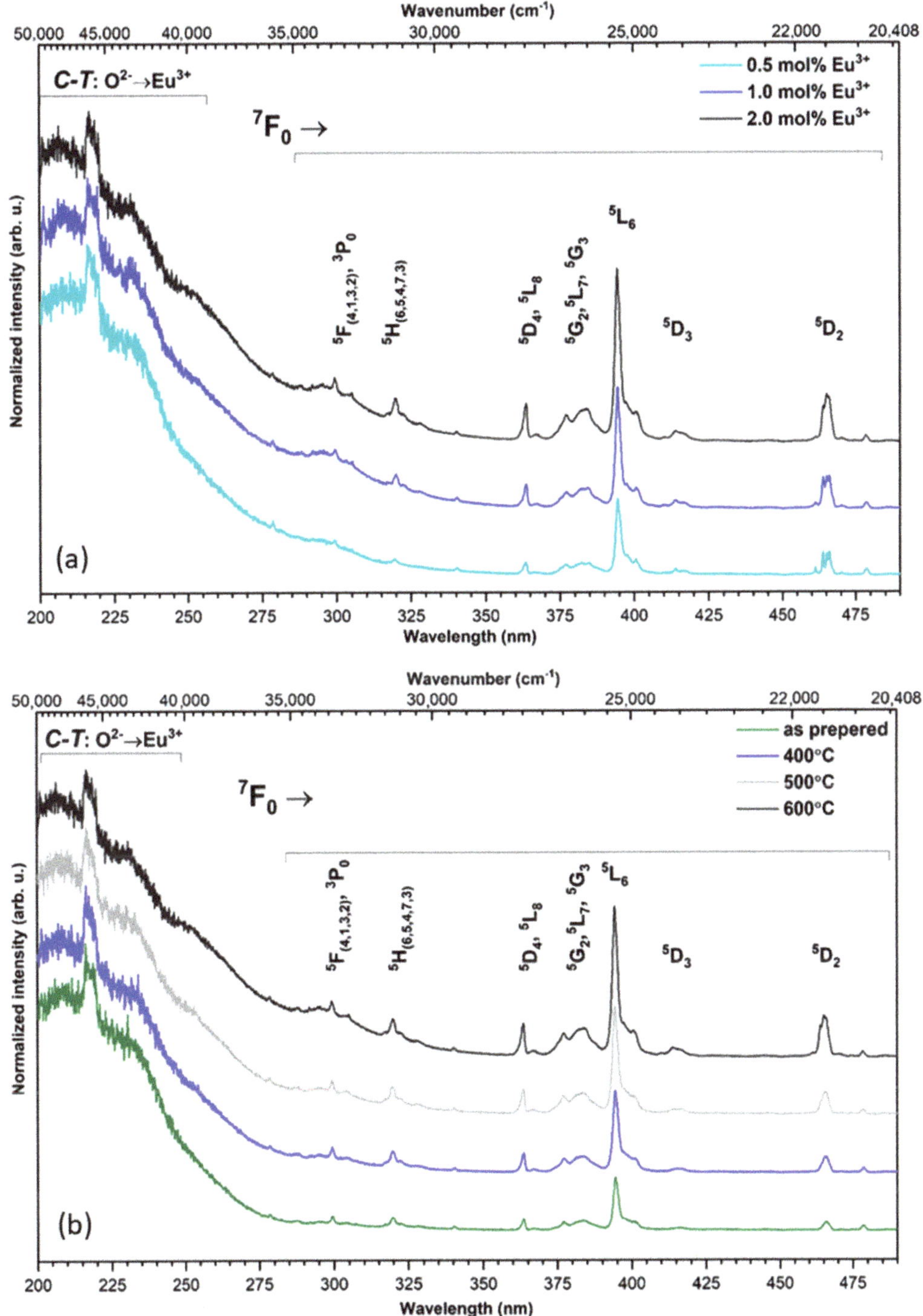

Figure 3. Excitation emission spectra of $Ca_{9.8-x}Sr_{0.2}Eu_x(PO_4)_2(SiO_4)_4(OH)_2$, where x = 0.5, 1.0, 2.0 mol%, sintered at 600 °C (**a**) as well as $Ca_{9.6}Sr_{0.2}Eu_{0.2}(PO_4)_2(SiO_4)_4(OH)_2$ as a function of sintering temperature (**b**).

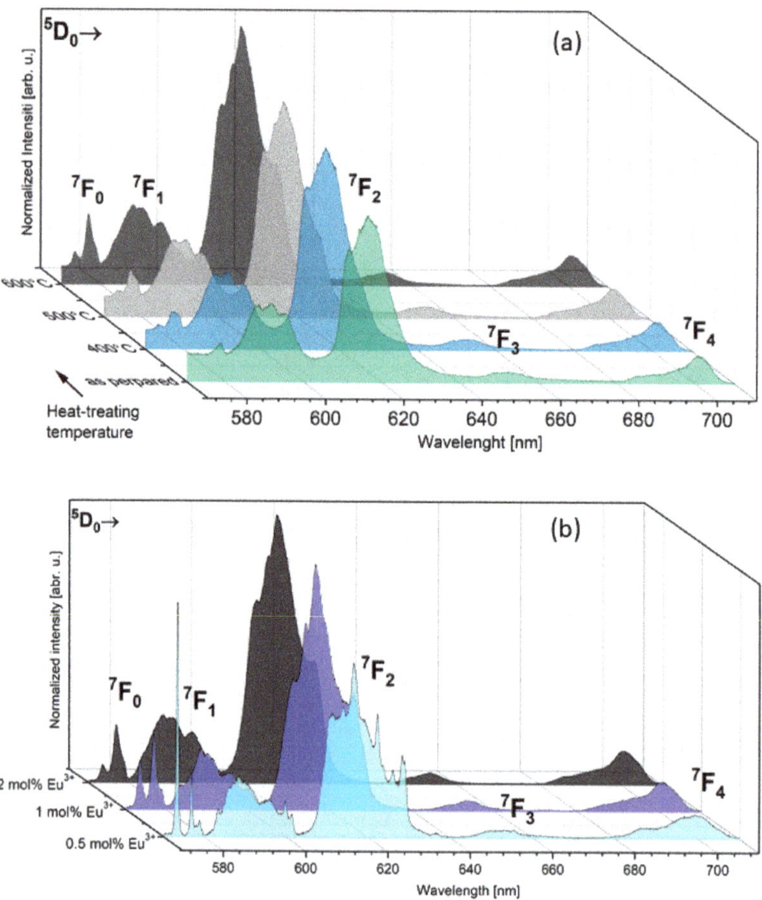

Figure 4. Emission spectra of $Ca_{9.6}Sr_{0.2}Eu_{0.2}(PO_4)_2(SiO_4)_4(OH)_2$ as a function of sintering temperature (**a**) and $Ca_{9.8-x}Sr_{0.2}Eu_x(PO_4)_2(SiO_4)_4(OH)_2$, where x = 0.5, 1.0, 2.0 mol%, sintered at 600 °C (**b**).

Emission spectra recorded for the investigated samples in the range of 570 to 580 nm are presented in Figure 5. In the emission spectra of 2 mol% Sr^{2+} and 0.5 mol% Eu^{3+} co-doped silicate-substituted hydroxyapatites, the transition attributed to $^5D_0 \rightarrow {}^7F_0$ presents some interesting features due to abnormally strong intensity. This fact can be related to an important perturbation of the symmetry of the dopant induced by the introduction of silicate groups. Moreover, the abnormally strong intensity of the $^5D_0 \rightarrow {}^7F_0$ transition has been reported in apatites such as oxyapatite $Ca_{10}(PO_4)_6O_2$, fluoroapatite $Ca_5(PO_4)_3F$, hydroxyapatite $Sr_{10}(PO_4)_6(OH)_2$, or silicophosphate apatite $Sr_5(PO4)_2SiO_4$, etc. This intense emission is attributed to the existence of the strong covalence of the Eu^{3+}-O^{2-} bond in the Ca(2) site in the apatite lattice [28–30].

In the other cases, the population of the Ca^{2+} position replaced by the Eu^{3+} ion could be related to the thermal diffusion process of dopant ions in the apatite structure. It is commonly known that in the case of as-prepared apatite materials, only the emission associated with one type of site with C_3 symmetry was observed, whereas with an increase in the calcination temperature, additional 0–0 peaks appeared [1,31,32]. The emission intensity ratio Ca(1)/Ca(2) has been calculated and results are presented in Table 1, showing values from 2.4 for as-prepared samples to 3.1 for sintered samples at 600 °C. Taking into account all results, one can note that in the silicate-substituted strontium-doped hydroxyapatite matrix, the Ca(1) site is more occupied by Eu^{3+} ions.

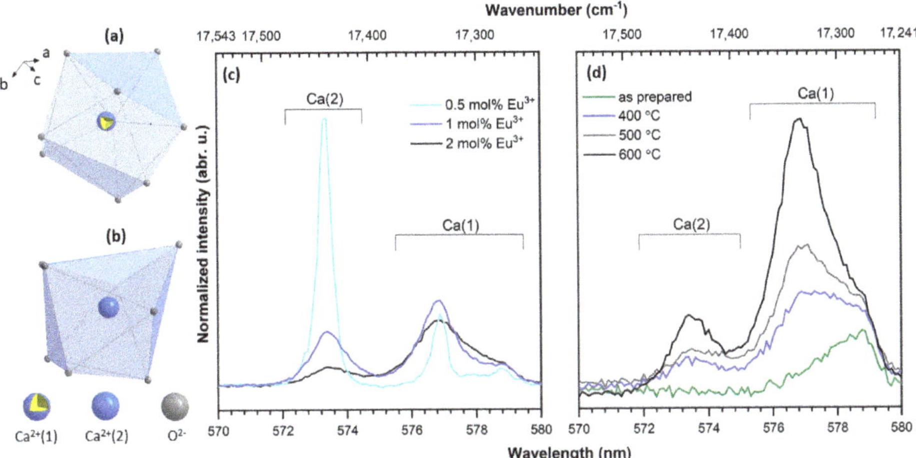

Figure 5. The projection of the coordination polyhedra of (**a**) Ca(1) and (**b**) Ca(2) cations, respectively. Emission spectra of (**c**) $Ca_{9.8-x}Sr_{0.2}Eu_x(PO_4)_2(SiO_4)_4(OH)_2$, where x = 0.5, 1.0, 2.0 mol%, sintered at 600 °C and (**d**) $Ca_{9.6}Sr_{0.2}Eu_{0.2}(PO_4)_2(SiO_4)_4(OH)_2$ as a function of sintering temperature, for the $^5D_0 \rightarrow {}^7F_0$ transition.

To obtain additional insight into the luminescence behavior of Eu^{3+} ions in silicate-substituted strontium co-doped apatite, the Judd–Ofelt theory was applied [33,34]. The results of the calculation of radiative (A_{rad}), non-radiative (A_{nrad}) and total (A_{tot}) processes, as well as intensity parameters (Ω_2, Ω_4), quantum efficiency (η) and asymmetry ratio (R), are presented in Table 1. These parameters were calculated based on emission spectra and decay profiles according to equations defined in previous reports [35,36].

A comparison of the Judd–Ofelt intensity parameters (Ω_2 and Ω_4) is made between the samples with different Eu^{3+} ion concentrations and sintering temperatures. An increase in the Ω_2 value is noted with an increase in the heat-treating temperature. This result indicates the increasingly hypersensitive character of the $^5D_0 \rightarrow {}^7F_2$ transition and the increasing polarization of the Eu^{3+} ion environment. Consequently, Eu^{3+} ions in the material heat-treated at the highest temperature are in a more polarizable environment. The influence of silicon group presence is analyzed in comparison with previous work by our group [37]. A decrease in the J-O Ω_2 parameter is noted from 6.683×10^{-20} cm^2 (see [37]) to 4.506×10^{-20} cm^2 (see Table 1) for $Ca_{9.7}Sr_{0.2}Eu_{0.1}(PO_4)_6(OH)_2$ and for $Ca_{9.7}Sr_{0.2}Eu_{0.1}(PO_4)_2(SiO_4)_4(OH)_2$, respectively. This observation can be related to the improvement of the Eu^{3+} cation polyhedral and to the decrease in the covalence character of the Eu^{3+}–O^{2-} bond in silicate-substituted apatite. In the case of the investigated samples, the $\Omega_2 > \Omega_4$ parameter suggests that Eu^{3+} ions are not located in the local symmetry of centrosymmetric character. The calculated results are in agreement with previous reports regarding apatite systems [1,24,35,37].

The highest quantum efficient (η) is observed for the 1 mol% Eu^{3+}-doped compound with the value of 40%. The quantum efficient is reduced by the increase as well as the decrease of optically active ion concentration to 37% and 34%, respectively. Samples heat-treated at higher temperatures demonstrate higher quantum efficient values, and the η parameter increases by the maximum of 15%.

The luminescence intensity ratio (R) of the electric dipole transition $^5D_0 \rightarrow {}^7F_2$ to the magnetic dipole transition $^5D_0 \rightarrow {}^7F_1$ has been calculated and the results are presented in Table 1. As expected, the sintering process and concentration of optical active ions have a significant influence on the symmetry of the Eu^{3+} ion environment in the apatite matrix.

The R factor increases with an increase in the heat-treating temperature, which indicates lower symmetry around Eu^{3+} ions and suggests that the covalence of Eu^{3+}-O^{2-} is higher. The R parameter increases for 0.5 mol% Eu^{3+} to 1.0 mol% Eu^{3+} from 3.1 to 4.5, respectively. Then, the opposite trend is observed, and the R parameter decreases from 4.5 to 4.1 for 1.0 mol% Eu^{3+} and 2.0 mol% Eu^{3+}.

Table 1. Decay rates of radiative (A_{rad}), non-radiative (A_{nrad}) and total (A_{tot}) processes of $^5D_0 \rightarrow {}^7F_J$ transitions, luminescence lifetimes (τ), intensity parameters (Ω_2, Ω_4), quantum efficiency (η) and asymmetry ratio (R) for investigated samples.

Sample	$Ca_{9.8-x}Sr_{0.2}Eu_x(PO_4)_2(SiO_4)_4(OH)_2$ 600 °C			$Ca_{9.6}Sr_{0.2}Eu_{0.2}(PO_4)_2(SiO_4)_4(OH)_2$			
	x = 0.5 mol% Eu^{3+}	x = 1.0 mol% Eu^{3+}	x = 2.0 mol% Eu^{3+}	as Prepared	400 °C	500 °C	600 °C
Ca(1)/Ca(2)	0.42	2.08	3.06	2.46	2.64	2.78	3.06
A_{rad} (s^{-1})	159.02	207.91	192.34	147.11	158.88	168.89	192.34
A_{nrad} (s^{-1})	306.85	308.56	327.17	509.69	475.04	435.80	327.17
A_{tot} (s^{-1})	465.87	516.48	519.51	656.80	633.92	604.69	519.51
τ (ms)	2.15	1.94	1.92	1.52	1.58	1.65	1.92
Ω_2 (10^{-20} cm^2)	4.084	5.898	5.377	3.678	4.166	4.506	5.377
Ω_4 (10^{-20} cm^2)	1.078	1.119	0.994	0.997	0.907	0.977	0.995
h (%)	34.13	40.26	37.02	22.40	25.06	27.93	37.02
R	2.85	4.11	3.75	2.57	2.91	3.14	3.75

Figure 6 presents the energy level diagram of Eu^{3+} corresponding to the detected excitation and emission spectra. As seen, Eu^{3+} ions were pumped to upper excited levels. The emission bands from the $^5D_1$, $^5D_2$ and $^5D_3$ levels are not observed at room temperature in the case of the investigated samples, which suggests a fast, non-radiative (NR), multiphonon relaxation from the excited state $^5L_6$ to the $^5D_0$ state.

3.5. Temperature-Dependent Emission

To further study the possible application under high temperature, the temperature-dependent emission spectra of the silicate-substituted hydroxyapatite co-doped with 2 mol% Sr^{2+} and 2 mol% Eu^{3+} were measured and are presented in Figure 7. Emissions were recorded in the range of 80 to 725 K and at an excitation wavelength of 375 nm into the $^5G_2$ level. It is seen that the emission intensity decreases clearly with an increase in the ambient temperature, but the decrease is not linear in the whole range of measured temperatures. For the most intense line corresponding to the $^5D_0 \rightarrow {}^7F_2$ transition, the linear relationship applies between 80 and 325 K (R^2 = 99.4%). Then, between 450 and 800 K, the decrease can be described by the exponential equation (see Figure S3). The line corresponding to the $^5D_0 \rightarrow {}^7F_0$ transition of the Ca(2) calcium site (573 nm) is completely eliminated at 350 K.

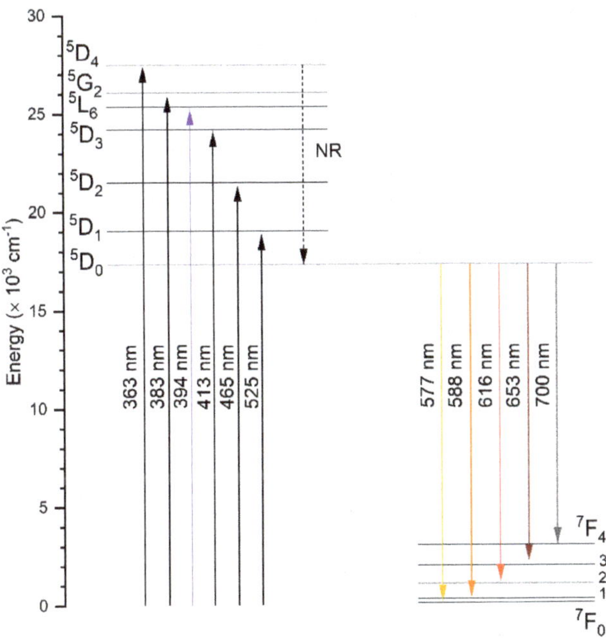

Figure 6. The simplified energy level scheme for Eu^{3+} ion in silicate-substituted strontium-doped hydroxyapatite.

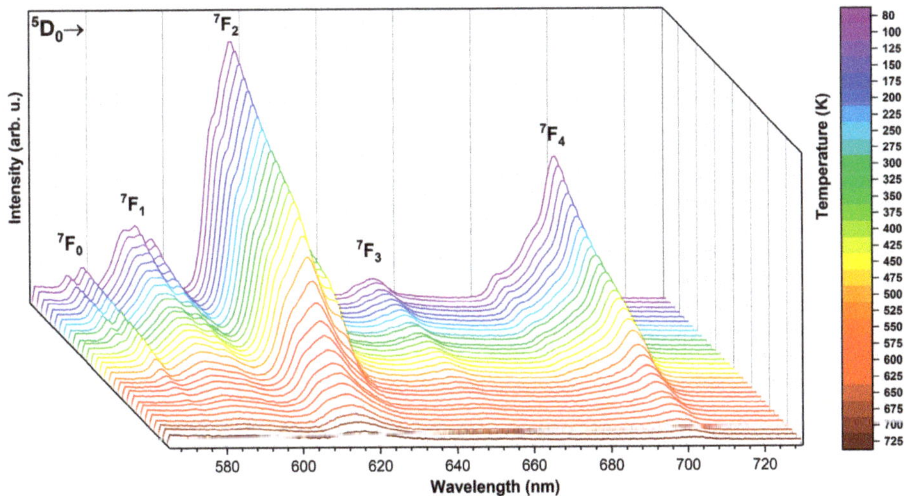

Figure 7. Temperature-dependent emission spectra of the $Ca_{9.6}Sr_{0.2}Eu_{0.2}(PO_4)_2(SiO_4)_4(OH)_2$, sintered at 600 °C.

In Figure 8 has been shown the Commission Internationale de l'Eclairage (CIE) 1931 chromaticity diagram for the $Ca_{9.6}Sr_{0.2}Eu_{0.2}(PO_4)_2(SiO_4)_4(OH)_2$ sample. The CIE color coordinates are listed in Table S3, which are calculated from the temperature-dependent emission spectra [38]. It has been reported that the emission color changed from reddish-orange to orange with the increasing of ambient temperature, but the reddish-orange emission was stable until 750 K. These results show that the Eu^{3+}-activated silicate-substituted apatites have the potential to be color-stable materials capable of operating at a wide range of ambient temperatures.

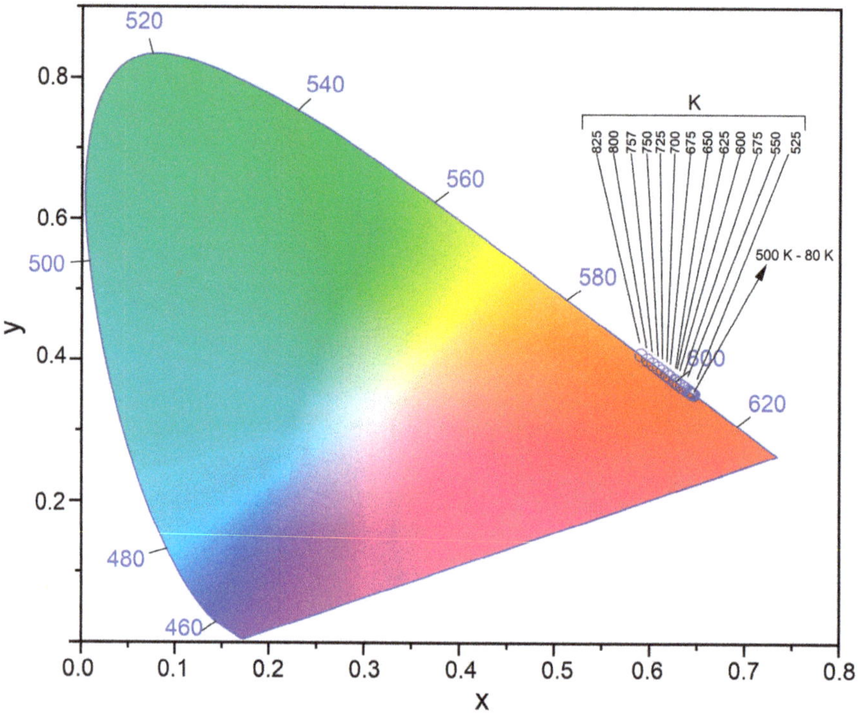

Figure 8. CIE 1931 chromaticity diagram of the $Ca_{9.6}Sr_{0.2}Eu_{0.2}(PO_4)_2(SiO_4)_4(OH)_2$ as a function of ambient temperature.

3.6. Decay Time

The luminescence kinetics corresponding to the $^5D_0 \rightarrow {}^7F_2$ transition were obtained at room temperature. As expected, all the recorded decays presented non-single exponential character. This phenomenon was consistent with the existence of two non-equivalent Eu^{3+} positions and, because of this fact, the effective emission lifetime was calculated by Equation (6). The recorded decays and calculated luminescence lifetimes (τ) are presented in Figure 9, as a function of Eu^{3+} concentration (a) and (b) as well as heat-treating temperature ((c) and (d)), respectively.

$$\tau_m = \frac{\int_0^\infty tI(t)dt}{\int_0^\infty I(t)dt} \cong \frac{\int_0^{t_{max}} tI(t)dt}{\int_0^{t_{max}} I(t)dt} \tag{6}$$

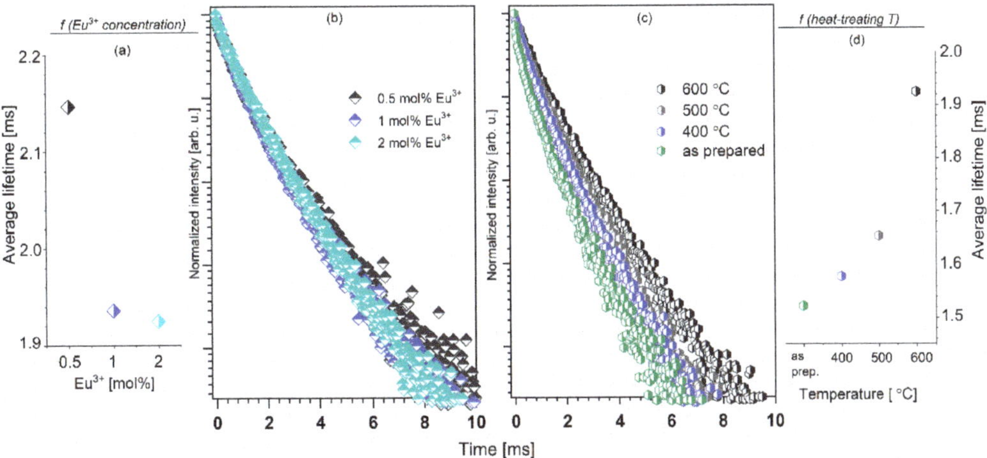

Figure 9. Calculated average lifetimes (**a**) as well as luminescence decay profiles (**b**) of $Ca_{9.8-x}Sr_{0.2}Eu_x(PO_4)_2(SiO_4)_4(OH)_2$, sintered at 600 °C, as a function of Eu^{3+} ion concentration. Luminescence decay profiles (**c**) and calculated average lifetimes (**d**) of the $Ca_{9.6}Sr_{0.2}Eu_{0.2}(PO_4)_2(SiO_4)_4(OH)_2$ as a function of sintering temperature.

4. Conclusions

In this study, it has been shown for the first time that a series of silicate-substituted hydroxyapatite co-doped with 2 mol% Sr^{2+} and Eu^{3+} ions in the range of 0.5–2.0 mol % in a ratio to the entire Ca^{2+} ion content were successfully synthesized by the hydrothermal method assisted with microwave and heat-treated. The average crystal sizes of the studied materials were in the range of 16–56 nm as calculated by the Rietveld method.

Attention was paid to the structural and spectroscopic properties related to a variable amount of Eu^{3+} ion concentration. The spectroscopic properties have shown for the sintered samples that Eu^{3+} ions occupied three independent crystallographic sites: one Ca(1) site with C_3 local symmetry and two Ca(2) sites with C_s local symmetry with *cis* and *trans* symmetry. The $^5D_0 \rightarrow {}^7F_2$ hypersensitive transition is the most intense for most obtained materials, excluding the sample co-doped with 0.5 mol% Eu^{3+} ion and sintered at 600 °C, where the most dominant is $^5D_0 \rightarrow {}^7F_0$ transition. Moreover, the charge compensation mechanism in the materials induced by the substitution of Eu^{3+} and Sr^{2+} ions into the silicate-substituted hydroxyapatite host lattice was rendered in the Kröger–Vink notation.

The luminescence decay times corresponding to the most intense $^5D_0 \rightarrow {}^7F_2$ transition were recorded. The luminescence kinetics was characterized by a non-exponential decay profile and was in the range of 2.15 ms (0.5 mol% Eu^{3+}) and 19.4 ms (1 mol% Eu^{3+}) to 1.9 ms (2 mol% Eu^{3+}) for the samples sintered at 600 °C. On the other hand, the decay times for the samples doped with 2 mol% Eu^{3+} as a function of sintered temperature were in the range of 15.2 to 1.92 ms. The typical modes of the vibrations of the silicate-substituted hydroxyapatite ion group were detected in the FT-IR spectra, and these included the OH^- group vibrations characteristic of the hydroxyapatite matrix. The simplified Judd–Ofelt theory was used for a detailed analysis of the luminescence spectra. The hydroxyapatite containing 1 mol% of Eu^{3+} ions was evaluated to be the most optically efficient material among all the studied silicate-substituted hydroxyapatites. The International Commission on Illumination (CIE) color coordinates showed that the emission color can be tuned by varying the ambient temperature. The emission color was changed from reddish-orange to orange.

Supplementary Materials: The following are available online at https://www.mdpi.com/2079-4991/11/1/27/s1, Figure S1: The representative SEM image (a) and EDS spectra (b) of the $Ca_{9.6}Sr_{0.2}Eu_{0.2}$

$(PO_4)_2(SiO_4)_4(OH)_2$ nanopawders. Figure S2: Representative results of the of the $Sr_{0.2}Eu_{0.2}Ca_{9.6}(PO_4)_2(SiO_4)_4(OH)_2$, obtained at 600 °C, Rietveld analysis (red—fitted diffraction; blue—differential pattern; column—reference phase peak position). Figure S3: Temperature-dependent emission intensity of the lines correspond to the listed transitions. Table S1. The amount of substrates used for synthesis of silicate-substituted hydroxyapatite co-doped wth Eu^{3+} and Sr^{2+}. Table S2: Unit cell parameters (a,c), cell volume (V), grain size as well as refine factor (R_W) for the $Ca_{10}(PO_4)_2(SiO_4)_4(OH)_2$ co-doped with 2 mol% Sr^{2+} and x mol/% Eu^{3+} ions (where x = 0.5–2). Table S3: The comparison of the CIE color coordinates (x,y) of $Ca_{9.6}Sr_{0.2}Eu_{0.2}(PO_4)_2(SiO_4)_4(OH)_2$ as a function of ambient temperature.

Author Contributions: R.J.W. conceived and designed the experiments and contributed reagents/materials/analysis tools and participated in funding acquisition in addition to analyzing all data; S.T. designed the experiments in addition to analyzing data. Both authors have read and agreed to the published version of the manuscript.

Funding: The authors would like to acknowledge financial support from the National Science Centre (NCN) within the project "Preparation and characterisation of biocomposites based on nanoapatites for theranostics" (No. UMO-2015/19/B/ST5/01330).

Institutional Review Board Statement: Not applicable.

Informed Consent Statement: Not applicable.

Data Availability Statement: Not applicable.

Acknowledgments: We are grateful to E. Bukowska for the help with XRD measurements and to D. Szymanski for SEM measurements.

Conflicts of Interest: The authors declare no conflict of interest.

References

1. Zawisza, K.; Strzep, A.; Wiglusz, R.J. Influence of annealing temperature on the spectroscopic properties of hydroxyapatite analogues doped with Eu^{3+}. *New J. Chem.* **2017**, *41*, 9990–9999. [CrossRef]
2. Guo, Y.; Moon, B.K.; Choi, B.C.; Jeong, J.H.; Kim, J.H.; Choi, H. Fluorescence properties with red-shift of Eu^{2+} emission in novel phosphor-silicate apatite $Sr_3LaNa(PO_4)_2SiO_4$ phosphors. *Ceram. Int.* **2016**, *42*, 18324–18332. [CrossRef]
3. Zawisza, K.; Wiglusz, R.J. Preferential site occupancy of Eu^{3+} ions in strontium hydroxyapatite nanocrystalline- $Sr_{10}(PO_4)_6(OH)_2$ -structural and spectroscopic characterisation. *Dalton Trans.* **2017**, *46*, 3265–3275. [CrossRef]
4. Pogosova, M.A.; Eliseev, A.A.; Kazin, P.E.; Azarmi, F. Synthesis, structure, luminescence, and color features of the Eu- and Cu-doped calcium apatite. *Dye Pigment.* **2017**, *141*, 209–216. [CrossRef]
5. Sokolnicki, J.; Zych, E. Synthesis and spectroscopic investigations of $Sr_2Y_8(SiO_4)_6O_2:Eu^{2+},Eu^{3+}$ phosphor for white LEDs. *J. Lumin.* **2015**, *158*, 65–69. [CrossRef]
6. Xia, Z.; Molokeev, M.S.; Im, W.B.; Unithrattil, S.; Liu, Q. Crystal structure and photoluminescence evolution of $La_5(S_xB_{1-x})(O_{13-x}N_x)$: Ce^{3+} solid solution phosphors. *J. Phys. Chem. C* **2015**, *119*, 9488–9495. [CrossRef]
7. Sobierajska, P.; Wiglusz, R.J. Influence of the grain sizes on Stokes and anti-Stokes fluorescence in the Yb^{3+} and Tb^{3+} ions co-doped nanocrystalline fluorapatite. *J. Alloy. Compd.* **2019**, *785*, 808–818. [CrossRef]
8. Liu, N.; Mei, L.; Liao, L.; Fu, J.; Yang, D. High thermal stability apatite phosphors $Ca_2La_8(SiO_4)_6O_2:Dy^{3+}/Sm^{3+}$ for white light emission: Synthesis, structure, luminescence properties and energy transfer. *Sci. Rep.* **2019**, *9*. [CrossRef]
9. Fleet, M.E.; Liu, X.; Pan, Y. Site preference of rare earth elements in hydroxyapatite $[Ca_{10}(PO_4)_6(OH)_2]$. *J. Solid State Chem.* **2000**, *149*, 391–398. [CrossRef]
10. Chouard, N.; Caurant, D.; Majérus, O.; Dussossoy, J.L.; Loiseau, P.; Grygiel, C.; Peuget, S. External irradiation with heavy ions of neodymium silicate apatite ceramics and glass-ceramics. *J. Nucl. Mater.* **2019**, *516*, 11–29. [CrossRef]
11. Zhu, G.; Shi, Y.; Mikami, M.; Shimomura, Y.; Wang, Y. Design, synthesis and characterization of a new apatite phosphor $Sr_4La_2Ca_4(PO_4)_6O_2:Ce^{3+}$ with long wavelength Ce^{3+} emission. *Opt. Mater. Express* **2013**, *3*, 229. [CrossRef]
12. Xia, Y.; Liu, Y.G.; Huang, Z.; Fang, M.; Molokeev, M.S.; Mei, L. $Ca_6La_4(SiO_4)_2(PO_4)_4O_2:Eu^{2+}$: A novel apatite green-emitting phosphor for near-ultraviolet excited w-LEDs. *J. Mater. Chem. C* **2016**, *4*, 4675–4683. [CrossRef]
13. Zhang, W.; Yu, M.; Wu, Z.; Wang, Y.; Zhang, P. White emission enhancement of $Ca_5(PO_4)_3Cl:Dy^{3+}$ phosphor with Li^+/Eu^{3+} co-doping for white light-emitting diodes. *J. Mater. Sci. Mater. Electron.* **2018**, *29*, 8224–8233. [CrossRef]
14. Gu, Z.; Zhang, X.; Li, L.; Wang, Q.; Yu, X.; Feng, T. Acceleration of segmental bone regeneration in a rabbit model by strontium-doped calcium polyphosphate scaffold through stimulating VEGF and bFGF secretion from osteoblasts. *Mater. Sci. Eng. C* **2013**, *33*, 274–281. [CrossRef] [PubMed]
15. Wiens, M.; Wang, X.; Schröder, H.C.; Kolb, U.; Schloßmacher, U.; Ushijima, H.; Müller, W.E.G. The role of biosilica in the osteoprotegerin/RANKL ratio in human osteoblast-like cells. *Biomaterials* **2010**, *31*, 7716–7725. [CrossRef]

16. Lin, K.; Xia, L.; Li, H.; Jiang, X.; Pan, H.; Xu, Y.; Lu, W.W.; Zhang, Z.; Chang, J. Enhanced osteoporotic bone regeneration by strontium-substituted calcium silicate bioactive ceramics. *Biomaterials* **2013**, *34*, 10028–10042. [CrossRef]
17. Kikuchi, M.; Yamazaki, A.; Otsuka, R.; Akao, M.; Aoki, H. Crystal structure of Sr-substituted hydroxyapatite synthesized by hydrothermal method. *J. Solid State Chem.* **1994**, *113*, 373–378. [CrossRef]
18. Rietveld, H.M. A profile refinement method for nuclear and magnetic structures. *J. Appl. Crystallogr.* **1969**, *2*, 65–71. [CrossRef]
19. Lutterotti, L.; Matthies, S.; Wenk, H.-R. MAUD: A friendly Java program for material analysis using diffraction. *IUCr Newsl. CPD* **1999**, *21*, 14–15. [CrossRef]
20. Pham, V.H.; Ngoc Trung, N. Luminescence of europium doped silicon-substituted hydroxyapatite nanobiophosphor via a coprecipitation method. *Mater. Lett.* **2014**, *136*, 359–361. [CrossRef]
21. Shannon, R.D. Revised effective ionic radii and systematic studies of interatomie distances in halides and chaleogenides. *Acta Cryst.* **1976**, *A32*, 751–767. [CrossRef]
22. Balan, E.; Delattre, S.; Roche, D.; Segalen, L.; Morin, G.; Guillaumet, M.; Blanchard, M.; Lazzeri, M.; Brouder, C.; Salje, E.K.H. Line-broadening effects in the powder infrared spectrum of apatite. *Phys. Chem. Miner.* **2011**, *38*, 111–122. [CrossRef]
23. Williams, R.L.; Hadley, M.J.; Jiang, P.J.; Rowson, N.A.; Mendes, P.M.; Rappoport, J.Z.; Grover, L.M. Thiol modification of silicon-substituted hydroxyapatite nanocrystals facilitates fluorescent labelling and visualisation of cellular internalisation. *J. Mater. Chem. B* **2013**, *1*, 4370–4378. [CrossRef] [PubMed]
24. Targonska, S.; Szyszka, K.; Rewak-Soroczynska, J.; Wiglusz, R.J. A new approach to spectroscopic and structural studies of the nano-sized silicate-substituted hydroxyapatite doped with Eu^{3+} ions. *Dalton Trans.* **2019**, *48*, 8303–8316. [CrossRef]
25. Wiglusz, R.J.; Bednarkiewicz, A.; Strek, W. Synthesis and optical properties of Eu^{3+} ion doped nanocrystalline hydroxya patites embedded in PMMA matrix. *J. Rare Earths* **2011**, *29*, 1111–1116. [CrossRef]
26. Marycz, K.; Sobierajska, P.; Smieszek, A.; Maredziak, M.; Wiglusz, K.; Wiglusz, R.J. Li^+ activated nanohydroxyapatite doped with Eu^{3+} ions enhances proliferative activity and viability of human stem progenitor cells of adipose tissue and olfactory ensheathing cells. Further perspective of nHAP:Li^+, Eu^{3+} application in theranostics. *Mater. Sci. Eng. C* **2017**, *78*, 151–162. [CrossRef]
27. Lima, T.A.R.M.; Valerio, M.E.G. X-ray absorption fine structure spectroscopy and photoluminescence study of multifunctional europium (III)-doped hydroxyapatite in the presence of cationic surfactant medium. *J. Lumin.* **2018**, *201*, 70–76. [CrossRef]
28. El Ouenzerfi, R.; Kbir-Ariguib, N.; Trabelsi-Ayedi, M.; Piriou, B. Spectroscopic study of Eu^{3+} in strontium hydroxyapatite $Sr_{10}(PO_4)_6(OH)_2$. *J. Lumin.* **1999**, *85*, 71–77. [CrossRef]
29. Huang, Y.; Gan, J.; Seo, H.J. Luminescence investigation of Eu-activated $Sr_5(PO_4)_2SiO_4$ phosphor by combustion synthesis. *J. Am. Ceram. Soc.* **2011**, *94*, 1143–1148. [CrossRef]
30. Yu, R.; Jeong, J.H.; Wang, Y.F. A novel Eu^{3+}- and self-activated vanadate phosphor of $Ca_4La(VO_4)_3O$ with oxyvanadate apatite structure. *J. Am. Ceram. Soc.* **2017**, *100*, 5649–5658. [CrossRef]
31. Doat, A.; Fanjul, M.; Pellé, F.; Hollande, E.; Lebugle, A. Europium-doped bioapatite: A new photostable biological probe, internalizable by human cells. *Biomaterials* **2003**, *24*, 3365–3371. [CrossRef]
32. Martin, P.; Carlot, G.; Chevarier, A.; Den-Auwer, C.; Panczer, G. Mechanisms involved in thermal diffusion of rare earth elements in apatite. *J. Nucl. Mater.* **1999**, *275*, 268–276. [CrossRef]
33. Judd, B.R. Optical absorption intensities of rare-earth ions. *Phys. Rev.* **1962**, *127*, 750–761. [CrossRef]
34. Ofelt, G.S. Intensities of crystal spectra of rare-earth ions. *J. Chem. Phys.* **1962**, *37*, 511–520. [CrossRef]
35. Kodaira, C.A.; Brito, H.F.; Malta, O.L.; Serra, O.A. Luminescence and energy transfer of the europium (III) tungstate obtained via the Pechini method. *J. Lumin.* **2003**, *101*, 11–21. [CrossRef]
36. Hreniak, D.; Strek, W.; Amami, J.; Guyot, Y.; Boulon, G.; Goutaudier, C.; Pazik, R. The size-effect on luminescence properties of $BaTiO_3$:Eu^{3+} nanocrystallites prepared by the sol-gel method. *J. Alloy. Compd.* **2004**, *380*, 348–351. [CrossRef]
37. Sobierajska, P.; Wiglusz, R.J. Influence of Li^+ ions on the physicochemical properties of nanocrystalline calcium–strontium hydroxyapatite doped with Eu^{3+} ions. *New J. Chem.* **2019**, *43*, 14908–14916. [CrossRef]
38. Shionoya, S.; Yen, W.; Yamamoto, H. *Phosphor Handbook*; Phosphor Research Society; CRC Press: Cleveland, OH, USA, 2007.

nanomaterials

Article

Mechanistic Insight of Sensing Hydrogen Phosphate in Aqueous Medium by Using Lanthanide(III)-Based Luminescent Probes

Jashobanta Sahoo [1,2,3], Santlal Jaiswar [4], Pabitra B. Chatterjee [2,5], Palani S. Subramanian [1,2,*] and Himanshu Sekhar Jena [6,*]

[1] Inorganic Materials and Catalysis Division, Central Salt and Marine Chemicals Research Institute (CSIR-CSMCRI), Bhavnagar, Gujarat 364 002, India; s.jashobanta@gmail.com
[2] Academy of Scientific and Innovative Research (AcSIR), CSIR-CSMCRI, Bhavnagar, Gujarat 364 002, India; pbchatterjee@csmcri.res.in
[3] Department of Chemistry, Hindol College, Khajuriakata, Higher Education Department, State Government of Odisha, Bhubaneswar, Odisha 751001, India
[4] Discipline of Marine Biotechnology and Ecology, CSIR-CSMCRI, Bhavnagar, Gujarat 364 002, India; santlal@csmcri.res.in
[5] Analytical Discipline and Centralized Instrument Facility, CSIR-CSMCRI, Bhavnagar, Gujarat 364 002, India
[6] Department of Chemistry, Ghent University, Krijgslaan 281-S3 B, 9000 Ghent, Belgium
* Correspondence: siva140@yahoo.co.in or siva@csmcri.org (P.S.S.); hsjena@gmail.com or Himanshu.jena@ugent.be (H.S.J.)

Citation: Sahoo, J.; Jaiswar, S.; Chatterjee, P.B.; Subramanian, P.S.; Jena, H.S. Mechanistic Insight of Sensing Hydrogen Phosphate in Aqueous Medium by Using Lanthanide(III)-Based Luminescent Probes. *Nanomaterials* 2021, 11, 53. https://doi.org/10.3390/nano11010053

Received: 26 October 2020
Accepted: 23 December 2020
Published: 28 December 2020

Publisher's Note: MDPI stays neutral with regard to jurisdictional claims in published maps and institutional affiliations.

Abstract: The development of synthetic lanthanide luminescent probes for selective sensing or binding anions in aqueous medium requires an understanding of how these anions interact with synthetic lanthanide probes. Synthetic lanthanide probes designed to differentiate anions in aqueous medium could underpin exciting new sensing tools for biomedical research and drug discovery. In this direction, we present three mononuclear lanthanide-based complexes, $EuLCl_3$ (**1**), $SmLCl_3$ (**2**), and $TbLCl_3$ (**3**), incorporating a hexadentate aminomethylpiperidine-based nitrogen-rich heterocyclic ligand **L** for sensing anion and establishing mechanistic insight on their binding activities in aqueous medium. All these complexes are meticulously studied for their preferential selectivities towards different anions such as HPO_4^{2-}, SO_4^{2-}, CH_3COO^-, I^-, Br^-, Cl^-, F^-, NO_3^-, CO_3^{2-}/HCO_3^-, and HSO_4^- at pH 7.4 in aqueous HEPES (2-[4-(2-hydroxyethyl)piperazin-1-yl]ethanesulfonic acid) buffer. Among the anions scanned, HPO_4^{2-} showed an excellent luminescence change with all three complexes. Job's plot and ESI-MS support the 1:2 association between the receptors and HPO_4^{2-}. Systematic spectrophotometric titrations of **1–3** against HPO_4^{2-} demonstrates that the emission intensities of **1** and **2** were enhanced slightly upon the addition of HPO_4^{2-} in the range 0.01–1 equiv and 0.01–2 equiv., respectively. Among the three complexes, complex **3** showed a steady quenching of luminescence throughout the titration of hydrogen phosphate. The lower and higher detection limits of HPO_4^{2-} by complexes **1** and **2** were determined as 0.1–4 mM and 0.4–3.2 mM, respectively, while complex **3** covered 0.2–100 µM. This concludes that all complexes demonstrated a high degree of sensitivity and selectivity towards HPO_4^{2-}.

Keywords: lanthanides; luminescence; nitrogen-rich ligand; phosphate sensing; quenching

1. Introduction

Inorganic phosphates, the charged anions of phosphoric acid such as $[H_2PO_4]^-$, $[HPO_4]^{2-}$, and $[PO_4]^{3-}$, are essential components during the synthesis of DNA/RNA and phospholipid membrane [1]. Further, their influence in the metabolic process in human, plant, and animal cells are inevitable. Sensing of phosphate draws special attention [2–13] due to its biological role as polyphosphate, and hyper- and hypophosphatemia in Chronic Kidney Disease (CKD) patients [14]; energy source through dephosphorylation [15] of

ATP, ADP, AMP, and PPi; and reverse polycondensation to form polyphosphates. Various methods were developed for the determination of phosphates in fertilizers, plants, natural waters, and other environmental samples [16–18]. Generally, serum phosphates are measured based on a photometric approach using ammonium phosphate, which forms a chromogenic complex with inorganic phosphates (Pi) [19]. However, the search for new receptors with selective response to phosphates remains active behind many challenges. Moreover, with phosphates being important bioanlytes [20–29], varieties of colorimetric sensors [30–35] and fluorosensors [36–38] were reported for their detection. Among these, luminescent lanthanide [20] complexes gained significant attention due to their potential applications in clinical diagnosis, biomarkers [39,40], MRI contrast agents [41–47], screening of drugs, etc. Parker et al. reported Eu(III) and Tb(III) tetra-azaphenylene complexes for the detection of phosphates in live cells [48,49]. It is important to note that the concentrations of phosphate vary significantly in inter- and intracellular environments of human cells, ranging from 0.15 to 1.3 mM [50–52]. Among the various lanthanide complexes reported so far in the literature, Eu-Tc [53,54] was recognized as an efficient probe for phosphates due to its lower detection limit (LOD = 3 μmolL^{-1}). Moreover, there are many intracellular processes, where the concentrations of phosphate vary among different subcellular compartments present therein [55]. Therefore, a highly sensitive and selective probe which can detect phosphate at a considerably low concentration is very much required to investigate such intracellular processes. In this context, recently, we have reported a set of europium(III) and terbium(III) complexes, incorporating different hexadentate ligands which showed highly selective and efficient recognition of inorganic phosphates and nucleoside phosphates [56,57]. In this direction and as a part of our ongoing research, herein, we report a series of relatively simple, cheap, and water-soluble Ln(III) complexes **1**, **2**, and **3** (Scheme 1) (Ln = Eu, Sm, and Tb, respectively) using an aminomethylpiperidine-functionalized 1,10-phenanthrolene-based nitrogenous heterocyclic ligand **L** as the metal chelator. The anion-sensing ability of these hydrophilic rare-earth complexes (**1–3**) was explored and found high selectivity and sensitivity for hydrogen phosphate ions in HEPES (2-[4-(2-hydroxyethyl)piperazin-1-yl]ethanesulfonic acid) buffer at pH 7.4. Moreover, a mechanistic insight into the anion binding behavior of complex **1** was also explored in this work.

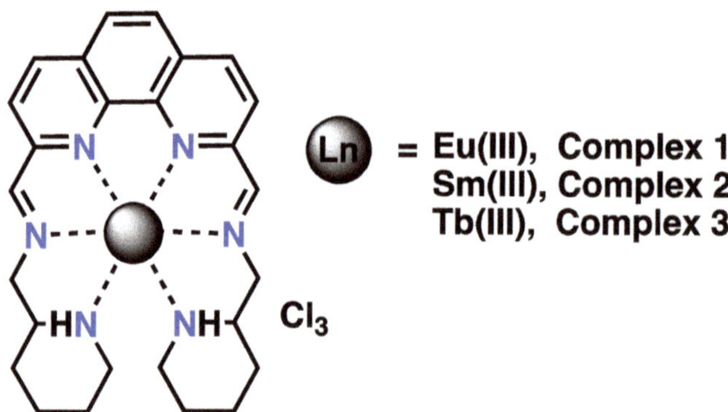

Scheme 1. 2,9-Bis(aminomethylpiperidine)-1,10-phenanthrolene (**L**) and its Ln(III) complexes **1**, **2**, and **3**.

2. Results and Discussion

Schiff base ligand **L** was obtained by condensing 2,9-dialdehyde 1,10-phenanthroline and 2-(aminomethyl) piperidine. The characteristic azomethine peak at 8.25δ in ^{1}H NMR (Figure S1) and ^{13}C (Figure S2), DEPT 135° NMR (Figure S3) in combination with MS spectrum (Figure S4), CD spectrum (Figure S5) and IR spectra (Figure S6), confirms the formation of **L**. Treating **L** with the respective LnCl$_3$ salt, the corresponding complexes

EuLCl$_3$(**1**), TbLCl$_3$(**2**), and SmLCl$_3$(**3**) were isolated as per the procedure. The formulation of each complex was confirmed from the ESI-MS analysis (Figure S7–S9). The emission spectra of these complexes were studied at 25 °C in aqueous HEPES buffer at pH 7.4 (Figure S10). All the complexes showed a significant red-shift of the emission spectra with respect to the emission profile of the free ligand **L**. Being luminescent in nature, we sought to investigate the excited state photophysical properties of **1–3** in the presence of various important anions. Complexes **1** and **3** showed characteristic luminescent bands at 614 nm and 545 nm, respectively, while complex **2** displayed two sensitive bands at 595 and 644 nm attributable for their metal centered emission. Although water functions as a luminescence quencher [58], all these complexes showed an intense luminescence in aqueous HEPES buffer at physiological pH at 25 °C. The effects of the addition of a range of anions such as hydrogen phosphate, sulfate, acetate, iodide, bromide, chloride, fluoride, nitrate, carbonate/bicarbonate, and bisulfate on the emission spectra of **1–3** is showed in Figure 1a–c. The bar diagrams, as insets in Figure 1a,c, show the changes in emission intensities of the hypersensitive peaks at 614 and 545 nm of complexes **1** and **3**, while Figure 1b depicts the changes in the ratio of the 644/595 nm hypersensitive peaks of complex **2** with the addition of various anions.

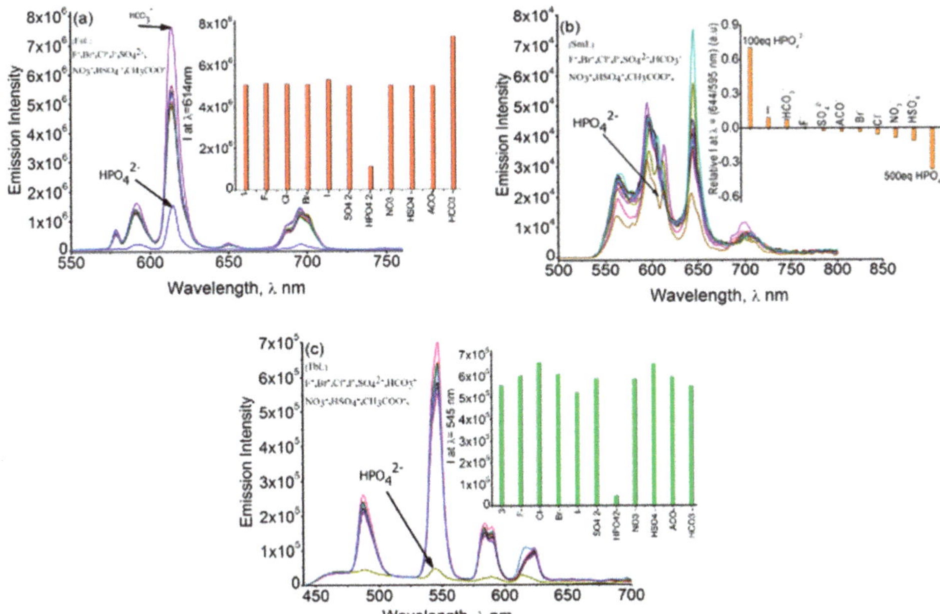

Figure 1. Emission spectra of (**a**) **1** (1 × 10^{-5} M), (**b**) **2** (2 × 10^{-5} M), and (**c**) **3** (4 × 10^{-5} M) upon the addition of various anions (100 equiv. for **1** and 10 equiv. for **3** in aqueous HEPES (2-[4-(2-hydroxyethyl)piperazin-1-yl]ethanesulfonic acid) buffer at pH = 7.4, λ_{exi} = 276 nm): in the case of **2**, spectra obtained after the addition of both 100 (cyano) and 500 equiv. (light brown) of HPO$_4$$^{2-}$ are overlaid. Insets: luminescence intensities of **1** and **3** at 614 and 545 nm, respectively, in the presence of different anions, while in the case of **2**, relative intensity ratios (644/595 nm) are plotted along the y-axis.

Among the anions scanned, complex **1** illustrated a significant emission change with hydrogen phosphate and bicarbonate ions. While the addition of HCO$_3$$^-$ showed 14.7% luminescence enhancement, phosphate in contrast leads to luminescence quenching by 29% of the emission intensity of **1** (Figure 1a). In Figure S11, the emission spectra of **1** against varying concentrations of phosphates (0–400 equiv.) are shown. The emission intensity was found to increase (2.6-fold) initially in the range 0.01–1 equiv. of HPO$_4$$^{2-}$ (Figure 2a).

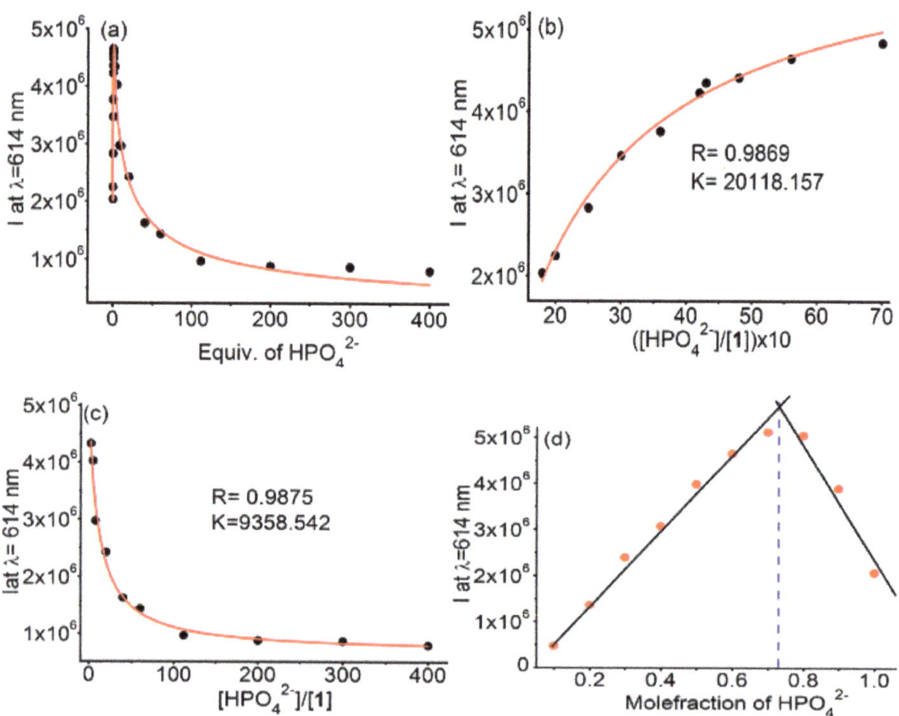

Figure 2. (**a**) Changes in emission maxima of **1** (1×10^{-5} M) upon gradual addition of HPO_4^{2-} in aqueous HEPES buffer at pH = 7.4; (**b**) nonlinear curve fitting of the titration data as a function of HPO_4^{2-} concentrations in the range 0.01–1 equiv. (luminescence enhancement part); (**c**) nonlinear curve fitting of the titration data as a function of HPO_4^{2-} concentrations in the range 1–400 equiv. (luminescence quenching part), in which a factor of 10 was multiplied with enhancement to maintain the same intesnsity for both quenching and enhancement; and (**d**) Job's plot analysis of mixtures of complex **1** with HPO_4^{2-} ($C_{complex\ 1} + C_{HPO4}^{2-} = 1.0\ \mu M$) in aqueous HEPES buffer pH 7.0, indicating 1:2 complex formation (λ_{emi} = 614 nm).

Upon further addition of HPO_4^{2-} to the reaction mixture, a gradual decrease in the luminescence, at 614 nm, of receptor **1** was observed (Figure 2a). The changes in the luminescence intensities, as displayed in Figure 2a, can be attributed to the two distinct behaviours of **1** against HPO_4^{2-}. Therefore, the spectrometric titration (Figure S11) has offered two association constants. Figure 2a also displayed that the luminescence intensity of the emission maximum decreased to a constant level after the addition of 4 mM of phosphate. Therefore, it is evident from Figure 2a that **1** can be used to sense a wide range of phosphate concentrations. The analytical limit of detection (LOD) [59–62] of **1** for phosphate was calculated as 0.1 μM. Since the sensing of hydrogen phosphate showed nonlinear fitting in Figure 2b,c with one enhanced and the other quenching the luminescence intensity, we applied the nonlinear fit data point results by following Equations (1) and (2), [63–66] respectively, providing the binding constants $K_1 = 2.0 \times 10^4$ M^{-1} (R = 0.9869), 1st part, attributable to luminescence enhancement and $K_2 = 0.94 \times 10^4$ M^{-1} (R = 0.9875), 2nd part, associated to luminescence quenching.

$$F = F_0 + \frac{F_{max} - F_0}{2} + \left\{ \left(1 + \frac{[M]}{C_L} + \frac{1}{C_L K}\right) - \sqrt{\left(1 + \frac{[M]}{C_L} + \frac{1}{C_L K}\right)^2 - 4\frac{[M]}{C_L}} \right\} \tag{1}$$

$$F = F_{max} + \frac{F_0 - F_{max}}{2} + \left\{ \left(1 + \frac{[M]}{C_L} + \frac{1}{C_L K}\right) - \sqrt{\left(1 + \frac{[M]}{C_L} + \frac{1}{C_L K}\right)^2 - 4\frac{[M]}{C_L}} \right\} \quad (2)$$

where F_0 is the luminescence intensity in the absence of hydrogen phosphate and F_{max} is the luminescent intensity in the presence of HPO_4^{2-}, and C_L and K are the concentration and binding constant of the complex, respectively. To find the association stoichiometry between complex **1** and HPO_4^{2-}, Job's plot was performed (Figure 2d), which established 1:2 binding stoichiometry, i.e., [**1**:HPO_4^{2-} = 1:2]. Further, a final confirmation regarding the abovementioned 1:2 stoichiometry was provided by ESI-MS (Figure S12), where a peak at $m/z = 855.23$ with 100% abundance was attributed to $[EuL(HPO_4)_2(H_2O) + 2Na^+] \cdot H_2O$ (calcd. $m/z = 855.11$).

Unlike hydrogen phosphate, HCO_3^- showed little enhancement in the emission of **1** (Figure 1a). The binding constant (K) was determined to be 1.2×10^3 M^{-1} from the spectrometric titrations of **1** against increasing concentrations of HCO_3^- (10 to 600 equiv.), as shown in Figure S13. Luminescence enhancement of **1**, observed in the addition of HCO_3^-, may occur due to chelate formation between the bicarbonate ion and europium (III) center by replacing the weakly bound inner sphere water molecules [67]. In Figure 1b, the luminescence response of **2** towards different anions is displayed. Among the four emission bands observed for **2**, the peaks at 595 nm ($^5G_{5/2} \rightarrow {}^6H_{7/2}$) and 644 nm ($^5G_{5/2} \rightarrow {}^6H_{9/2}$) were found to be hypersensitive [68–70].

Spectrophotometric titrations of **2** (Figure S14 and Figure 3a) against varying amounts of HPO_4^{2-} ranging from 0.01 to 800 equiv. illustrated similar patterns as observed earlier in case of **1**. Interestingly, the initial luminescent enhancement of **2** was found up to 18 equiv. of phosphate addition and further increases in phosphate concentrations quench the luminescence of the resulting solution. Applying nonlinear fitting of the data points (Figure 3b,c), the respective binding constants (K_1 and K_2) were calculated to be 2.1×10^4 M^{-1} (R = 0.9828), 1st part of luminescence enhancement and 2.9×10^3 M^{-1} (R = 0.9858), 2nd part of luminescence quenching. The LOD was calculated to be 0.4 µM, a little higher than that observed for **1**. To derive the complex **2** to phosphate ratio, Job's plot was performed, which clearly indicated 1:2 stoichiometry (Figure 3d). The respective positive ion ESI-MS (Figure S15) also confirmed the proposed 1:2 composition by depicting a molecular ion peak at $m/z = 862.27$ attributable to the formation of $[Sm(L-2H)(HPO_4)_2 + 4Na^+]$ (calcd. $m/z = 862.05$).

Screening complex **3** towards various anions (10 equiv.), shown in Figure 1c, illustrates again an excellent luminescent probe for HPO_4^{2-} with superior selectivity and sensitivity. Upon the addition of HPO_4^{2-}, the emission intensity of **3** was reduced to 8%. Systematic spectrophotometric titration with an increasing concentration of hydrogen phosphate ions in the range 0.01–5 equiv. was performed (Figure 4a and Figure S16). Unlike, **1** and **2**, the luminescence intensity of **3** demonstrated a steady quenching process. The luminescence intensity was quenched continuously from the very beginning of HPO_4^{2-} addition and at the 100 µM HPO_4^{2-} concentration; the emission quenched completely and remained constant thereafter (Figure 4a). The LOD for complex **3** was derived as 0.2 µM. The binding constant was determined from nonlinear data fitting using the Equation (2), and the respective association constant (K) was found to be 7.0×10^4 M^{-1} (R = 0.9870) (Figure 4b). A molecular ion peak at $m/z = 803.56$ (Figure S17) can be attributed to the generation of $[TbL(HPO_4)_2 + Na^+ + H^+]$ (calcd. $m/z = 803.45$) in solution. Thus, all three complexes are established to bind phosphates in 1:2 stoichiometries.

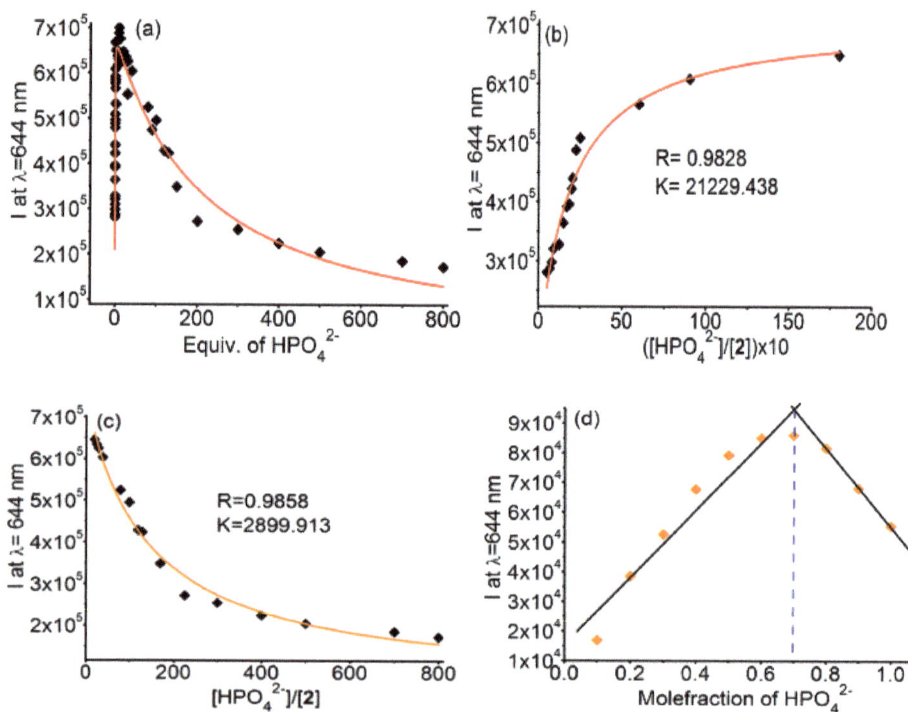

Figure 3. (**a**) Changes in emission maxima of **2** (4 × 10^{-5} M) upon gradual addition of HPO$_4$$^{2-}$ in aqueous HEPES buffer at pH = 7.4; (**b**) nonlinear curve fitting of the titration data as a function of HPO$_4$$^{2-}$ concentrations in the range 0.01–18 equiv. (luminescence enhancement part); (**c**) nonlinear curve fitting of the titration data as a function of HPO$_4$$^{2-}$ concentrations in the range 20–800 equiv. (luminescence quenching part); and (**d**) Job's plot of complex **2** with HPO$_4$$^{2-}$ ($C_{complex\ 2}$ + $C_{HPO4}$$^{2-}$ = 1.0 μM) in aqueous HEPES buffer pH 7.0 showing 1:2 complex formation (λ_{emi} = 644 nm).

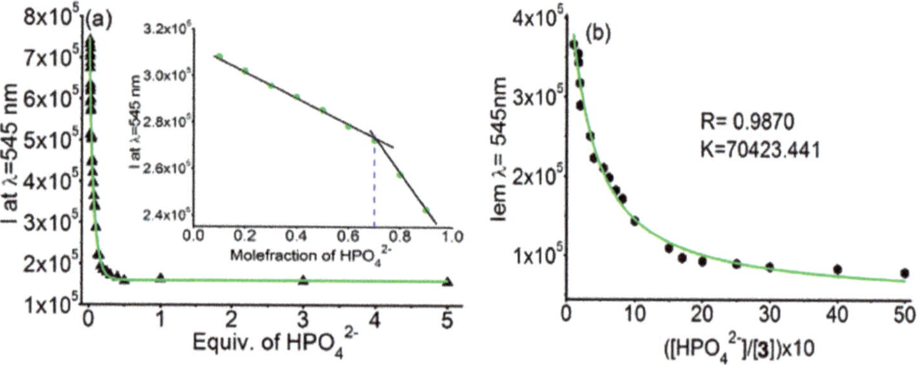

Figure 4. (**a**) Change in the emission spectra of **3** (2 × 10^{-5} M) upon the addition of HPO$_4$$^{2-}$ in aqueous HEPES buffer at pH = 7.4: the inset shows the Job's plot of mixtures of complex **3** with HPO$_4$$^{2-}$ ($C_{complex\ 3}$ + $C_{HPO4}$$^{2-}$ = 1.0 μM) in aqueous HEPES buffer pH 7.4 showing 1:2 complex formation. (**b**) Nonlinear curve fitting of luminescence intensities of **3** (2 × 10^{-5} M) as a function of HPO$_4$$^{2-}$ concentrations (λ_{emi} = 545 nm).

Ligand **L** with its hexadentate nature fulfils six coordination sites of Eu(III), Sm(III), and Tb(III) in their respective complexes **1**, **2**, and **3**. The remaining sites at the metal centers were calculated by measuring the hydration states [11,71] (denoted hereafter by "q") of **1–3**, adapting Equation (3). Based on the experimental results, the respective inner-sphere hydration numbers calculated for **1–3** are compiled in Table 1. Accordingly, complexes **1** and **2** are found to possess four coordinated water molecules while complex **3** accommodates only three water molecules, presumably due to the smaller ionic radius of Tb(III).

$$q_{corr} = A' \Delta k_{corr} \; [where \; \Delta k_{corr} = (k_{H2O} - k_{D2O})] \tag{3}$$

whereas k_{H2O} and k_{D2O} are radiative rate constants in H_2O and D_2O solvent, A' is a proportionality constant signifying the sensitivity of the lanthanide ion to vibronic quenching by OH oscillators, and q_{corr} is the hydration state, i.e., number of solvent molecules attached to a metal center.

Table 1. Excited state lifetime measurements of **1–3** in H_2O and D_2O at pH 7.4 (HEPES buffer).

Complex	τH_2O (ms)	τD_2O (ms)	q_{corr} [a]	Coordination Number
1	0.22	1.56	4.38	10
2	0.15	1.11	3.70 *	10
3	0.47	0.73	3.50	9

[a] q_{corr} values were determined by adapting $A' = 1.2$ ms (Eu^{3+}) and 5 ms (Tb^{3+}) and $\Delta k_{corr} = -0.25$ ms^{-1} (Eu^{3+}) and -0.06 ms^{-1} (Tb^{3+}). * For Sm(III), since Δk_{corr} values are not available in the literature, the q_{corr} value for complex **2** is calculated without applying the correction.

To understand the underlying mechanism behind the titration profiles of **1** and **2** against the hydrogen phosphate ions (Figures 2a and 3a), we performed time-resolved luminescence decay studies (Figures S18–S21) and also calculated the quantum yield of each complexes (Table S1). As a representative example lifetime was determined for **1** in the absence and presence of phosphate ions at different stoichiometry and compared with lifetime of the complex **1**, a laser excitation source of 276 nm was used, and the decay luminescence pattern was monitored at 614 nm. Aqueous (H_2O as well as D_2O) HEPES buffer solutions of **1** were used for these studies. In the absence of HPO_4^{2-}, luminescence profiles for **1** could be best fitted to single exponential decay traces with lifetime values $\tau = 0.22$ ms (H_2O) and 1.56 ms (D_2O) ($\kappa^2 = 1.16$, 1.18) (Figure S19, Table 1). Upon the addition of one equivalent HPO_4^{2-} to these two solutions, the luminescence decay profiles could be best fitted to $\tau = 0.41$ ms (H_2O) and 1.72 ms (D_2O) ($\kappa^2 = 1.17$, 1.13) (Table 2). Thus, the decrease in the hydration state of **1** from $q = 4$ to $q = 2$ upon the addition of 1 equivalent of HPO_4^{2-} (i.e., upon 1:1 association) was obvious to understand. After the addition of another equivalent of HPO_4^{2-} to this mixture, the lifetime values were found to be 0.60 ms (H_2O) and 1.89 ms (D_2O) ($\kappa^2 = 1.20$, 1.18) (Table 2), therefore revealing that hydration state $q = 1$, i.e., one coordinated water molecules present at 1:2 binding ratio between **1** and HPO_4^{2-}. The gradual addition of excess HPO_4^{2-} did not change the hydration state of the resulting species in solution further, i.e., $q = 1$ (Table 2). Based on the results summarized in Table 2, a plausible mechanism of phosphate's interaction with complex **1** is schematically represented in Figure 5. The initial little luminescence enhancement of **1** upon one equivalent HPO_4^{2-} addition possibly arose due to the replacement of two coordinated water molecules by the incoming phosphate group, which normally functions as a strong chelating species [24]. It is noteworthy to mention that such an effect has already been observed in the case of bicarbonate [72]. The addition of a second equivalent of HPO_4^{2-} to this 1:1 mixture resulted in the displacement of one more coordinated water molecule from the metal center (also supported by Figure S12) and shows quenching of luminescence. Possibly, steric crowding played an important role here by forcing the phosphate group to form a hydrogen bond with the piperidine NH moiety of ligand **L**, causing an energy mismatch between the lowest triplet state (T_1) of **L** and the excited state of the Eu(III). Therefore, the energy transfer process from the ligand L to europium(III) terminated and hence resulted in quenching of the luminescence process. However, at higher concentrations, the emission intensity reduces completely, which can be ascribed to the leaching of lanthanides from the complexes in the presence of more strongly coordinating phosphates ions.

Table 2. Summary of the changes in the lifetimes of complex **1** in the presence of different equivalent HPO_4^{2-} in aqueous media (H_2O as well as D_2O).

Species	$1 + HPO_4^{2-}$ (1:1)	$1 + HPO_4^{2-}$ (1:2)	$1 + HPO_4^{2-}$ (1:10)
τ (H_2O) (ms)	0.41	0.60	0.62
τ (D_2O) (ms)	1.72	1.90	1.91
q_{corr}	1.98	1.12	1.06

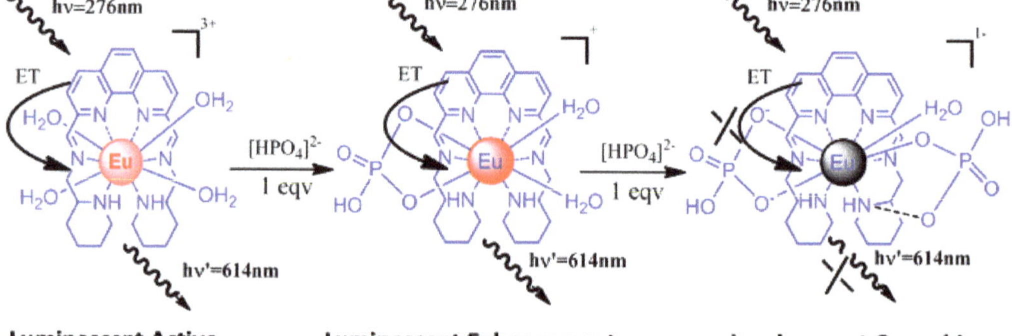

Figure 5. Proposed mechanistic pathway for successive HPO_4^{2-} binding with **1** in aqueous medium.

Although a significant number of luminescent complexes are applied in bio-imaging with excitation range below 300 nm [73,74], the presented complexes with excitation at 276 nm (i.e., in the UV region) limit their usage in in vivo bio-imaging. However, for in vitro conditions, the complexes are expected to be significant.

3. Conclusions

A series of mononuclear Ln(III) complexes (**1–3**) based on aminomethylpiperidine functionalized 1,10-phenanthrolene-based nitrogen-rich hexadentate heterocyclic ligand **L** has been reported. All these rare-earth complexes showed red-shifted metal-centered luminescence. The excited state photophysical properties of these complexes were explored to find their specific recognition affinity towards various important anions. The selective sensing of **1–3** for hydrogen phosphates over other anions is remarkable. Systematic spectrophotometric analysis demonstrates that, in the case of **1** and **2**, the emission intensities were increased slightly at the very beginning of phosphate addition (up to 1 and 2 equivalents, respectively) and finally decreased to a plateau at high phosphate concentrations (at mM level). The limits of detection (LOD) fall in the range 0.1–0.4 µM. Luminescence decay studies revealed that successive replacement of weakly bound coordinated water molecules from Eu(III) and Sm(III) probably caused the initial emission enhancement of these two complexes upon hydrogen phosphate addition. However, the addition of excess hydrogen phosphate causing steric crowding at the metal site and possible hydrogen bond formation between the piperidine NH group and phosphate might have created an energy mismatch between the lowest triplet state (T_1) of **L** and the excited state of the Eu(III), which in turn resulted in termination of the energy transfer between the *o*-phenanthrolene moiety and Ln(III).

4. Experimental Section

4.1. Materials and Methods

All chemicals were purchased from Aldrich. Sodium salts of all anions were used in this study. Elemental analyses of the complexes were carried out by using a vario Micro cube from Elementar. IR spectra were recorded from KBr pellets (1% *w/w*) on a Perkin–Elmer spectrum GX FTIR spectrophotometer. Electronic spectra were recorded on a Shimadzu UV 3600 spectrophotometer and scanned in the range 200–800 nm. The mass-spectrometric analysis was performed by using the positive ESI technique on a Waters Q-ToF Micromass spectrometer in CH_3OH. NMR spectra were recorded on a Bruker *Avance* 500 MHz FT-NMR spectrometer. The chemical shifts (δ) for proton resonances are reported in ppm relative to the internal standard TMS (Tetramethylsilane). The CD spectra were recorded by using a JASCO 815 spectrometer. Milli-Q water was used as a solvent. pH measurements were carried out using an ORION VERSA STAR pH meter. Emission spectra were recorded using an Edinburgh Instruments model Xe-900, and all the spectra recorded are reported hereafter applying emission correction. The slit sizes for emission and excitation were adjusted as 3.0/3.0 nm. For Job plot analysis (continuous variations method), a series of samples were prepared with a constant sum of concentrations at 1.0 μM but with varying concentrations of complex and hydrogen phosphate. The luminescence spectra were recorded for each sample with λ_{ex} = 276 nm for all these complexes. The maximum luminescence intensity was plotted versus the mole fraction of the corresponding hydrogen phosphate. For determination of the maximum, the ascending and descending segments of the curve were fitted to linear lines, respectively, and the intercept of both lines denotes the maximum and thus the stoichiometry of the complex.

Synthesis of ligand L. 4(H$_2$O): 1,9-Diformyl-1,10-phenanthroline (0.001 mmol, 0.200 g) (17) was dissolved in 50 mL of CH_3OH. To this methanolic solution, 2-(aminomethyl) piperidine (0.002 mmol, 0.184 g) was added drop by drop. This reaction mixture was stirred continuously for 48 h at 50 °C. During this, the color of the reaction mixture changed to red-brown, indicating the formation of a Schiff base. The solvent was removed under vacuum, and the resultant orange-red powder was isolated. Yield. 70%. -^{1}H NMR (CDCl$_3$, 500 MHz): δ = 8.26, 8.24 (*dd*, *J* = 3 Hz, 2H), 7.86, 784, 7.82(*t*, *J* = 10 Hz, 4H), 7.77(*s*, 2H), 3.55(*brs*, 2H), 3.27(*m*, 2H), 3.09, 3.07, 3.05, 3.03 (*q*, *J* = 10 Hz,2H), 2.83, 2.81, 2.79 (*t*, *J* = 12 Hz, 2H), 2.50 (*m*, 2H), 2.24, 2.22, 2.20 (*t*, *J* = 11 Hz, 2H), 1.93–1.83(*dd*, *J* = 12 Hz, 4H), 1.60–1.50(*m*, 6H), 1.34–1.31 (*m*, 2H). ^{13}C NMR, (CDCl$_3$, 125. MHz) δ = 160.22, 145.27, 136.89, 128.66, 126.30, 122.75, 122.64, 83.14, 83.06, 64.20, 50.66, 48.99, 28.64, 24.91, 23.95. DEPT-135°. 131.59, 121.00, 117.44, 117.34, 77.85, 58.93 (CH, UP), 45.40, 43.72, 23.37, 19.64, 18.68 (CH$_2$, DOWN). IR (KBr): v cm^{-1} = 3418 (br), 1616 (s), 1598 (s), 1370 (s). UV vis (CH$_3$OH, nm (ε, M^{-1} cm^{-1})): λ = 275 (32170), 234(31530); ESI[MS]$^+$ in methanol: *m/z* (calcd (found)) 429.28 (429.69 for L+H$^+$; 100% abundance); 451.26 (451.69 for L+Na$^+$; 90% abundance). Elemental data: Calc (found) for C$_{26}$H$_{32}$N$_6$.4H$_2$O: C 62.38 (62.44), H 8.05 (7.74), N 16.79 (16.22)%.

4.2. Synthesis of Complexes

General synthetic procedure for compounds 1–3. The methanolic solution of the ligand **L** (0.001 mmol) and LnCl$_3$ salt (0.001 mmol) was mixed together and allowed for constant stirring at room temperature for 4 h. After completion of the reaction, the solution was evaporated by rotary and the solid was further dried under vacuum.

Complex 1. Yield: 75%. IR (KBr): v cm^{-1} = 3398 (br), 1623 (s), 1458 (m), 1432 (m) 1400 (s), UV vis (HEPES Buffer, pH 7.4, nm (ε, M^{-1} cm^{-1})): λ = 237 (21,392), 285 (20,212); ESI-[MS]$^+$ in methanol: *m/z* calcd(found) 687.10 (687.12) for ([EuL(Cl)$_3$+H$^+$]; 65% abundance), 651.13(651.15) for ([EuL(Cl)$_2$]$^+$; 100% abundance). Elemental data: Calc (found) for C$_{26}$H$_{70}$Cl$_3$EuN$_6$O$_{19}$, C 30.34(30.54), H 6.86 (6.51), N 8.17 (8.24)%.

Complex 2. Yield: 68%. IR (KBr): v cm^{-1} = 3436 (br), 1629 (s), 1459 (m), 1431 (m) 1386 (s), -UV vis (HEPES Buffer, pH 7.4, nm (ε, M^{-1} cm^{-1})): λ = 236 (16,825), 287 (14,175); ESI-[MS]$^+$ in methanol: m/z calcd(found) 706.27 (706.21) for ([Sm(**L-2H).4**(CH$_3$OH)]$^+$; 100% abundance); 770.33 (770.27) for ([Sm**L-2H).6**(CH$_3$OH)]$^+$; 100% abundance). Elemental data: Calc (found) for C$_{26}$H$_{66}$Cl$_3$N$_6$O$_{17}$Sm: C 31.49(31.11), H 6.71(6.65), N 8.48(8.54)%.

Complex 3. Yield 65%. IR (KBr): v cm^{-1} = 3432 (br), 1627(s), 1459 (m), 1432 (m) 1390 (s). -UV vis (HEPES Buffer, pH 7.4, nm (ε, M^{-1} cm^{-1})): λ = 236 (215,725), 285 (13,917); ESI-[MS]$^+$ in methanol: m/z calcd(found) 693.11.(693.09) for ([TbL(Cl)$_3$+H$^+$]; 90% abundance); 657.13 (657.12) for ([TbL(Cl)$_2$]$^+$; 100% abundance). Elemental data: Calc (found) for C$_{26}$H$_{70}$Cl$_3$N$_6$O$_{19}$Tb: C 30.14(29.68), H 6.81(7.05), N 8.11(8.34)%.

Detection limit (DL) calculation:

$$DL = CL \times ET \tag{4}$$

where DL = detection limit, CL = Concentration of complex, and ET = Equivalent of Titrant at which change was observed. Here, the titrant is phosphate.

Supplementary Materials: The following are available online at https://www.mdpi.com/2079-4991/11/1/53/s1, Figure S1: ^{1}H NMR of **L** in CDCl$_3$, Figure S2: ^{13}C NMR of **L** in CDCl$_3$, Figure S3: DEPT-135° NMR of **L** in CDCl$_3$, Figure S4: ESI-MS Spectrum of **L**, Figure S5:CD spectra of Ligand **L** in CHCl$_3$, Figure S6: IR-Spectra of **L**, **1**, **2**, and **3**, Figure S7: ESI-MS Spectrum of **1**, Figure S8: ESI-MS Spectrum of **2**, Figure S9: ESI-MS Spectrum of **3**, Figure S10: Normalization Spectra, Figure S11: Emission Curve of **1** against HPO$_4^{2-}$, Figure S12: ESI-MS spectrum of [**1**]:2[HPO$_4^{2-}$], Figure S13: Non-linear fit curve of **1** against HCO$_3^{-}$, Figure S14: Emission curve of **2** against HPO$_4^{2-}$, Figure S15: ESI-MS spectrum of [**2**]:2[HPO$_4^{2-}$], Figure S16: Emission curve of **3** against HPO$_4^{2-}$, Figure S17: ESI-MS spectra of [**3**]:2[HPO$_4^{2-}$], Figure S18: (a) UV-vis spectra of ligand **L** and its complexes **1**, **2** and **3** and (b) possible energy transfer, Figure S19: Excited state lifetime of complex **1** with HPO$_4^{2-}$ with 1:1 ratio, Figure S20: Excited state lifetime of complex **1** with HPO$_4^{2-}$ with 1:2 ratio, Figure S21: Excited state lifetime of complex **1** with HPO$_4^{2-}$ with 1:10 ratio, Table S1: Quantum yield calculation for complex **1**, **2** and **3**.

Author Contributions: Concept, J.S., P.S.S.; methodology, J.S., P.S.S.; experiments and data collections, J.S., S.J., P.B.C.; validation, J.S., P.S.S., H.S.J.; writing—original draft, J.S., P.S.S., H.S.J.; formatting, H.S.J.; funding acquisition, H.S.J. All authors have read and agreed to the published version of the manuscript.

Funding: This research received no external funding.

Institutional Review Board Statement: Not applicable.

Informed Consent Statement: Not applicable.

Data Availability Statement: Data used to support the findings of this study are included within the article and supporting material.

Acknowledgments: The manuscript has been assigned CSMCRI communication number CSIR-CSMCRI 193/2015. J.S. acknowledges CSIR, Govt. of India, New Delhi for the CSIR-SRF award. The members of Analytical Division and Centralized Instrument Facility (AD&CIF) of CSIR-CSMCRI are gratefully acknowledged for their analytical supports. H.S.J. thanks FWO [PEGASUS]^2 Marie Sklodowska-Curie grant agreement No. 665501 for the incoming Post-doctoral Fellowship.

Conflicts of Interest: The authors declare no conflict of interest.

References

1. Saenger, W. *Principles of Nucleic Acid Structure*; Springer: New York, NY, USA, 1984; ISBN 0-387-90762-9.
2. Massue, J.; Quinn, S.J.; Gunnlaugsson, T. Lanthanide Luminescent Displacement Assays: The Sensing of Phosphate Anions Using Eu(III)−Cyclen-Conjugated Gold Nanoparticles in Aqueous Solution. *J. Am. Chem. Soc.* **2008**, *130*, 6900–6901. [CrossRef]
3. Bowler, M.W.; Cliff, M.J.; Waltho, J.P.; Blackburn, G.M. Why did Nature select phosphate for its dominant roles in biology? *New. J. Chem.* **2010**, *34*, 784–794. [CrossRef]
4. Vullev, V.I.; Jones, G. Photoinduced charge transfer in helical polypeptides. *Res. Chem. Intermed.* **2002**, *28*, 795–815. [CrossRef]
5. Barela, T.D.; Sherry, A.D. A simple, one-step fluorometric method for determination of nanomolar concentrations of terbium. *Anal. Biochem.* **1976**, *71*, 351–352. [CrossRef]

6.	Cable, M.L.; Kirby, J.P.; Levine, D.J.; Manary, M.J.; Gray, H.B.; Ponce, A. Detection of Bacterial Spores with Lanthanide—Macrocycle Binary Complexes. *J. Am. Chem. Soc.* **2009**, *131*, 9562–9570. [CrossRef] [PubMed]

7.	Caffrey, D.F.; Gunnlaugsson, T. Displacement assay detection by a dimeric lanthanide luminescent ternary Tb(iii)–cyclen complex: High selectivity for phosphate and nitrate anions. *Dalton Trans.* **2014**, *43*, 17964–17970. [CrossRef] [PubMed]

8.	Li, S.-H.; Yuan, W.-T.; Zhu, C.-Q.; Xu, J.-G. Species-differentiable sensing of phosphate-containing anions in neutral aqueous solution based on coordinatively unsaturated lanthanide complex probes. *Anal. Biochem.* **2004**, *331*, 235–242. [CrossRef] [PubMed]

9.	Coates, J.; Gay, E.; Sammes, P.G. Anion effects on the luminescence of europium complexes. *Dyes Pigm.* **1997**, *34*, 195–205. [CrossRef]

10.	Zyryanov, G.V.; Palacios, M.A.; Anzenbacher, P., Jr. Rational design of a fluorescence-turn-on sensor array for phosphates in blood serum. *Angew. Chem. Int. Ed.* **2007**, *46*, 7849–7852. [CrossRef]

11.	Xu, W.; Zhou, Y.; Huang, D.; Su, M.; Wang, K.; Xiang, M.; Hong, M. Luminescent sensing profiles based on anion-responsive lanthanide(iii) quinolinecarboxylate materials: Solid-state structures, photophysical properties, and anionic species recognition. *J. Mater. Chem. C* **2015**, *3*, 2003–2015. [CrossRef]

12.	Wang, Y.-W.; Liu, S.-B.; Yang, Y.-L.; Wang, P.-Z.; Zhang, A.-J.; Peng, Y. A Terbium(III)-Complex-Based On–Off Fluorescent Chemosensor for Phosphate Anions in Aqueous Solution and Its Application in Molecular Logic Gates. *ACS Appl. Mater. Interfaces* **2015**, *7*, 4415–4422. [CrossRef] [PubMed]

13.	Xu, H.; Cao, C.-S.; Zhao, B. A water-stable lanthanide-organic framework as a recyclable luminescent probe for detecting pollutant phosphorus anions. *Chem.Commun.* **2015**, *51*, 10280–10283. [CrossRef] [PubMed]

14.	Toussaint, N.D.; Pedagogos, E.; Tan, S.-J.; Badve, S.V.; Hawley, C.M.; Perkovic, V.; Elder, G.J. Phosphate in early chronic kidney disease: Associations with clinical outcomes and a target to reduce cardiovascular risk. *Nephrology* **2012**, *17*, 433–444. [CrossRef] [PubMed]

15.	Chana, T.O.; Zhang, J.; Rodeck, U.; Pascal, J.M.; Armen, R.S.; Spring, M.; Dumitru, C.D.; Myers, V.; Li, X.; Cheung, J.Y.; et al. Resistance of Akt kinases to dephosphorylation through ATP-dependent conformational plasticity. *Proc. Natl. Acad. Sci. USA* **2011**, *108*, E1120–E1127. [CrossRef] [PubMed]

16.	Pradhan, S.M.; Pokhrel, R. Spectrophotometric determination of phosphate in sugarcane juice, fertilizer, detergent and water samples by molybdenum blue method. *Sci. World* **2013**, *11*, 58–62. [CrossRef]

17.	Kipngetich, T.E.; Hillary, M.; Swamy, T.A. Determination of levels of phosphates and sulphates in domestic water from three selected springs in Nandi County, Kenya. *Int. J. Pharm. Life Sci.* **2013**, *4*, 2828–2833.

18.	Sahu, S.K.; Ajmal, P.Y.; Bhangare, R.C.; Tiwari, M.; Pandit, G.G. Natural radioactivity assessment of a phosphate fertilizer plant area. *J. Radiat. Res. Appl. Sci.* **2014**, *7*, 123–128. [CrossRef]

19.	Gaddaum, J.H. The Estimation of Phosphorus in Blood. *Biochem. J.* **1926**, *20*, 1204–1207. [CrossRef]

20.	Bünzli, J.-C.G. Lanthanide Luminescence for Biomedical Analyses and Imaging. *Chem. Rev.* **2010**, *110*, 2729–2755. [CrossRef]

21.	McCleskey, S.C.; Griffin, M.J.; Schneider, S.E.; McDevitt, J.T.; Anslyn, E.V. Differential Receptors Create Patterns Diagnostic for ATP and GTP. *J. Am. Chem. Soc.* **2003**, *125*, 1114–1115. [CrossRef]

22.	Wang, X.; Chang, H.; Xie, J.; Zhao, B.; Liu, B.; Xu, S.; Pei, W.; Ren, N.; Huang, L.; Huang, W. Recent developments in lanthanide-based luminescent probes. *Coord. Chem. Rev.* **2014**, *273–274*, 201–212. [CrossRef]

23.	Hong, J.; Pei, D.; Guo, X. Quantum dot-Eu3+ conjugate as a luminescence turn-on sensor for ultrasensitive detection of nucleoside triphosphates. *Talanta* **2012**, *99*, 939–943. [CrossRef] [PubMed]

24.	Weitz, E.A.; Chang, J.Y.; Rosenfield, A.H.; Morrow, E.A.; Pierre, V.C. The basis for the molecular recognition and the selective time-gated luminescence detection of ATP and GTP by a lanthanide complex. *Chem. Sci.* **2013**, *4*, 4052–4060. [CrossRef]

25.	Mizukami, S.; Agano, T.; Urano, Y.; Odani, A.; Kikuchi, K. A Fluorescent Anion Sensor That Works in Neutral Aqueous Solution for Bioanalytical Application. *J. Am. Chem. Soc.* **2002**, *124*, 3920–3925. [CrossRef] [PubMed]

26.	Schneider, S.E.; O'Neil, S.N.; Anslyn, E.V. Coupling Rational Design with Libraries Leads to the Production of an ATP Selective Chemosensor. *J. Am. Chem. Soc.* **2000**, *122*, 542–543. [CrossRef]

27.	Epstein, D.M.; Chappell, L.L.; Khalili, H.; Supkowski, R.M. Eu(III) Macrocyclic Complexes Promote Cleavage of and Bind to Models for the 5′-Cap of mRNA. Effect of Pendent Group and a Second Metal Ion. *Inorg. Chem.* **2000**, *39*, 2130–2134. [CrossRef]

28.	Shao, N.; Jin, J.; Wang, G.; Zhang, Y.; Yang, R.; Yuan, J. Europium(iii) complex-based luminescent sensing probes for multi-phosphate anions: Modulating selectivity by ligand choice. *Chem. Commun.* **2008**, 1127–1129. [CrossRef]

29.	Butler, S.J. Ratiometric Detection of Adenosine Triphosphate (ATP) in Water and Real-Time Monitoring of Apyrase Activity with a Tripodal Zinc Complex. *Chem. Eur. J.* **2014**, *20*, 15768–15774. [CrossRef]

30.	Rao, A.S.; Kim, D.; Nam, H.; Jo, H.; Kim, K.H.; Ban, C.; Ahn, K.H. A turn-on two-photonfluorescent probe for ATP and ADP. *Chem. Commun.* **2012**, *48*, 3206–3208.

31.	Kaur, J.; Singh, P. ATP selective acridone based fluorescent probes for monitoring of metabolic events. *Chem. Commun.* **2011**, *47*, 4472–4474. [CrossRef]

32.	Moro, A.J.; Cywinski, P.J.; Korsten, S.; Mohr, G.J. An ATP fluorescent chemosensor based on a Zn(ii)-complexed dipicolylaminereceptor coupled with a naphthalimidechromophore. *Chem. Commun.* **2010**, *46*, 1085–1087. [CrossRef] [PubMed]

33.	Mahato, P.; Ghosh, A.; Mishra, S.K.; Shrivastav, A.; Mishra, S.; Das, A. Zn(II) based colorimetric sensor for ATP and its use as a viable staining agent in pure aqueous media of pH 7.2. *Chem. Commun.* **2010**, *46*, 9134–9136. [CrossRef] [PubMed]

34. Ngo, H.T.; Liu, X.; Jolliffe, K.A. Anion recognition and sensing with Zn(ii)–dipicolylamine complexes. *Chem. Soc. Rev.* **2012**, *41*, 4928–4965. [CrossRef] [PubMed]

35. Ojida, A.; Takashima, I.; Kohira, T.; Nonaka, H.; Hamachi, I. Turn-On Fluorescence Sensing of Nucleoside Polyphosphates Using a Xanthene-Based Zn(II) Complex Chemosensor. *J. Am. Chem. Soc.* **2008**, *130*, 12095–12101. [CrossRef] [PubMed]

36. Zhou, Y.; Xu, Z.; Yoon, J. Fluorescent and colorimetric chemosensors for detection of nucleotides, FAD and NADH: Highlighted research during 2004–2010. *Chem. Soc. Rev.* **2011**, *40*, 2222–2235. [CrossRef]

37. Santos-Figueroa, E.L.; Moragues, M.E.; Climent, E.; Agostini, A.; Martınez-Manez, R.; Sancenon, F. Chromogenic and fluorogenic chemosensors and reagents for anions. A comprehensive review of the years 2010–2011. *Chem. Soc. Rev.* **2013**, *42*, 3489–3613. [CrossRef]

38. Xu, Z.; Singh, N.J.; Lim, J.; Pan, J.; Kim, H.N.; Park, S.; Kim, K.S.; Yoon, J. Unique Sandwich Stacking of Pyrene-Adenine-Pyrene for Selective and Ratiometric Fluorescent Sensing of ATP at Physiological pH. *J. Am. Chem. Soc.* **2009**, *131*, 15528–15533. [CrossRef]

39. Montgomery, C.P.; Murray, B.S.; New, E.J.; Pal, R.; Parker, D. Cell-Penetrating Metal Complex Optical Probes: Targeted and Responsive Systems Based on Lanthanide Luminescence. *Acc. Chem. Res.* **2009**, *42*, 925–937. [CrossRef]

40. Bünzli, J.-C.G.; Comby, S.; Chauvin, A.-S.; Vandevyver, C.D.B. New opportunities for lanthanide luminescence. *J. Rare. Earths.* **2007**, *25*, 257–274.

41. Comby, S.; Surender, E.M.; Kotova, O.; Truman, L.K.; Molloy, J.K.; Gunnalaugsson, T. Lanthanide-Functionalized Nanoparticles as MRI and Luminescent Probes for Sensing and/or Imaging Applications. *Inorg. Chem.* **2014**, *53*, 1867–1879. [CrossRef]

42. Major, J.L.; Meade, T.J. Bioresponsive, Cell-Penetrating, and Multimeric MR Contrast Agents. *Acc. Chem. Res.* **2009**, *42*, 893–903. [CrossRef] [PubMed]

43. Werner, E.J.; Datta, A.; Jocher, C.J.; Raymond, K.N. High-relaxivity MRI contrast agents: Where coordination chemistry meets medical imaging. *Angew. Chem. Int. Ed.* **2008**, *47*, 8568–8580. [CrossRef] [PubMed]

44. Caravan, P.; Ellison, J.J.; McMurry, T.J.; Lauffer, R.B. Gadolinium(III) Chelates as MRI Contrast Agents: Structure, Dynamics, and Applications. *Chem. Rev.* **1999**, *99*, 2293–2352. [CrossRef] [PubMed]

45. Caravan, P. Protein-Targeted Gadolinium-Based Magnetic Resonance Imaging (MRI) Contrast Agents: Design and Mechanism of Action. *Acc. Chem. Res.* **2009**, *42*, 851–862. [CrossRef]

46. Bottrill, M.; Kwok, L.; Long, N.J. Lanthanides in magnetic resonance imaging. *Chem. Soc. Rev.* **2006**, *35*, 557–571. [CrossRef]

47. Doble, D.M.J.; Melchior, M.; O'Sullivan, B.; Siering, C.; Xu, J.; Pierre, V.C.; Raymond, K.N. Toward Optimized High-Relaxivity MRI Agents: The Effect of Ligand Basicity on the Thermodynamic Stability of Hexadentate Hydroxypyridonate/Catecholate Gadolinium(III) Complexes. *Inorg. Chem.* **2003**, *42*, 4930–4937. [CrossRef]

48. New, E.J.; Parker, D.; Peacock, R.D. Spin state tuning of non-heme iron-catalyzed hydrocarbon oxidations: Participation of FeIII–OOH and FeV[double bond, length as m-dash]O intermediates. *Dalton Trans.* **2009**, *38*, 672–679. [CrossRef]

49. Smith, D.G.; Mcmahon, B.K.; Pal, R.; Parker, D. Live cell imaging of lysosomal pH changes with pH responsive ratiometric lanthanide probes. *Chem. Commun.* **2012**, *48*, 8520–8522. [CrossRef]

50. Katz, L.A.; Swain, J.A.; Portman, M.A.; Balaban, R.S. Intracellular pH and inorganic phosphate content of heart in vivo: A 31P-NMR study. *Am. J. Physiol. Heart Circ. Physiol.* **1988**, *255*, H189–H196. [CrossRef]

51. Heipertz, R.; Eickhoff, K.; Karstens, K.H. Cerebrospinal fluid concentrations of magnesium and inorganic phophate in epilepsy. *J. Neurol. Sci.* **1979**, *41*, 55–60. [CrossRef]

52. Butler, S.J.; Parker, D. Anion binding in water at lanthanide centres: From structure and selectivity to signalling and sensing. *Chem. Soc. Rev.* **2015**, *42*, 1652–1666. [CrossRef] [PubMed]

53. Schferling, M.; Wolfbeis, O.S. Europium Tetracycline as a Luminescent Probe for Nucleoside Phosphates and Its Application to the Determination of Kinase Activity. *Chem. Eur. J.* **2007**, *13*, 4342–4349. [CrossRef]

54. Duerkop, A.; Turel, M.; Lobnik, A.; Wolfbeis, O.S. Microtiter plate assay for phosphate using a europium–tetracycline complex as a sensitive luminescent probe. *Anal. Chim. Acta.* **2006**, *555*, 292–298. [CrossRef]

55. Bergwitz, C.; Jüppner, H. Phosphate sensing. *Adv. Chronic Kidney Dis.* **2011**, *18*, 132–144. [CrossRef] [PubMed]

56. Nadella, S.; Selvakumar, P.M.; Suresh, E.; Subramanian, P.S.; Albrecht, M.; Giese, M.; Fröhlich, R. Lanthanide(III) Complexes of Bis-semicarbazone and Bis-imine-Substituted Phenanthroline Ligands: Solid-State Structures, Photophysical Properties, and Anion Sensing. *Chem. Eur. J.* **2012**, *18*, 16784–16792. [CrossRef]

57. Nadella, S.; Sahoo, J.; Subramanian, P.S.; Sahu, A.; Mishra, S.; Albrecht, M. Sensing of Phosphates by Using Luminescent EuIII and TbIII Complexes: Application to the Microalgal Cell Chlorella vulgaris. *Chem. Eur. J.* **2014**, *20*, 6047–6053. [CrossRef] [PubMed]

58. Shavaleev, N.M.; Scopelliti, R.; Gumy, F.; Bunzli, J.-C.G. Benzothiazole- and Benzoxazole-Substituted Pyridine-2-Carboxylates as Efficient Sensitizers of Europium Luminescence. *Inorg. Chem.* **2009**, *48*, 6178–6191. [CrossRef] [PubMed]

59. Kumar, M.; Kumar, N.; Bhalla, V. Highly selective fluorescent probe for detection and visualization of palladium ions in mixed aqueous media. *RSC Adv.* **2013**, *3*, 1097–1102. [CrossRef]

60. Long, G.L.; Winefordner, J.D. Limit of Detection A Closer Look at the IUPAC Definition. *Anal. Chem.* **1983**, *55*, 712A–724A.

61. Gole, B.; Bar, A.K.; Mukherjee, P.S. Modification of Extended Open Frameworks with Fluorescent Tags for Sensing Explosives: Competition between Size Selectivity and Electron Deficiency. *Chem. Eur. J.* **2014**, *20*, 2276–2291. [CrossRef]

62. Gole, B.; Bar, A.K.; Mukherjee, P.S. Multicomponent Assembly of Fluorescent-Tag Functionalized Ligands in Metal–Organic Frameworks for Sensing Explosives. *Chem. Eur. J.* **2014**, *20*, 13321–13336. [CrossRef]

63. Thordarson, P. Determining association constants from titration experiments in supramolecular chemistry. *Chem. Soc. Rev.* **2011**, *40*, 1305–1323. [CrossRef] [PubMed]
64. Raju, M.; Patel, T.J.; Nair, R.R.; Chatterjee, P.B. Xanthurenic acid: A natural ionophore with high selectivity and sensitivity for potassium ions in an aqueous solution. *New J. Chem.* **2016**, *40*, 1930–1934. [CrossRef]
65. Raju, M.; Nair, R.R.; Raval, I.H.; Haldar, S.; Chatterjee, P.B. Reporting a new siderophore based Ca2+ selective chemosensor that works as a staining agent in the live organism Artemia. *Analyst* **2015**, *140*, 7799–7809. [CrossRef] [PubMed]
66. Li, P.; Zhou, X.; Huang, R.; Yang, L.X.; Tang, X.; Dou, W.; Zhao, Q.; Liu, W. A highly fluorescent chemosensor for Zn2+ and the recognition research on distinguishing Zn^{2+} from Cd^{2+}. *Dalton Trans.* **2014**, *43*, 706–713. [CrossRef] [PubMed]
67. Gusev, A.N.; Hasegawa, M.; Nishchymenko, G.A.F.; Shul'gin, V.B.; Meshkova, S.; Doga, P.; Linert, W. Ln(iii) complexes of a bis(5-(pyridine-2-yl)-1,2,4-triazol-3-yl)methaneligand: Synthesis, structure and fluorescent properties. *Dalton Trans.* **2013**, *42*, 6936–6943. [CrossRef] [PubMed]
68. Horrocks, W.D., Jr.; Sudnick, D.R. Lanthanide ion probes of structure in biology. Laser-induced luminescence decay constants provide a direct measure of the number of metal-coordinated water molecules. *J. Am. Chem. Soc.* **1979**, *101*, 334–340.
69. Beeby, A.; Clarkson, I.M.; Dickins, R.S.; Faulkner, S.; Parker, D.; Royle, L.; de Sousa, A.S.; Williams, J.A.G.; Woods, M. Non-radiative deactivation of the excited states of europium, terbium and ytterbium complexes by proximate energy-matched OH, NH and CH oscillators: An improved luminescence method for establishing solution hydration states. *J. Chem. Soc. Perkin Trans.* **1999**, *2*, 493–503. [CrossRef]
70. Zhang, P.; Kimura, T.; Yoshida, Z. Luminescence Study on the Inner-Sphere Hydration Number of Lanthanide(III) Ions in Neutral Organo-Phosphorus Complexes. *Solvent Extr. Ion. Exch.* **2004**, *22*, 933–945. [CrossRef]
71. Dos Santos, C.M.G.; Gunnlaugsson, T. Exploring the luminescent sensing of anions by the use of an urea functionalised 1,10-phenanthroline (phen)-based (3:1) Eu(III) complex. *Supramol. Chem.* **2009**, *21*, 173–180. [CrossRef]
72. Bridou, L.; Nielsen, L.G.; Sørensen, T.J. Using europium(III) complex of 1,4,7,10-tetraazacyclododecane-1,4,7-triacedic acid Eu.DO3A as a luminescent sensor for bicarbonate. *J. Rare Earths* **2020**, *38*, 498–505. [CrossRef]
73. McMahon, B.; Mauer, P.; McCoy, C.P.; Lee, T.C.; Gunnlaugsson, T. Selective Imaging of Damaged Bone Structure (Microcracks) Using a Targeting Supramolecular Eu(III) Complex As a Lanthanide Luminescent Contrast Agent. *J. Am. Chem. Soc.* **2009**, *131*, 17542–17543. [CrossRef] [PubMed]
74. Law, G.-L.; Man, C.; Parker, D.; Walton, J.W. Observation of the selective staining of chromosomal DNA in dividing cells using a luminescent terbium(iii) complex. *Chem. Commun.* **2010**, *46*, 2391–2393. [CrossRef] [PubMed]

Article

Quenching of the Eu^{3+} Luminescence by Cu^{2+} Ions in the Nanosized Hydroxyapatite Designed for Future Bio-Detection

Katarzyna Szyszka [1,*], Sara Targońska [1], Agnieszka Lewińska [2], Adam Watras [1] and Rafal J. Wiglusz [1,3,*]

[1] Institute of Low Temperature and Structure Research, PAS, Okolna 2, 50-422 Wroclaw, Poland; s.targonska@intibs.pl (S.T.); a.watras@intibs.pl (A.W.)
[2] Faculty of Chemistry, University of Wroclaw, Joliot-Curie 14, 50-383 Wroclaw, Poland; agnieszka.lewinska@chem.uni.wroc.pl
[3] International Institute of Translational Medicine, Jesionowa 11 St., 55–124 Malin, Poland
* Correspondence: k.szyszka@intibs.pl (K.S.); r.wiglusz@intibs.pl (R.J.W.); Tel.: +48-71-3954-274 (K.S.); +48-71-3954-159 (R.J.W.)

Abstract: The hydroxyapatite nanopowders of the Eu^{3+}-doped, Cu^{2+}-doped, and Eu^{3+}/Cu^{2+}-co-doped $Ca_{10}(PO_4)_6(OH)_2$ were prepared by a microwave-assisted hydrothermal method. The structural and morphological properties of the products were investigated by X-ray powder diffraction (XRD), transmission electron microscopy techniques (TEM), and infrared spectroscopy (FT-IR). The average crystal size and the unit cell parameters were calculated by a Rietveld refinement tool. The absorption, emission excitation, emission, and luminescence decay time were recorded and studied in detail. The $^5D_0 \rightarrow {}^7F_2$ transition is the most intense transition. The Eu^{3+} ions occupied two independent crystallographic sites in these materials exhibited in emission spectra: one Ca(1) site with C_3 symmetry and one Ca(2) sites with C_s symmetry. The Eu^{3+} emission is strongly quenched by Cu^{2+} ions, and the luminescence decay time is much shorter in the case of Eu^{3+}/Cu^{2+} co-doped materials than in Eu^{3+}-doped materials. The luminescence quenching mechanism as well as the schematic energy level diagram showing the Eu^{3+} emission quenching mechanism using Cu^{2+} ions are proposed. The electron paramagnetic resonance (EPR) technique revealed the existence of at least two different coordination environments for copper(II) ion.

Keywords: apatite; europium ions; cooper ions; photoluminescence spectroscopy; EPR spectroscopy

Citation: Szyszka, K.; Targońska, S.; Lewińska, A.; Watras, A.; Wiglusz, R.J. Quenching of the Eu^{3+} Luminescence by Cu^{2+} Ions in the Nanosized Hydroxyapatite Designed for Future Bio-Detection. *Nanomaterials* **2021**, *11*, 464. https://doi.org/10.3390/nano11020464

Academic Editor: Jiangshan Chen

Received: 31 December 2020
Accepted: 8 February 2021
Published: 11 February 2021

Publisher's Note: MDPI stays neutral with regard to jurisdictional claims in published maps and institutional affiliations.

1. Introduction

Apatite-type materials can be applied in many industrial fields, e.g., as sorbents, biocompatible and biodegradable materials for bone and teeth reconstruction, catalysts, materials for the wastewater treatment, fertilizers, and luminescent materials [1,2]. Hydroxyapatite ($Ca_{10}(PO_4)_6(OH)_2$—abbr. HAp) is used in medicine as a bone implant material due to its biocompatibility, bioactivity, and similarity to bone mineral [3,4]. However, it is still widely investigated in order to improve its properties by obtaining appropriate grain size, morphology, mechanical strength, and solubility and by adding some dopants that are, e.g., naturally built into bone apatite, ions possessing antibacterial properties, or ions enabling bio-imaging [5,6]. Infections after grafting of bone implant material are a serious problem in surgery. The idea is that doping with antibacterial ions into the grafted biomaterial will prevent bacterial biofilm formation and infection development. Inorganic antibacterial agents possess advantages such as stability and safety. The antibacterial agents include ions such as copper, silver, and zinc [7–9], and several studies have shown that they can play important roles in the prevention or minimization of initial bacterial adhesion [10,11]. Metal ions can react with microbial membrane, causing structural changes and permeability. Then, Ag^+ and Cu^{2+} ions have the ability to complex anions such as $-NH_2$, $-S-S-$, and $-CONH-$ of the proteins or enzymes in the bacterial cells. It provides bacterial DNA and RNA damage and inhibits proliferation [7,11,12]. Copper is an essential microelement that is involved in many metabolic processes that taken place in human bodies [7,11,13,14]. However, copper ions may have potentially toxic effects at higher amount in human beings due to their ability to generate ROSs (Reactive Oxygen Species). On the other hand, recent studies have shown that copper-doped apatite-type materials are very promising as a new kind of ow-toxic pigment that can be used in the paint and varnish industry [1,15–18].

Lanthanide(III)-doped nanomaterials are promising candidates for fluorescent bio-labels due to their stable luminescence over time, high photochemical stability, sharp emission peak, low levels of photobleaching, and toxicity compared with organic fluorophores. Eu^{3+} ions are structural and luminescence probe-sensitive to changes in the local environment around the ion. Furthermore, the luminescence of Eu^{3+} ions are identified by a narrow emission band and long lifetimes of the excited state [19,20].

Apatite is a big family of compounds, and it is widely investigated due to its outstanding properties such as good biocompatibility or possibility to be doped with different ions in a broad concentration range for applications in the industry, in medicine, etc. There are a lot of papers focusing on apatite synthesis [21–24]; its doping with antibacterial ions [9,10,25]; as well as its doping with luminescence ions such as Eu^{3+} [5,19,26,27], Tb^{3+} [28,29], Eu^{3+}/Tb^{3+} [30], Er^{3+}/Yb^{3+}, or Eu^{3+}/Cu^{2+} [17]. Moreover, there is a lot of research focusing on anion-substituted apatite such as silicate [31,32], vanadate [33], borate [34], or carbonate [35].

In the presented work, the synthesis, structural, morphological, and luminescence properties of Eu^{3+}/Cu^{2+} co-doped HAp were investigated attentively. To the best of our knowledge, this is the first time that the quenching mechanism in this system has been elucidated.

2. Materials and Methods

2.1. Synthesis

The $Ca_{10}(PO_4)_6(OH)_2$ nanopowders doped with Eu^{3+} and Cu^{2+} ions were synthesized by a microwave-assisted hydrothermal method. The starting materials used were $CaCO_3$ (99.0%, Alfa Aesar, Karlsruhe, Germany), $NH_4H_2PO_4$ (99.0%, Fluka, Bucharest, Romania), Eu_2O_3 (99.99%, Alfa Aesar, Karlsruhe, Germany), $Cu(NO_3)_2 \cdot 2.5H_2O$ (98.0–102.0%, Alfa Aesar, Karlsruhe, Germany), and $NH_3 \cdot H_2O$ (99%, Avantor, Gliwice, Poland) as a pH regulation reagent. The concentration of dopants was calculated based on inductively coupled plasma-optical emission spectrometer (ICP-OES) results. The concentrations of europium ions were 0.5 mol%, 1 mol%, 2 mol%, and 5 mol%, and the concentrations of the cooper ions were 2 mol% and 5 mol% to the overall molar content of calcium cations. First, the stoichiometric amounts of $CaCO_3$ as well as Eu_2O_3 were separately digested in excess of HNO_3 (suprapur Merck, Darmstadt, Germany) to obtain water-soluble nitrates. The obtained europium nitrate hydrate was recrystallized three times to remove excess HNO_3. Then, the stoichiometric amount of $Eu(NO_3)_3$ was dissolved in deionized water, and then, the $Cu(NO_3)_2$ was added to the stoichiometric amount of calcium nitrate. After this, $NH_4H_2PO_4$ was added to the abovementioned mixture and the pH value was adjusted to 9 by ammonia. The suspension was transferred to a Teflon vessel and was placed into the microwave reactor (ERTEC MV 02-02, Wrocław, Poland). The reaction system was heat-treated at 280 °C for 90 min under autogenous pressure of 60 atm. The obtained product was washed several times with deionized water and dried at 70 °C for 24 h.

2.2. Powder Characterization

The crystal structure and phase purity were studied using a PANalytical X'Pert Pro diffractometer (Malvern Panalytical Ltd., Malvern, UK) equipped with Ni-filtered Cu Kα radiation (V = 40 kV and I = 30 mA). The recorded X-ray powder diffraction patterns (XRD) were compared with the reference standard of hexagonal calcium hydroxyapatite ($P6_3/m$) from the Inorganic Crystal Structure Database (ICSD-2866) and analyzed. Rietveld structural refinement was performed with the aid of a Maud program (version 2.93) (University of Trento-Italy, Department of Industrial Engineering, Trento, Italy) [36,37] based on the apatite hexagonal crystal structure with better approximation and indexing of the Crystallographic Information File (CIF). The quality of structural refinement was checked by R-values (R_w, R_{wnb}, R_{all}, R_{nb}, and σ), which were followed to get a structural refinement with better quality and reliability.

The morphology was investigated by high-resolution transmission electron microscopy (HRTEM) using a Philips CM-20 SuperTwin microscope (Eindhoven, The Netherlands), operating at 200 kV. The specimen for the HRTEM measurement was obtained by dispersing a small amount of powder in methanol and by putting a droplet of the suspension onto a copper microscope grid covered with carbon.

Fourier transform infrared spectra were measured using a Thermo Scientific Nicolet iS50 FT-IR spectrometer (Waltham, MA, USA) in the range of 4000–400 cm^{-1} at 295 K. Absorption spectra were recorded with an Agilent Cary 5000 spectrophotometer, employing a spectral bandwidth (SBW) of 0.1 nm in the visible and ultraviolet range and of 0.7 nm in the infrared. The spectra were recorded at room temperature.

The excitation spectra were recorded with the aid of an FLS980 Fluorescence Spectrometer (Edinburgh Instruments, Kirkton Campus, UK) equipped with 450 W Xenon lamp. The excitation of 300 mm focal length monochromator was in Czerny–Turner configuration and the excitation arm was supplied with holographic grating of 1800 lines/mm grating blazed at 250 nm. The excitation spectra were corrected to the excitation source intensity. The emission spectra were measured by using a Hamamatsu PMA-12 photonic multichannel analyzer (Hamamatsu, Hamamatsu City, Japan) equipped with BT-CCD line (Hamamatsu, Hamamatsu City, Japan). As an excitation source, a pulsed 266 nm line of Nd:YAG laser (3rd harmonic; LOTIS TII, Minsk, Belarus) was chosen (f = 10 Hz, t < 10 ns). The detection setup was calibrated and had a flat response for the whole working range (350–1100 nm). The measurements were carried out at 300 K.

The time-resolved luminescence spectrum was obtained by recording decay curves during changing observed wavelength and by creating a two-dimensional map (i.e., intensity vs. time and wavelength). It was recorded by an in-house developed software that controlled the equipment. A Dongwoo Optron DM711 monochromator (Hoean-Daero, Opo-Eup, Gyeonggi-Do, Korea) with a focal length of 750 mm was used to select the observed wavelength, while luminescence decay curves were acquired with a Hamamatsu R3896 photomultiplier (Hamamatsu, Hamamatsu City, Japan) connected to a digital Tektronix MDO 4054B oscilloscope (Bracknell, UK). An optical parametric oscillator Opotek Opolette 355 LD (Carlsbad, CA, USA) emitting 5 ns pulses was used as an excitation source.

The luminescence kinetics were measured by using a Jobin-Yvon THR1000 monochromator (HORIBA Jobin-Yvon, Palaiseu, France) equipped with a Hamamatsu R928 photomultiplier (Hamamatsu, Hamamatsu City, Japan) as a detector and a LeCroy WaveSurfer as a digital oscilloscope (Teledyne LeCroy, Chestnut Ridge, NY, USA). As an excitation source, a pulsed 266 nm line from an Nd:YAG laser was used. The luminescence kinetics were monitored at 618 nm according to the most intense electric dipole transition ($^5D_0 \rightarrow {}^7F_2$), and the effective emission lifetimes were calculated using the following equation:

$$\tau_m = \frac{\int_0^\infty t I(t)\,dt}{\int_0^\infty I(t)\,dt} \cong \frac{\int_0^{t_{max}} t I(t)\,dt}{\int_0^{t_{max}} I(t)\,dt} \tag{1}$$

where $I(t)$ is the luminescence intensity at time t corrected for the background and the integrals are calculated over the range of $0 < t < t_{max}$, where $t_{max} \gg \tau_m$.

The effective content of elements was determined by using an Agilent 720 bench-top optical emission spectrometer with inductively coupled Ar plasma (Ar-ICP-OES) and was corrected to an effective value. The ICP standard solutions were used to record the calibration curves to determine the Ca^{2+}, P^{5+}, Cu^{2+}, and Eu^{3+} ion content. The samples for elemental analysis were prepared by digesting in the pure HNO_3 acid (65% suprapur Merck).

The electron paramagnetic resonance (EPR) spectra were measured at 295 K and 77 K using a Bruker Elexsys 500 CW-EPR (Bruker GmbH, Rheinstetten, Germany) spectrometer operating at the X-band frequency ($\approx$9.7 GHz), equipped with frequency counter (E 41 FC) and NMR teslameter (ER 036TM). The spectra were measured with a modulation frequency of 100 kHz, microwave power of 10 mW, modulation amplitude of 10 G, time constant of 40 ms, and a conversion time of 160 ms. The first derivative of the absorption power was recorded as a function of the magnetic field value. An analysis of the EPR spectra was carried out using the WinEPR software package, version 1.26b (Bruker WinEPR GmbH, Rheinstetten, Germany).

3. Results and Discussion

3.1. Structural Analysis

The structural characterization of the HAp nanocrystals doped with xEu^{3+} (where x = 0.5, 1, and 3 mol%) and co-doped with xEu^{3+} and yCu^{2+} (where x = 0.5, 1, and 4 mol% and y = 0.5 and 1 mol%) was carried out by powder X-Ray diffraction measurements as a function of doping ion concentration (see Figure 1). Detectable crystallinity and pure hexagonal phase corresponding to the reference standard (ICSD—180315 [38]) were observed. Only in the case of the 4 mol% Eu^{3+}/0.5 mol% Cu^{2+}:HAp, an extra peak at 29.5° of 2θ was observed (assigned as asterisk in Figure 1).

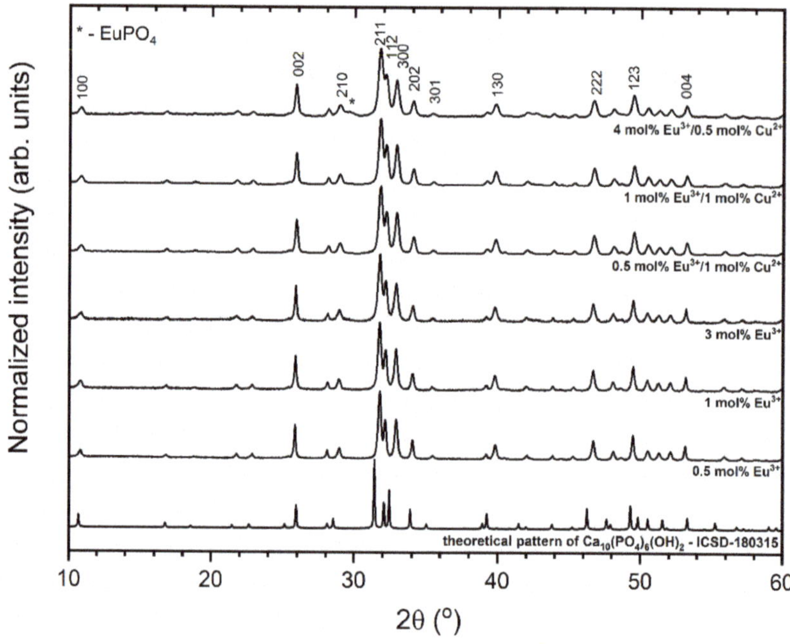

Figure 1. X-ray powder diffraction patterns of the Eu^{3+}-doped and Eu^{3+}/Cu^{2+} co-doped $Ca_{10}(PO_4)_6(OH)_2$.

Structural refinement was performed to obtain the unit cell parameters and the average grain sizes of synthesized materials. Hexagonal phase formation and the successful incorporation of Eu^{3+} and Cu^{2+} ions were verified. The theoretical fit with the observed XRD pattern was found to be in good agreement, which indicated the success of the Rietveld refinement method (see Figure 2). More details are displayed in Table 1. As can be seen, it was possible to observe an increase in the cell volume and a parameters with the increase in Eu^{3+} ion concentration in single-doped materials, which was caused by a smaller ionic radii of the dopant (Ca^{2+} (coordination number—CN9), 1.18 Å; Eu^{3+} (CN9), 1.12 Å; Ca^{2+} (CN7), 1.06 Å; and Eu^{3+} (CN7), 1.01 Å) [39]. Moreover, shrinkage of the average grain size with an increase in the Eu^{3+} ion concentration in the host lattice was observed. In the case of co-doped materials, no straightforward dopant concentration dependence on cell parameters (a, c, and V) or average grain size was observed.

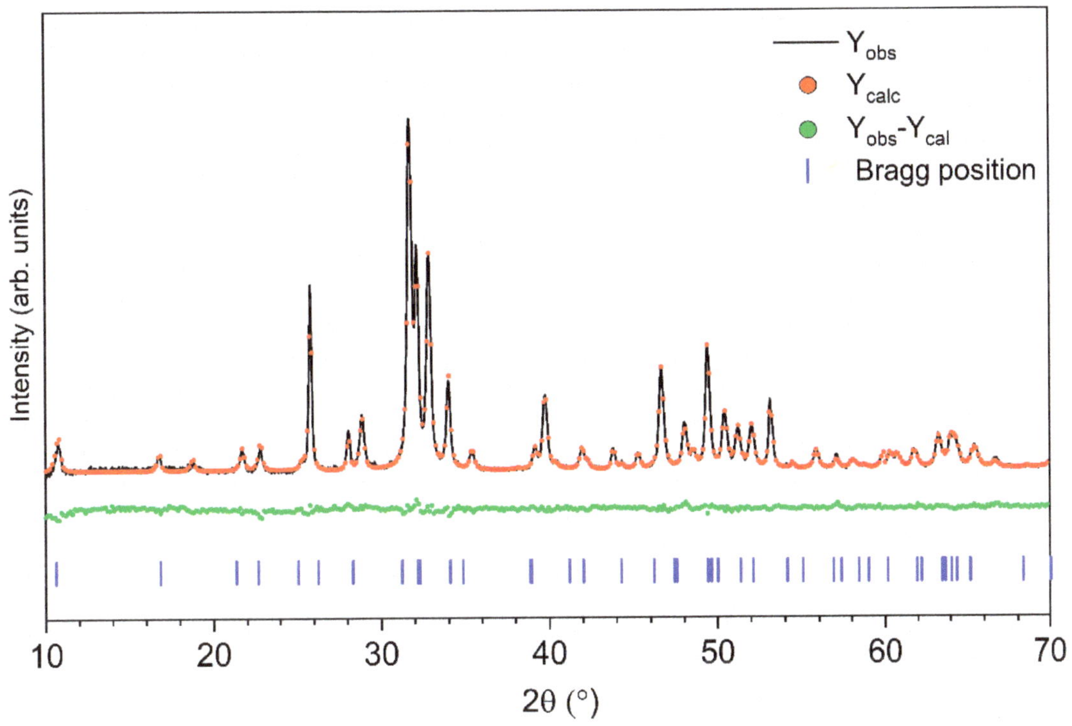

Figure 2. Representative results for the 0.5 mol% Eu^{3+}/1 mol% Cu^{2+}:$Ca_{10}(PO_4)_6(OH)_2$:, Rietveld analysis (red—fitted diffraction, green—differential pattern, and blue column—reference phase peak position).

Table 1. Unit cell parameters (a and c), cell volume (V), grain size, as well as refine factor (R_W) for the Eu^{3+}-doped and Eu^{3+}/Cu^{2+} co-doped $Ca_{10}(PO_4)_6(OH)_2$.

Sample	a (Å)	c (Å)	V (Å³)	Size (nm)	R_W (%)
single crystal	9.424(4)	6.879(4)	529.09(44)	–	–
doped with x mol% Eu^{3+}					
0.5 mol% Eu^{3+}	9.4139(6)	6.8905(0)	528.83(54)	55.6(2)	3.0
1 mol% Eu^{3+}	9.4259(6)	6.8881(3)	530.19(10)	45.5(9)	2.8
3 mol% Eu^{3+}	9.4276(8)	6.8891(5)	530.26(80)	38.0(1)	3.3
co-doped with x mol% Eu^{3+} and y mol% Cu^{2+}					
0.5 mol% Eu^{3+}/1 mol% Cu^{2+}	9.4300(2)	6.8875(1)	530.41(48)	42.0(2)	3.1
1 mol% Eu^{3+}/1 mol% Cu^{2+}	9.4264(5)	6.8858(9)	529.87(91)	41.2(4)	2.7
4 mol% Eu^{3+}/0.5 mol% Cu^{2+}	9.4288(2)	6.8893(5)	530.41(84)	58.2(0)	3.2

The morphology of the calcium hydroxyapatite was investigated by HRTEM. Nanoparticles are crystalline in nature and elongated, as can be seen in Figure 3. The particle size distribution is relatively wide, and the mean grain sizes of particle is in the range between 60 and 120 nm in length and about 40 nm in width.

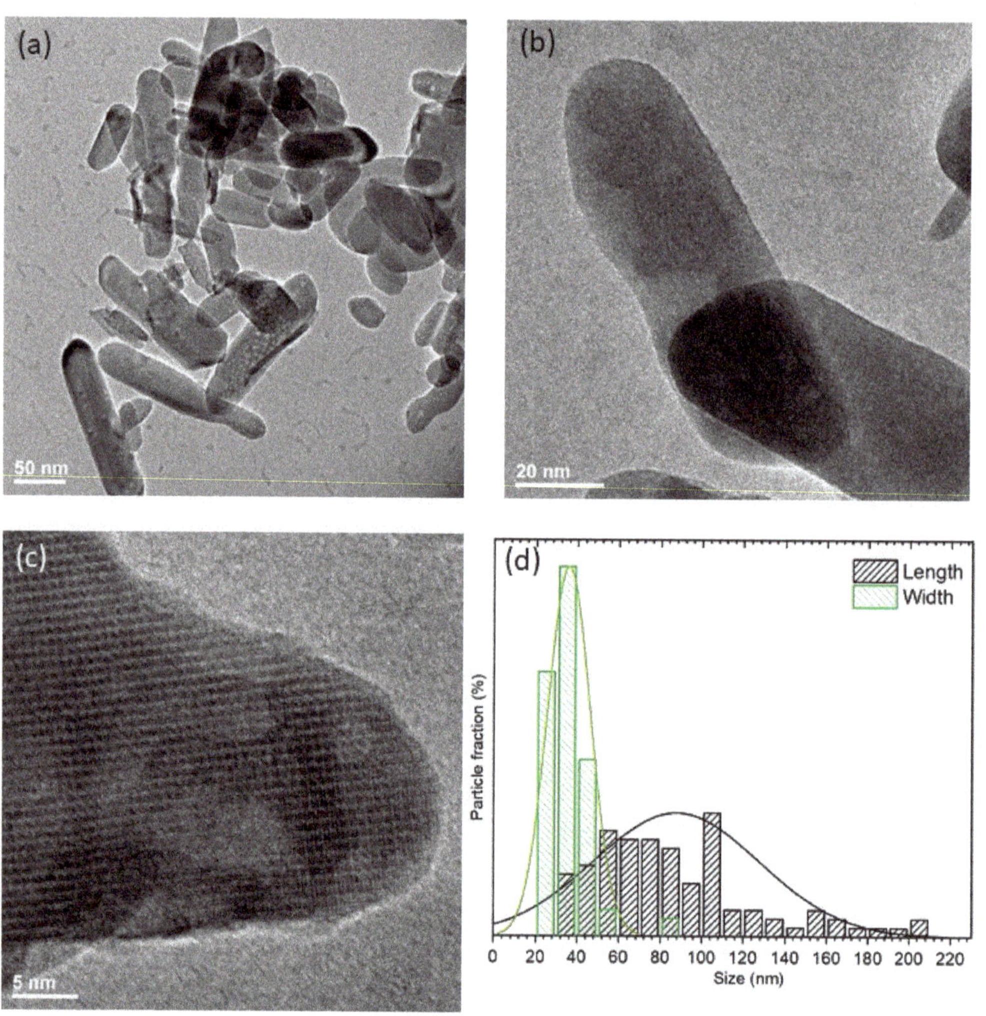

Figure 3. Representative TEM images (**a–c**) and particle size distribution (**d**) of the 1 mol% Eu^{3+}:$Ca_{10}(PO_4)_6(OH)_2$.

The infrared spectra of the copper-doped, europium-doped, and co-doped hydroxyapatite materials are presented in Figure 4. The most intense peaks are the triply degenerated antisymmetric stretching bands of phosphate groups $v_3(PO_4^{3-})$ located at 1044.5 cm^{-1} and 1097.8 cm^{-1}. The peaks observed at 566.0 cm^{-1} and 603.1 cm^{-1} correspond to the triply degenerated $v_4(PO_4^{3-})$ vibrations. The peaks at 963.0 cm^{-1} are assigned to the non-degenerated symmetric stretching $v_1(PO_4^{3-})$ band. Two peaks corresponding to OH$^-$ group at 3571.5 cm^{-1} and 633.5 cm^{-1} are observed on the infrared spectra. The existence of these peaks clearly confirms the hydroxyapatite structure with a hydroxyl group in the host lattice. The broad bands between 3690 and 3290 cm^{-1} were connected with H_2O vibration.

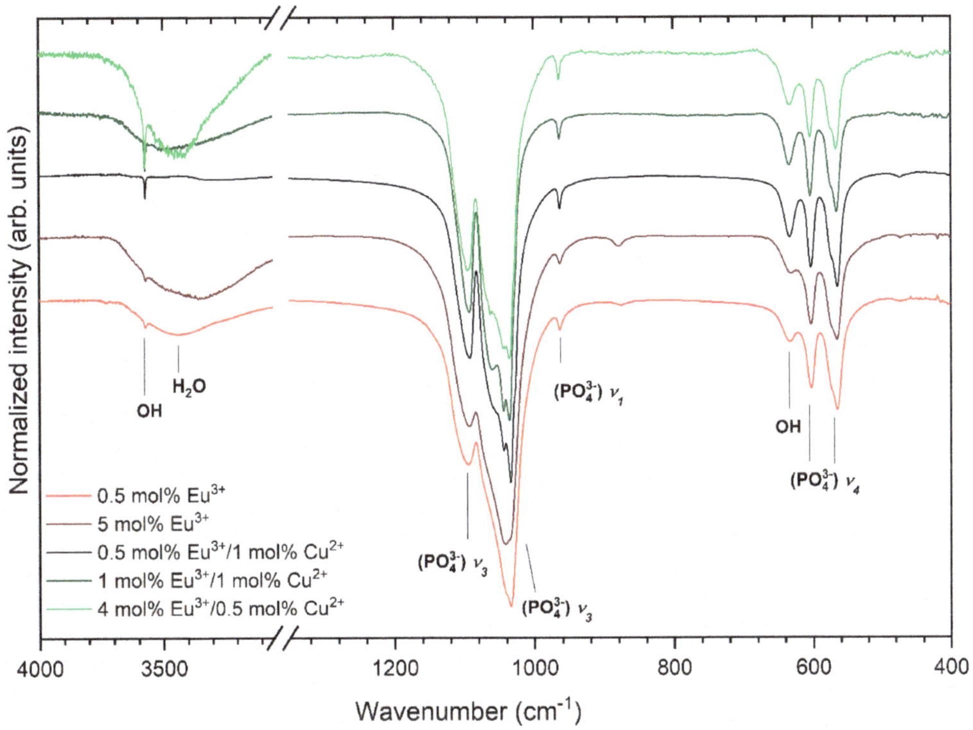

Figure 4. Infrared spectra of the Eu^{3+}-doped and Eu^{3+}/Cu^{2+} co-doped $Ca_{10}(PO_4)_6(OH)_2$.

3.2. Absorption, Excitation, and Emission Spectra

The absorption spectra of the pure, copper-doped, europium-doped, and co-doped hydroxyapatite nanopowders were recorded in the visible range from 350 nm to 800 nm at room temperature (see Figure 5). The pure hydroxyapatite matrix is transparent for these wavelengths. The copper-doped materials absorbed the blue radiation in the range from 350 nm to 420 nm. All materials are relatively transparent for the radiation from 450 nm to 550 nm of the wavelength. The copper-doped materials absorbed the radiation from 550 nm to 800 nm, and the absorption coefficient increases with the increase in wavelength. The broad absorption band is attributed to the $^2E \rightarrow ^2T_2$ intra-configurational (d-d) transition of the Cu^{2+} ions [40,41]. In the absorption spectra, the peaks related to the 4f-4f transitions of Eu^{3+} ions are observed. These peaks are attributed to the following transitions: the $^7F_0 \rightarrow ^5D_4, ^5L_8$ at 362 nm and $^7F_0 \rightarrow ^5G_6, ^5L_7, ^5G_3$ at 376 nm. The $^7F_0 \rightarrow ^5L_6$ transition with a maximum at 394 nm was observed in the case of europium and the copper co-doped materials. This transition is the most intense f-f transition of Eu^{3+} ions.

The excitation emission spectra, which were recorded at room temperature by monitoring the intense red emission at 618 nm ($^5D_0 \rightarrow ^7F_2$), of investigated materials are presented in Figure 6. The representative excitation spectra of the 1 mol% Eu^{3+}:HAp and 1 mol% $Eu^{3+}/1$ mol% Cu^{2+}:HAp are presented. As demonstrated, the excitation spectra consisted of visible intra-configurational 4f-4f transitions with sharp lines characteristic of Eu^{3+} ions. Particularly, these narrow bands located at around 320, 363, 383, 395, 416, and 466 nm originated from the $^7F_0 \rightarrow ^5H_J$; $^7F_0 \rightarrow ^5D_4, ^5L_8$; $^7F_0 \rightarrow G_2, ^5L_7, ^5G_3$; $^7F_0 \rightarrow ^5L_6$; $^7F_0 \rightarrow ^5D_3$ transitions of Eu^{3+} ions, respectively [31,35]. The absorption peak of the $^7F_0 \rightarrow ^5H_J$ transition at 320 nm indicates that the energy band-gap of HAp is considerably larger than that in, e.g., $Eu_2Ti_2O_7$ oxide [42], in which the $^5H_{3,6}$-related transition peak is completely masked by the charge transfer band due to its lower energy band-gap nature. The f-f electron transitions are weakly affected by the crystal field; thus, their positions remain almost steady due to good isolations of lanthanide's f orbitals by an external shell [19,31,43]. As can be seen, the intensity of the emission excitation spectra is much lower in the case of the co-doped material than in that single doped with Eu^{3+} in the HAp host.

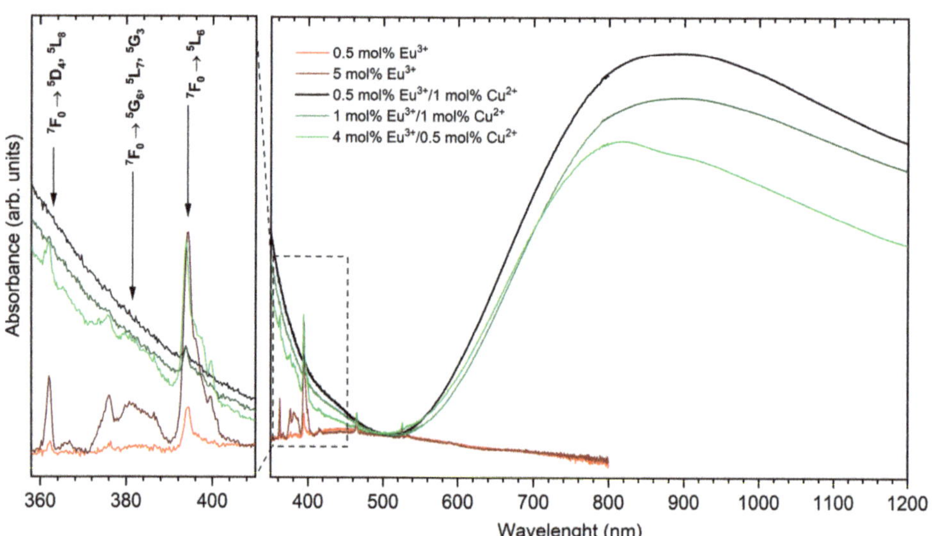

Figure 5. The absorption spectra of the Eu^{3+}-doped and Eu^{3+}/Cu^{2+}-co-doped $Ca_{10}(PO_4)_6(OH)_2$.

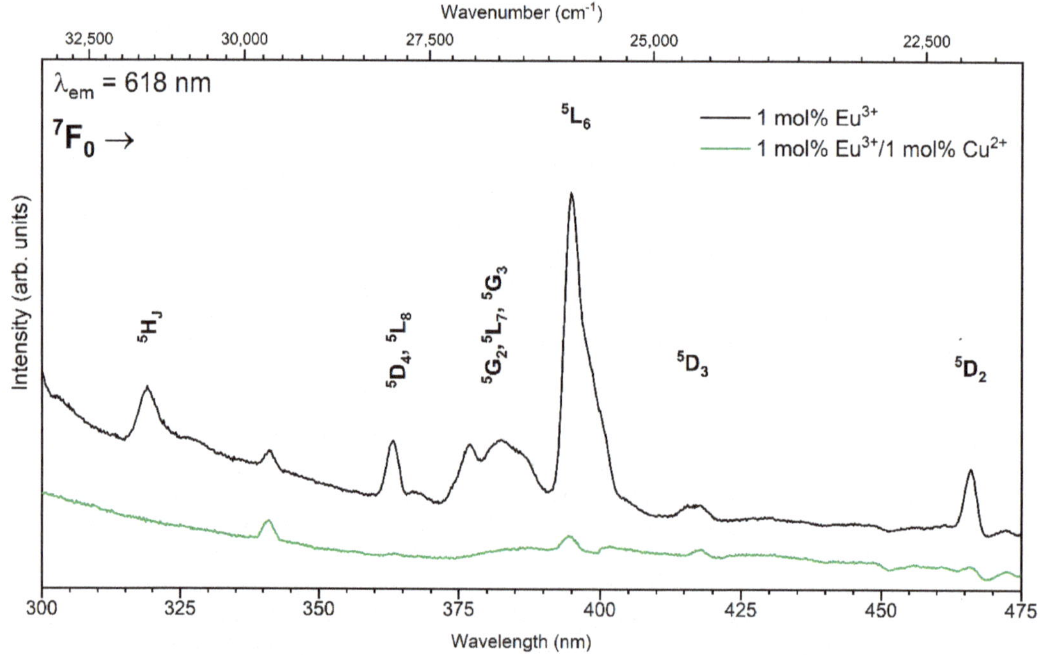

Figure 6. The excitation spectra of the 1 mol% Eu^{3+} and 1 mol% $Eu^{3+}/1$ mol% Cu^{2+}-co-doped $Ca_{10}(PO_4)_6(OH)_2$.

The spectroscopic properties of Eu^{3+} ions allow us to receive vital information about the symmetry of the Eu^{3+} ions surrounding the crystal lattice; the amount of crystallographic positions; and therefore, potential sites of substitution, structural changes occurring in the matrix caused by external factors, etc. The emission spectra of the Eu^{3+} ions consist of characteristic bands present in the red region of the electromagnetic radiation assigned to the electron transitions developing in the 4f-4f shell of Eu^{3+} ions. The $^5D_0 \rightarrow {}^7F_{0,1,2}$ transitions are the most important in analysis, particularly in correlation with the structural properties. The $^5D_0 \rightarrow {}^7F_0$ transition can provide direct information about the number of crystallographic sites occupied by Eu^{3+} ions in the host lattice [19].

The emission spectra of y mol% Eu^{3+}:HAp (where y—0.5, 1, and 3 mol%) and x mol% Eu^{3+}/y mol% Cu^{2+}:HAp (where x—0.5, 1, and 4 mol% and y—0.5 and 1 mol%) were measured by excitation wavelength at 266 nm at room temperature and are shown in Figure 7. The spectra were normalized to the $^5D_0 \rightarrow {}^7F_1$ magnetic dipole transition. The emission spectra were dominated by an intense red emission band situated at about 618 nm corresponding to the $^5D_0 \rightarrow {}^7F_2$ transition of Eu^{3+} ions. Meanwhile, four weaker emission bands peaking at around 578, 589, 652, and 698 nm were also detected and ascribed to the $^5D_0 \rightarrow {}^7F_0$, $^5D_0 \rightarrow {}^7F_1$, $^5D_0 \rightarrow {}^7F_3$, and $^5D_0 \rightarrow {}^7F_4$ transitions of Eu^{3+} ions, respectively [31,35]. The presence of a $^5D_0 \rightarrow {}^7F_0$ transition confirms that europium ions are located in a low-symmetry environment. Furthermore, the number of lines directly indicate the number of occupied crystallographic positions in the investigated lattice by Eu^{3+} ions. In the apatite molecule, ten calcium atoms are found in two non-equal crystallographic positions, in agreement with the results showed in Figure 7.

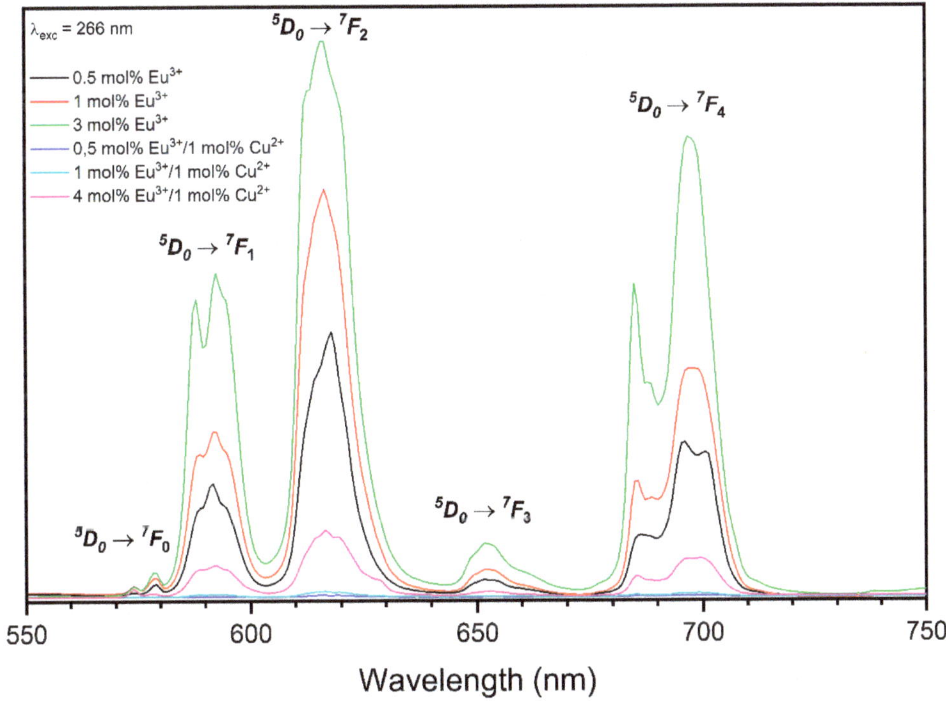

Figure 7. The emission spectra of the Eu^{3+}-doped and Eu^{3+}/Cu^{2+} co-doped $Ca_{10}(PO_4)_6(OH)_2$.

In Figure 8, the time-resolved emission spectrum of the 1 mol% Eu^{3+}/1 mol% Cu^{2+}:HAp is presented.

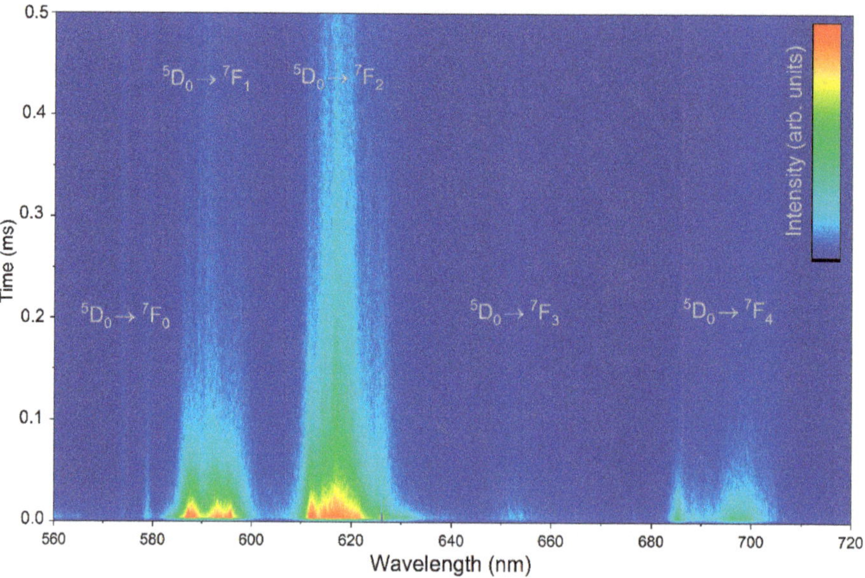

Figure 8. Representative emission spectra map of the 1 mol% Eu^{3+}/1 mol% Cu^{2+}:$Ca_{10}(PO_4)_6(OH)_2$.

3.3. Decay Profiles

The luminescence decay curves were registered and analyzed for the synthesized materials to determine the comprehensive characteristics of the luminescence properties. The decay curves presented in Figure 9 are not single-exponential, which is compatible with the presence of nonequivalent crystallographic sites of Eu^{3+} ions accordingly. The lifetimes values were calculated as the effective emission decay time by using Equation (1). The average lifetimes obtained for the sample single-doped by Eu^{3+} are equal to 0.93; 0.82, and 0.91 for concentrations of optically active ions at 0.5, 1.0, and 3.0 mol%, respectively.

The emission kinetic of Eu^{3+} ions strongly depends on the presence of Cu^{2+} ions, which effectively quenched the $^5D_0$ level. The average lifetime obtained for co-doped materials is much shorter than that for Eu^{3+}-doped materials, and the decay time values are estimated at about 0.33, 0.22, and 0.23 ms for the 0.5 mol% Eu^{3+}/1 mol% Cu^{2+}, 1 mol% Eu^{3+}/1 mol% Cu^{2+}, and 4 mol% Eu^{3+}/0.5 mol% Cu^{2+} co-doped materials, respectively. The emission quenching of the Eu^{3+} ions may be interpreted as nonradiative energy transfer between the Eu^{3+} and Cu^{2+} ions. The efficiency of energy transfer was estimated by Equation (2) for pairs of materials: single-doped with Eu^{3+} ions and co-doped with Eu^{3+} and Cu^{2+} ions with the same concentration of Eu^{3+} ions. The calculated lifetimes of Eu^{3+} ions (donor) in the absence and presence of Cu^{2+} ions (acceptor) are used [40,44]. The results of energy transfer efficiency are presented in Table 2.

$$\eta_{Eu^{3+} \to Cu^{2+}} = 1 - \left(\frac{\tau_{Eu^{3+} \to Cu^{2+}}}{\tau_{Eu^{3+}}} \right) \qquad (2)$$

Table 2. The average lifetime of Eu^{3+}-doped (τ_{Eu}), Eu^{3+}/1 mol% Cu^{2+} ($\tau_{Eu \to Cu}$) co-doped $Ca_{10}(PO_4)_6(OH)_2$ and energy transfer efficiency ($\eta_{Eu \to Cu}$).

	τ_{Eu} (ms)	$\tau_{Eu \to Cu}$ (ms)	$\eta_{Eu \to Cu}$ (%)
0.5 mol% Eu^{3+}	0.93	0.33	65
1 mol% Eu^{3+}	0.82	0.22	73

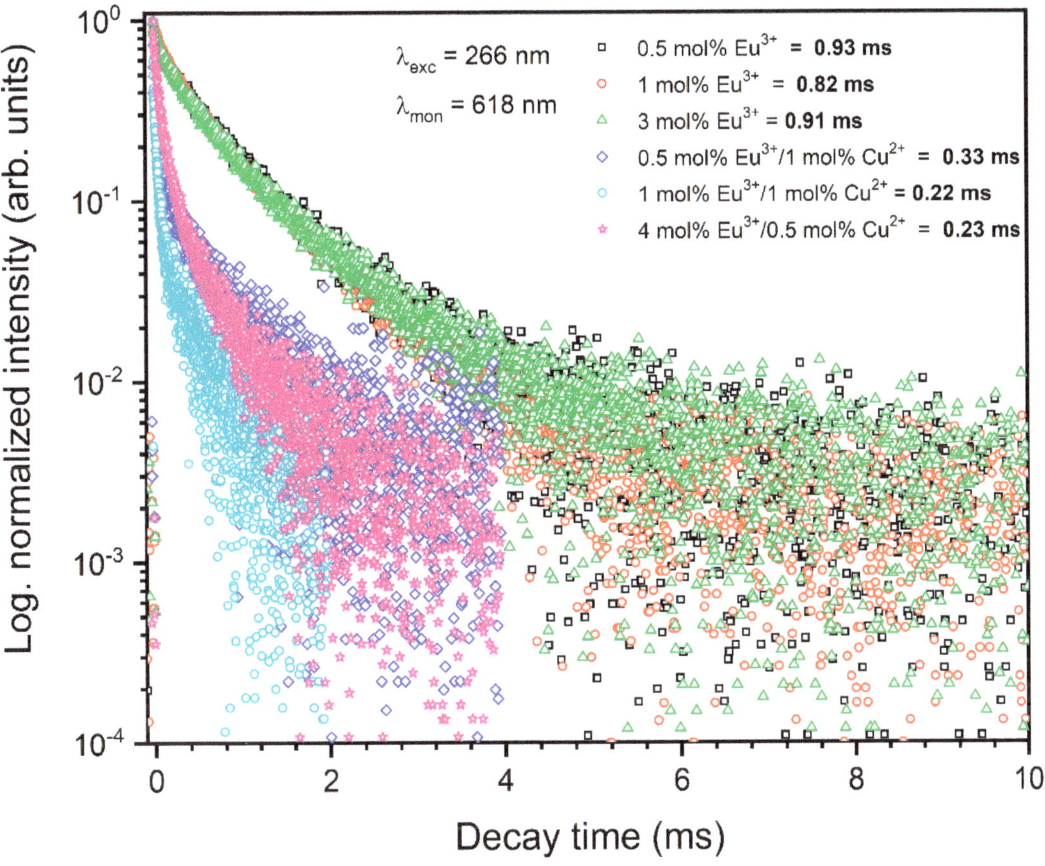

Figure 9. Decay times of the Eu^{3+}-doped and Eu^{3+}/Cu^{2+} co-doped $Ca_{10}(PO_4)_6(OH)_2$.

The obtained efficiency is equal to 65 and 73% for 0.5 mol% Eu^{2+}/1 mol% Cu^{2+} and 1 mol% Eu^{2+}/1 mol% Cu^{3+} co-doped HAp, respectively. The observed luminescence properties of europium ions in apatite lattice in the presence of copper ions are dominated by emission quenching of Eu^{3+} by Cu^{2+} ions. With an increase in Eu^{3+} concentration, the efficiency of quenching grew. This would suggest that the relatively huge probability of $Eu^{3+} \rightarrow Cu^{2+}$ nonradiative energy transfer probably behaves by electric dipole interaction [40,44].

The simplified energy level diagram of Eu^{3+} and Cu^{2+} ions was proposed and is shown in Figure 10 in order to explain the quenching mechanism occurring in hydroxyapatite co-doped with Eu^{3+} and Cu^{2+} ions. When the materials were excited by 266 nm wavelength, a charge transfer transition $O^{2-} \rightarrow Eu^{3+}$ occurred. Then, nonradiative relaxation to the $^5D_0$ first excited state was performed. From this state, the energy can be relaxed in two ways: by radiative transition to the ground state of Eu^{3+} ions ($^7F_{0-6}$) or by energy transfer to the $^2T_{2g}$ energy level of Cu^{2+} and then nonradiative relaxation to the ground state of Cu^{2+} ions. These two manners of energy relaxation compete among themselves, and doping with Cu^{2+} ions causes Eu^{3+} emission quenching.

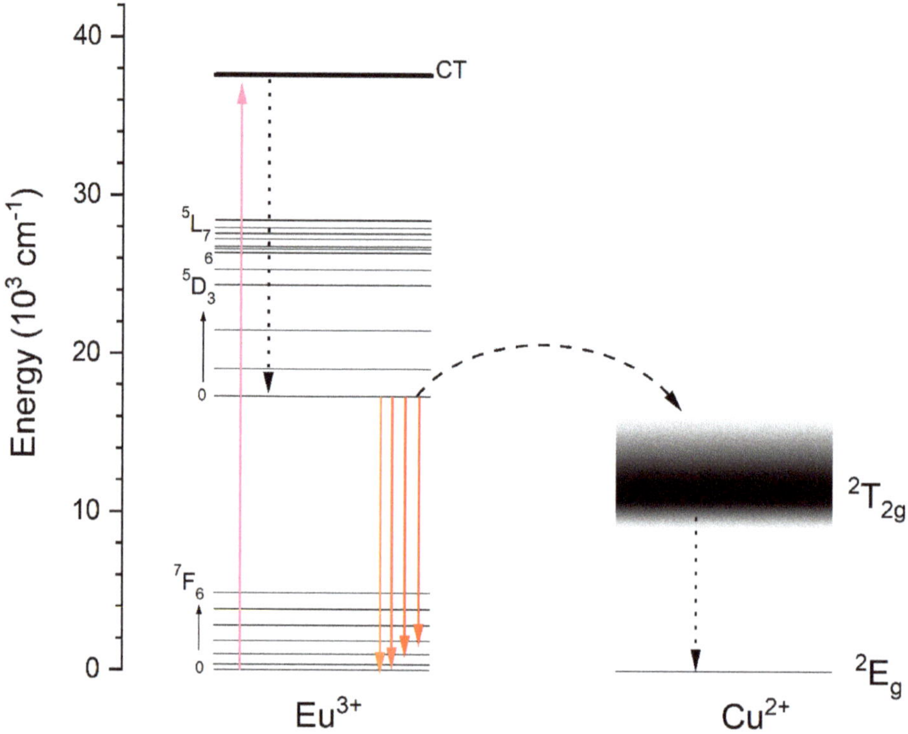

Figure 10. Simplified energy level scheme of Eu^{3+} and Cu^{2+} explaining quenching of Eu^{3+} ion emission.

3.4. The EPR Spectra Analysis

The EPR spectroscopy is especially predisposed to identifying the structural properties of paramagnetic compounds. The unpaired electron interacts (couples) with the nuclear spin (I) to form a 2I + 1 line hyperfine structure centered on g and spaced with the distance quantified by the hyperfine coupling parameter A. The coupling between the nuclear and electron spins becomes stronger as the A parameter becomes larger. The combination of g and A parameters can be utilized to differentiate between electron environments of ion.

There are two distinct Ca coordination sites in the HAp unit cell, that is the Ca(1) site with the Ca^{2+} ion surrounded by 9 oxygen atoms from 6 PO_4^{3-} groups and the Ca(2) site with the Ca^{2+} ion surrounded by 7 oxygen atoms from the 5 PO_4^{3-} and 1 OH^- anions. The Ca^{2+} ions in both coordination sites can be replaced by Eu^{3+} and Cu^{2+} ions [45]. The EPR properties of trivalent europium (Eu^{3+}) is relatively little because it is a non-Kramer ion, and its EPR spectrum should be silent because of the short spin-lattice relaxation time [46]. Therefore, in the EPR spectra recorded for the samples, only signals due to Cu^{2+} ions are observed.

The spectra recorded at room temperature and at 77 K (Figure 11) are anisotropic as a consequence of the Jahn–Teller effect operating for the d^9 electron configuration of Cu^{2+} ions that leads to considerable departure from a regular symmetry of the coordination sphere. The spectra reveal a weakly resolved hyperfine interaction between the spins of unpaired electrons and copper nuclei (I = 3/2), which for powder spectrum suggest a large distance between paramagnetic centers (Cu^{2+} ions).

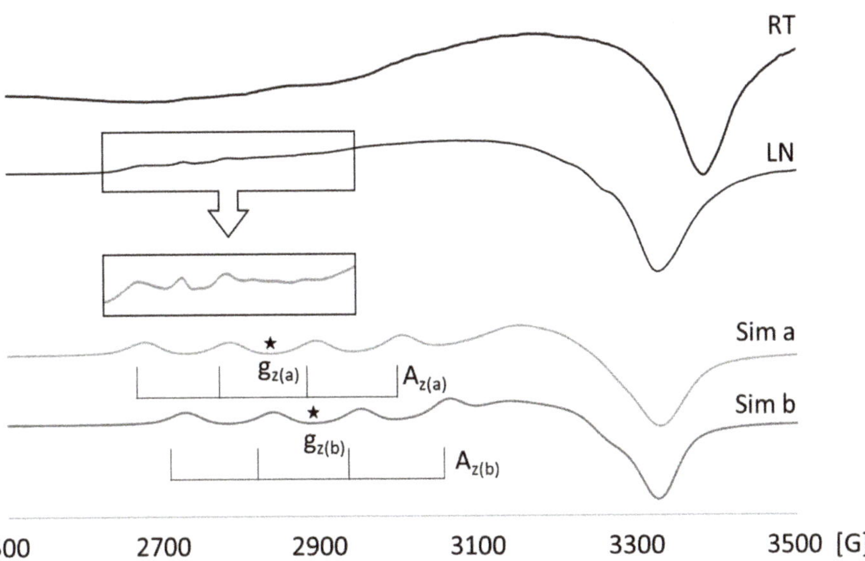

Figure 11. Experimental and simulated electron paramagnetic resonance (EPR) spectra of the 1 mol% Eu^{3+}/1 mol% Cu^{2+}:$Ca_{10}(PO_4)_6(OH)_2$.

Spectral analysis revealed the existence of at least two different coordination environments for copper(II) ions. The EPR spectrum of the 1 mol% Eu^{3+}/1 mol% Cu^{2+} co-doped HAp can be decomposed into two superimposed resonance signals due to two different Cu(II) coordination sites, hereinafter referred as "a" and "b". This stays in line with the fact that there are two distinct Ca coordination sites in HAp in which Cu(II) can be doped. From the simulation of the spectrum recorded at 77 K, the estimated parameters are $g_{z(a)}$ = 2.41, $g_{z(b)}$ = 2.37, g_y = 2.11, and g_x = 2.08 with $A_{z(a)}$ = $A_{z(b)}$ = 110 G. However, the accuracy of these parameters is inevitably limited due to the fact that the Cu(II) signals are superimposed.

The observed EPR parameters contrast their counterparts determined for synthetic hydroxyapatite doped with Cu^{2+}, in which also two different Cu(II) coordination sites were identified: g_z = 2.485, g_y = 2.17, g_x = 2.08, and A_z = 52 G for one coordination site and $g_{z(a)}$ = 2.420, g_y = 2.17, g_x = 2.08, and A_z = 92 G for the second [47]. This difference in g_z and A_z parameters clearly stems from the structural divergence between the 1 mol% Eu^{3+}/1 mol% Cu^{2+}:HAp and SHA. At the same time, the g_z and A_z parameters for 1 mol% Eu^{3+}/1 mol% Cu^{2+}:HAp are similar to the ones reported for copper(II) ions bonded to lattice oxygens in montmorillonite$((Cu(AlO)_n(H_2O)_{4-n})_x)$: g_z and A_z in the ranges 2.37–2.41 and 100–140 G, respectively [48]. Therefore, the geometry of Cu(II) coordination sites in 1 mol% Eu^{3+}/1 mol% Cu^{2+}:HAp are expected to structurally resemble $((Cu(AlO)_n(H_2O)_{4-n})_x)$.

Trends were found that enabled the Cu(II) EPR parameters to be correlated to the copper(II) ligands and the overall charge of the complexes [49–52]. According to these general trends, the g_z and A_z parameters for 1 mol% Eu^{3+}/1 mol% Cu^{2+}:HAp are characteristic of positively charged Cu–O complexes [49,50]. This fact indicates that, in the crystal lattice of the 1 mol% Eu^{3+}/1 mol% Cu^{2+}:HAp, the negative charge of the PO_4^{3-} and OH^- anions are primarily neutralized by remained Ca^{2+} cations. Moreover, the observed difference between g_z values found for two Cu coordination sites can be used to determine their possible assignment to Ca(1) and Ca(2). The increase in g_z is associated with the rise in positive charges for the Cu–O complex. Hence, the higher value of $g_{z(a)}$ indicates that this Cu(II) ion is surrounded by a lower number of oxygen atoms, which are the primary carriers of a negative charge, and therefore should be labeled as Cu(II) ion doped into the Ca(2) site.

4. Conclusions

The pure crystal hydroxyapatite powder doped co-doped with Eu^{3+} and Cu^{2+} ions was successfully synthesized by a microwave-assisted hydrothermal method that was confirmed by the X-ray powder diffraction method. The nanometric size of the obtained materials was confirmed by Rietveld refinement and TEM techniques. In the absorption spectra, the transitions occurring in Eu^{3+} as well as Cu^{2+} ions were observed. In the emission spectra, the typical transition of Eu^{3+} ions ($^5D_0 \rightarrow {}^7F_J$) were observed and the $^5D_0 \rightarrow {}^7F_2$ transition is the most intense. The $^5D_0 \rightarrow {}^7F_0$ transition consists of two lines, which means that the Eu^{3+} ions are localized in two independent crystallographic sites: in Ca(1) with C_3 point symmetry and in Ca(2) with C_s symmetry. The emission decay times of Eu^{3+}/Cu^{2+}:HAp are much shorter than the decay times of Eu^{3+}:HAp, which indicates that the Eu^{3+} emission is quenched by the Cu^{2+} ions. The simplified energy level diagram was proposed, and the quenching mechanism was explained. Based on the EPR measurement, the existence of at least two different coordinations surrounding copper(II) ions was detected.

Author Contributions: Conceptualization, K.S. and R.J.W.; methodology, K.S.; investigation, K.S., S.T., and A.L.; data curation, K.S. and S.T.; writing—original draft preparation, K.S.; writing—review and editing, K.S., S.T., A.L., A.W., and R.J.W.; visualization, K.S.; supervision, K.S. and R.J.W.; project administration, R.J.W.; funding acquisition, R.J.W. All authors have read and agreed to the published version of the manuscript.

Funding: This research was funded by the National Science Centre (NCN) within the projects "Preparation and characterisation of biocomposites based on nanoapatites for theranostics" (No. UMO-2015/19/B/ST5/01330) and "Elaboration and characteristics of biocomposites with anti-virulent and anti-bacterial properties against Pseudomonas aeruginosa" (No. UMO-2016/21/B/NZ6/01157).

Institutional Review Board Statement: Not applicable.

Informed Consent Statement: Not applicable.

Data Availability Statement: Not applicable.

Acknowledgments: We are grateful to E. Bukowska for the XRD measurements and to J. Komar for help with the emission spectra map measurements.

Conflicts of Interest: The authors declare no conflict of interest.

References

1. Pogosova, M.A.; González, L.V. Influence of anion substitution on the crystal structure and color properties of copper-doped strontium hydroxyapatite. *Ceram. Int.* **2018**, *44*, 20140–20147. [CrossRef]
2. Zawisza, K.; Sobierajska, P.; Renaudin, G.; Nedelec, J.-M.; Wiglusz, R.J. Effects of crystalline growth on structural and luminescence properties of $Ca_{(10-3x)}Eu_{2x}(PO_4)_6F_2$ nanoparticles fabricated by using a microwave driven hydrothermal process. *CrystEngComm* **2017**, *19*, 6936–6949. [CrossRef]
3. Bal, Z.; Kaito, T.; Korkusuz, F.; Yoshikawa, H. Bone regeneration with hydroxyapatite-based biomaterials. *Emergent Mater.* **2019**. [CrossRef]
4. Pajor, K.; Pajchel, L.; Kolmas, J. Hydroxyapatite and fluorapatite in conservative dentistry and oral implantology-a review. *Materials (Basel)* **2019**, *12*, 2683. [CrossRef]
5. Neacsu, I.A.; Stoica, A.E.; Vasile, B.S.; Andronescu, E. Luminescent Hydroxyapatite Doped with Rare Earth Elements for Biomedical Applications. *Nanomaterials* **2019**, *9*, 239. [CrossRef]
6. Szyszka, K.; Rewak-Soroczynska, J.; Dorotkiewicz-Jach, A.; Ledwa, K.A.; Piecuch, A.; Giersig, M.; Drulis-Kawa, Z.; Wiglusz, R.J. Structural modification of nanohydroxyapatite $Ca_{10}(PO_4)_6(OH)_2$ related to Eu^{3+} and Sr^{2+} ions doping and its spectroscopic and antimicrobial properties. *J. Inorg. Biochem.* **2020**, *203*, 110884. [CrossRef]
7. Shanmugam, S.; Gopal, B. Copper substituted hydroxyapatite and fluorapatite: Synthesis, characterization and antimicrobial properties. *Ceram. Int.* **2014**, *40*, 15655–15662. [CrossRef]

8.   Wiglusz, R.J.; Drulis-Kawa, Z.; Pazik, R.; Zawisza, K.; Dorotkiewicz-Jach, A.; Roszkowiak, J.; Nedelec, J.M. Multifunctional lanthanide and silver ion co-doped nano-chlorapatites with combined spectroscopic and antimicrobial properties. *Dalt. Trans.* **2015**, *44*, 6918–6925. [CrossRef]

9.   Zawisza, K.; Sobierajska, P.; Nowak, N.; Kedziora, A.; Korzekwa, K.; Pozniak, B.; Tikhomirov, M.; Miller, J.; Mrowczynska, L.; Wiglusz, R.J. Preparation and preliminary evaluation of bio-nanocomposites based on hydroxyapatites with antibacterial properties against anaerobic bacteria. *Mater. Sci. Eng. C* **2020**, *106*, 110295. [CrossRef]

10. Riaz, M.; Zia, R.; Ijaz, A.; Hussain, T.; Mohsin, M.; Malik, A. Synthesis of monophasic Ag doped hydroxyapatite and evaluation of antibacterial activity. *Mater. Sci. Eng. C* **2018**, *90*, 308–313. [CrossRef] [PubMed]

11. Stanić, V.; Dimitrijević, S.; Antić-Stanković, J.; Mitrić, M.; Jokić, B.; Plećaš, I.B.; Raičević, S. Synthesis, characterization and antimicrobial activity of copper and zinc-doped hydroxyapatite nanopowders. *Appl. Surf. Sci.* **2010**, *256*, 6083–6089. [CrossRef]

12. Kim, T.N.; Feng, Q.L.; Kim, J.O.; Wu, J.; Wang, H.; Chen, G.C.; Cui, F.Z. Antimicrobial effects of metal ions (Ag^+, Cu^{2+}, Zn^{2+}) in hydroxyapatite. *J. Mater. Sci. Mater. Med.* **1998**, *9*, 129–134. [CrossRef]

13. Kolmas, J.; Groszyk, E.; Kwiatkowska-Rózycka, D. Substituted hydroxyapatites with antibacterial properties. *Biomed Res. Int.* **2014**, *ID 178123*, 1–15. [CrossRef]

14. Huang, Y.; Hao, M.; Nian, X.; Qiao, H.; Zhang, X.; Zhang, X.; Song, G.; Guo, J.; Pang, X.; Zhang, H. Strontium and copper co-substituted hydroxyapatite-based coatings with improved antibacterial activity and cytocompatibility fabricated by electrodeposition. *Ceram. Int.* **2016**, *42*, 11876–11888. [CrossRef]

15. Pogosova, M.A.; Kasin, P.E.; Tretyakov, Y.D.; Jansen, M. Synthesis, structural features, and color of calcium-yttrium hydroxyapatite with copper ions in hexagonal channels. *Russ. J. Inorg. Chem.* **2013**, *58*, 381–386. [CrossRef]

16. Kazin, P.E.; Zykin, M.A.; Zubavichus, Y.V.; Magdysyuk, O.V.; Dinnebier, R.E.; Jansen, M. Identification of the chromophore in the apatite pigment [$Sr_{10}(PO_4)_6(Cu_xOH_{1-x-y})_2$]: Linear OCuO- featuring a resonance raman effect, an extreme magnetic anisotropy, and slow spin relaxation. *Chem. A Eur. J.* **2014**, *20*, 165–178. [CrossRef] [PubMed]

17. Pogosova, M.A.; Eliseev, A.A.; Kazin, P.E.; Azarmi, F. Synthesis, structure, luminescence, and color features of the Eu- and Cu-doped calcium apatite. *Dye. Pigment.* **2017**, *141*, 209–216. [CrossRef]

18. Kazin, P.E.; Zykin, M.A.; Schnelle, W.; Felser, C.; Jansen, M. Rich diversity of single-ion magnet features in the linearOCu^{III}O- ion confined in the hexagonal channels of alkaline-earth phosphate apatites. *Chem. Commun.* **2014**, *50*, 9325–9328. [CrossRef]

19. Zawisza, K.; Strzep, A.; Wiglusz, R.J. Influence of annealing temperature on the spectroscopic properties of hydroxyapatite analogues doped with Eu^{3+}. *New J. Chem.* **2017**, *41*, 9990–9999. [CrossRef]

20. Syamchand, S.S.; Sony, G. Europium enabled luminescent nanoparticles for biomedical applications. *J. Lumin.* **2015**, *165*, 190–215. [CrossRef]

21. Hassan, M.N.; Mahmoud, M.M.; El-Fattah, A.A.; Kandil, S. Microwave-assisted preparation of Nano-hydroxyapatite for bone substitutes. *Ceram. Int.* **2016**, *42*, 3725–3744. [CrossRef]

22. Chen, J.; Liu, J.; Deng, H.; Yao, S.; Wang, Y. Regulatory synthesis and characterization of hydroxyapatite nanocrystals by a microwave-assisted hydrothermal method. *Ceram. Int.* **2020**, *46*, 2185–2193. [CrossRef]

23. Jabalera, Y.; Oltolina, F.; Prat, M.; Jimenez-Lopez, C.; Fernández-Sánchez, J.F.; Choquesillo-Lazarte, D.; Gómez-Morales, J. Eu-doped citrate-coated carbonated apatite luminescent nanoprobes for drug delivery. *Nanomaterials* **2020**, *10*, 199. [CrossRef] [PubMed]

24. Mondal, S.; Hoang, G.; Manivasagan, P.; Kim, H.; Oh, J. Nanostructured hollow hydroxyapatite fabrication by carbon templating for enhanced drug delivery and biomedical applications. *Ceram. Int.* **2019**, *45*, 17081–17093. [CrossRef]

25. Targonska, S.; Rewak-Soroczynska, J.; Piecuch, A.; Paluch, E.; Szymanski, D.; Wiglusz, R.J. Preparation of a New Biocomposite Designed for Cartilage Grafting with Antibiofilm Activity. *ACS Omega* **2020**, *5*, 24546–24557. [CrossRef]

26. Targonska, S.; Wiglusz, R.J. Investigation of Physicochemical Properties of the Structurally Modified Nanosized Silicate-Substituted Hydroxyapatite Co-Doped with Eu^{3+} and Sr^{2+} Ions. *Nanomaterials* **2021**, *11*.

27. Escudero, A.; Calvo, M.E.; Rivera-Fernández, S.; de la Fuente, J.M.; Ocaña, M. Microwave-Assisted Synthesis of Biocompatible Europium-Doped Calcium Hydroxyapatite and Fluoroapatite Luminescent Nanospindles Functionalized with Poly(acrylic acid). *Langmuir* **2013**, *29*, 1985–1994. [CrossRef]

28. Wang, C.; Jeong, K.-J.; Kim, J.; Kang, S.W.; Kang, J.; Han, I.H.; Lee, I.-W.; Oh, S.-J.; Lee, J. Emission-tunable probes using terbium(III)-doped self-activated luminescent hydroxyapatite for in vitro bioimaging. *J. Colloid Interface Sci.* **2021**, *581*, 21–30. [CrossRef]

29. Muresan, L.E.; Perhaita, I.; Prodan, D.; Borodi, G. Studies on terbium doped apatite phosphors prepared by precipitation under microwave conditions. *J. Alloys Compd.* **2018**, *755*, 135–146. [CrossRef]

30. Ma, B.; Zhang, S.; Qiu, J.; Li, J.; Sang, Y.; Xia, H.; Jiang, H.; Claverie, J.; Liu, H. Eu/Tb codoped spindle-shaped fluorinated hydroxyapatite nanoparticles for dual-color cell imaging. *Nanoscale* **2016**, *8*, 11580–11587. [CrossRef]

31. Targonska, S.; Szyszka, K.; Rewak-Soroczynska, J.; Wiglusz, R.J. A new approach to spectroscopic and structural studies of the nano-sized silicate-substituted hydroxyapatite doped with Eu^{3+} ions. *Dalt. Trans.* **2019**. [CrossRef] [PubMed]

32. Priyadarshini, B.; Vijayalakshmi, U. Development of cerium and silicon co-doped hydroxyapatite nanopowder and its in vitro biological studies for bone regeneration applications. *Adv. Powder Technol.* **2018**, *29*, 2792–2803. [CrossRef]

33. Yoder, C.H.; Havlusch, M.D.; Dudrick, R.N.; Schermerhorn, J.T.; Tran, L.K.; Deymier, A.C. The synthesis of phosphate and vanadate apatites using an aqueous one-step method. *Polyhedron* **2017**, *127*, 403–409. [CrossRef]

34. Nakagawa, D.; Nakamura, M.; Nagai, S.; Aizawa, M. Fabrications of boron-containing apatite ceramics via ultrasonic spray-pyrolysis route and their responses to immunocytes. *J. Mater. Sci. Mater. Med.* **2020**, *31*. [CrossRef]
35. Gómez-Morales, J.; Verdugo-Escamilla, C.; Fernández-Penas, R.; Parra-Milla, C.M.; Drouet, C.; Maube-Bosc, F.; Oltolina, F.; Prat, M.; Fernández-Sánchez, J.F. Luminescent biomimetic citrate-coated europium-doped carbonated apatite nanoparticles for use in bioimaging: Physico-chemistry and cytocompatibility. *RSC Adv.* **2018**, *8*, 2385–2397. [CrossRef]
36. Rietveld, H.M. A profile refinement method for nuclear and magnetic structures. *J. Appl. Crystallogr.* **1969**, *2*, 65–71. [CrossRef]
37. Lutterotti, L.; Matthies, S.; Wenk, H.-R. MAUD: A friendly Java program for Material Analysis Using Diffraction. *IUCr Newsl. CPD* **1999**, *21*, 14–15. [CrossRef]
38. Balan, E.; Delattre, S.; Roche, D.; Segalen, L.; Morin, G.; Guillaumet, M.; Blanchard, M.; Lazzeri, M.; Brouder, C.; Salje, E.K.H. Line-broadening effects in the powder infrared spectrum of apatite. *Phys Chem Miner.* **2011**, *38*, 111–122. [CrossRef]
39. Shannon, R.D. Revised effective ionic radii and systematic studies of interatomic distances in halides and chalcogenides. *Acta Crystallogr.* **1976**, *A32*, 751–767. [CrossRef]
40. Jiménez, J.A. Photoluminescence of Eu^{3+}-doped glasses with Cu^{2+} impurities. *Spectrochim. Acta Part A Mol. Biomol. Spectrosc.* **2015**, *145*, 482–486. [CrossRef]
41. Lakshmi Reddy, S.; Endo, T.; Siva Reddy, G. Electronic (Absorption) Spectra of 3d Transition Metal Complexes. In *Advanced Aspects of Spectroscopy*; Farrukh, M.A., Ed.; IntechOpen: London, UK, 2012; pp. 3–48.
42. Orihashi, T.; Nakamura, T.; Adachi, S. Synthesis and Unique Photoluminescence Properties of $Eu_2Ti_2O_7$ and Eu_2TiO_5. *J. Am. Ceram. Soc.* **2016**, *99*, 3039–3046. [CrossRef]
43. Wang, S.; Wei, W.; Zhou, X.; Li, Y.; Hua, G.; Li, Z.; Wang, D.; Mao, Z.; Zhang, Z.; Ying, Z. Comparative study of the luminescence properties of $Ca_{2+x}La_{8-x}(SiO_4)_{6-x}(PO_4)_xO_2$:$Eu^{3+}$ ($x = 0, 2$) red phosphors. *J. Lumin.* **2020**, *221*, 1170743. [CrossRef]
44. Jiménez, J.A. Eu^{3+} amidst ionic copper in glass: Enhancement through energy transfer from Cu^+, or quenching by Cu^{2+}? *Spectrochim. Acta Part A Mol. Biomol. Spectrosc.* **2017**, *173*, 979–985. [CrossRef]
45. Cawthray, J.F.; Creagh, A.L.; Haynes, C.A.; Orvig, C. Ion exchange in hydroxyapatite with lanthanides. *Inorg. Chem.* **2015**, *54*, 1440–1445. [CrossRef]
46. Yong, G.; Chunshan, S. The spectral properties of BaB_8O_{13}:Eu, Tb phosphors. *J. Phys. Chem. Solids* **1996**, *57*, 1303–1306. [CrossRef]
47. Sutter, B.; Wasowicz, T.; Howard, T.; Hossner, L.R.; Ming, D.W. Characterization of Iron, Manganese, and Copper Synthetic Hydroxyapatites by Electron Paramagnetic Resonance Spectroscopy. *Soil Sci. Soc. Am. J.* **2002**, *66*, 1359–1366. [CrossRef]
48. Bahranowski, K.; Dula, R.; Labanowska, M.; Serwicka, E.M. ESR study of Cu centers supported on Al-, Ti-, and Zr-pillared montmorillonite clays. *Appl. Spectrosc.* **1996**, *50*, 1439–1445. [CrossRef]
49. Peisach, J.; Blumberg, W. Analysis of EPR Copper structural implications derived from the analysis of EPR spectra of natural and artificial Cu-proteins. *Arch Biochem Biophys* **1974**, *165*, 691–708. [CrossRef]
50. Kumar, S.; Sharma, R.P.; Venugopalan, P.; Witwicki, M.; Ferretti, V. Synthesis, characterization, single crystal X-ray structure, EPR and theoretical studies of a new hybrid inorganic-organic compound $[Cu(Hdien)_2(H_2O)_2](pnb)_4 \cdot 4H_2O$ and its structural comparison with related $[Cu(en)_2(H_2O)_2](pnb)2$. *J. Mol. Struct.* **2016**, *1123*, 124–132. [CrossRef]
51. Maślewski, P.; Wyrzykowski, D.; Witwicki, M.; Dołęga, A. Histaminol and Its Complexes with Copper(II) – Studies in Solid State and Solution. *Eur. J. Inorg. Chem.* **2018**, *2018*, 1399–1408. [CrossRef]
52. Kumar, S.; Pal Sharma, R.; Venugopalan, P.; Gondil, V.S.; Chhibber, S.; Aree, T.; Witwicki, M.; Ferretti, V. Hybrid inorganic-organic complexes: Synthesis, spectroscopic characterization, single crystal X-ray structure determination and antimicrobial activities of three copper(II)-diethylenetriamine-p-nitrobenzoate complexes. *Inorganica Chim. Acta* **2018**, *469*, 288–297. [CrossRef]

MDPI

St. Alban-Anlage 66

4052 Basel

Switzerland

Tel. +41 61 683 77 34

Fax +41 61 302 89 18

www.mdpi.com

Nanomaterials Editorial Office

E-mail: nanomaterials@mdpi.com

www.mdpi.com/journal/nanomaterials